CB Disarmament Negotiations, 1920–1970

SIPRI

Stockholm International Peace Research Institute

SIPRI is an independent institute for research into problems of peace and conflict, with particular attention to the problems of disarmament and arms regulation. It was established in 1966 to commemorate Sweden's 150 years of unbroken peace.

The financing is provided by the Swedish Government. The staff, the Governing Board, and the Scientific Council are international.

The Board and Scientific Council are not responsible for the views expressed in the publications of the Institute.

Governing Board

Professor Gunnar Myrdal, Chairman (Sweden)

Professor Hilding Eek, Vice Chairman (Sweden)

Academician Ivan Málek (Czechoslovakia)

Professor Leo Mates (Yugoslavia)

Professor Robert Neild (United Kingdom)

Professor Bert Röling (Holland)

Professor John Sanness (Norway)

The Director

Director

Dr. Frank Barnaby (United Kingdom)

SIPRI

Stockholm International Peace Research Institute
Sveavägen 166, S-113 46 Stockholm, Sweden

Cable: Peaceresearch, Stockholm   Telephone: 08-15 19 00

THE PROBLEM OF CHEMICAL AND BIOLOGICAL WARFARE

A study of the historical, technical, military, legal
and political aspects of CBW,
and possible disarmament measures

VOLUME IV
CB Disarmament Negotiations, 1920–1970

SIPRI

Stockholm International Peace Research Institute

Paul Elek Limited
London

Copyright © 1971 by SIPRI
Sveavägen 166, 113 46 Stockholm, Sweden

Distributed in
the United Kingdom and
Commonwealth
by Paul Elek Limited
At the Ibex
54–58 Caledonian Road
London N1 9RN

SBN 391-00203-1

Library of Congress Catalog
Card Number: 76-169793

KING ALFRED'S COLLEGE
WINCHESTER

341.735
STO

11602

Printed in Sweden by
Almqvist & Wiksells Boktryckeri AB, Uppsala 1971

Contents of the Study

Volume I. The Rise of CB Weapons

A description of the main lines of development in the technology underlying CBW and in the constraints affecting use of CB weapons. The period covered is approximately 1914–1945, although more recent developments in CW technology are also described. In addition, the volume includes an account of all instances known to SIPRI when CB weapons have been used in war, or when their use has been alleged; in this case the time-span is 1914–1970.

Volume II. CB Weapons Today

A description based on the open literature of the present state of CBW technology and of national CBW programmes and policies. It also includes a discussion of the attractions and liabilities of CB weapons, and of the consequences, intentional or unintentional, that might follow their use.

Volume III. CBW and International Law

A description of the legal limitations on the use of CB weapons. It comprises discussions of the field of application of the Geneva Protocol, particularly as regards non-lethal chemical weapons and anti-plant agents, of the existence, development and scope of the prohibition of CBW provided by the customary law of war, and of the application to CBW of general principles of the law of war. It also reviews the juristic works in this field.

Volume IV. CB Disarmament Negotiations, 1920–1970

A review of the activities of the League of Nations and United Nations in extending and reinforcing the prohibition concerning CB weapons, including a report of recent negotiations for international CB disarmament. The volume also contains an account of those instances when formal complaints of the use of CB weapons have been made to the two world organizations.

Volume V. The Prevention of CBW

A discussion of possible measures that might be adopted to prevent future CBW. The volume describes steps that might be taken to strengthen the legal prohibition of CBW, and the problems and possibilities, including those of verification, involved in the negotiation of CB disarmament.

Volume VI. Technical Aspects of Early Warning and Verification

A technical account of SIPRI research on methods of early warning and identification of biological warfare agents, together with a description of two experimental SIPRI projects on CB verification. The first project concerns the non-production of BW agents and involved visits to biological laboratories in several countries; the second concerns the non-production of organophosphorus CW agents and summarizes the results of a symposium.

Preface

The birth of this study of chemical and biological warfare can be traced back to 1964, when a group of microbiologists who were concerned about the problems of biological warfare started meeting under the auspices of Pugwash. After some meetings it became evident that there was need for more intense study than could be achieved through occasional gatherings of people who were busy with other work. In 1966–67 SIPRI, which was then starting up, decided to take on the task of making a major review of biological warfare. The study was soon extended to cover chemical warfare as well. It was found impossible to discuss one without the other. The two have traditionally been grouped together in law, in military organization, in political debate and in the public mind.

The aim of the study is to provide a comprehensive survey of all aspects of chemical and biological warfare and of the problems of outlawing it more effectively. It is hoped that the study will be of value to politicians, their advisers, disarmament negotiators, scientists and to laymen who are interested in the problem.

The authors of the report have come from a number of disciplines— microbiology, chemistry, economics, international law, medicine, physics and sociology and soldiery—and from many countries. It would be too much to claim that all the authors had come to share one precisely defined set of values in their approach to the problem. Some came to the problem because they were concerned that the advance of science in their field should not be twisted to military uses; others because they had taken a scholarly interest in the law or history of CBW; others because they had particular experience of military or technical aspects of it. What is true is that, after working together for a period of years, they have all come to share a sober concern about the potential dangers of CBW.

In reviewing the issues for policy (in Volume V) the aim has been not to produce a set of recommendations or a plan for action but to analyse the main factors influencing national policies and international negotiations over CBW, to indicate alternative courses of action as they emerge from the analysis, and to present as clearly as possible the perspective on the problem at which an international team of people working for a period of years on neutral soil has arrived.

At an early stage it was necessary to face the question whether, if we

Preface

assembled a lot of information on CBW and published all that we thought was relevant, we would risk contributing dangerously to the proliferation of these weapons. This proposition was rejected on the grounds that the service we could do by improving the level of public discussion was greater than any disservice we might do by transmitting dangerous knowledge. Secrecy in a field like this serves mostly to keep the public in ignorance. Governments find things out for themselves.

While the study has been in progress there has been an increase in public discussion of the subject. A group of experts appointed by the Secretary-General of the United Nations has produced a report on *Chemical and Bacteriological (Biological) Weapons and the Effects of Their Possible Use*. In the United States a rising tide of concern about CBW has given rise to Congressional hearings; a policy review, commissioned by the President, has led to the unilateral renunciation by the United States Government of biological weapons and to the decision to renounce first use of chemical weapons and to seek ratification of the Geneva Protocol. At the United Nations and at the Disarmament Committee in Geneva, CBW has received a lot of attention, culminating in current negotiations over a biological disarmament treaty.

In response to an invitation from the UN Secretary-General, early drafts of parts of this study were circulated to his group of experts in February 1969. These drafts were also made available to the World Health Organization for the preparation of its own submission to the UN group of experts; this submission, together with the subsequent WHO publication based upon it, *Health Aspects of Chemical and Biological Weapons,* was prepared by a group of consultants that included Julian Perry Robinson from SIPRI.

Provisional editions of parts of this study were issued in February 1970.

The authors are conscious of the problem of avoiding biases. A disproportionate part of the information we have used comes from the United States. This is partly because the United States has been very active in the field of chemical and biological warfare in the post-war period. It is also because the United States is much more open with information than most other countries.

Since this is a team work and since, like most studies of this size, it grew and changed shape and changed hands in some degree as it went along, it is not easy to attribute responsibility for its preparation. The authorship of each part is indicated at the start of it, but these attributions do not convey the whole story. The team of people who produced the study met together often, shared material, exchanged ideas, reviewed each others' drafts in greater or lesser degree, and so on. So it is a corporate product, and those who wrote the final drafts sometimes had the benefit of working papers, earlier drafts, ideas or material provided by others.

At first, Rolf Björnerstedt was briefly in charge of the study. After an interval, Carl-Göran Hedén took over. When he had to return to the Karolinska Institute—from which he has continued to give us his advice and help—I assumed responsibility for the project. The other members of the team have been Anders Boserup, who from the earliest stages has found time to come frequently from Copenhagen to help on the project, Jozef Goldblat, Milton Leitenberg, Theodor Nemec, Julian Perry Robinson and Hans von Schreeb. Åke Ljunggren was a member of the team in Stockholm in the early stages of the project. Sven Hirdman joined in at the later stages.

The work on rapid detection of the use of biological warfare agents (Volume VI) was undertaken separately from the main study by Konstantin Sinyak, who came from the Soviet Union to work at the Karolinska Institute in Stockholm, and Åke Ljunggren, who went from Sweden to work at the Microbiological Institute in Prague. Both worked in close contact with Carl-Göran Hedén who contributed a study on automation. We are indebted to the two host institutes for the facilities and help they generously provided.

It is usually wrong to single out one person from a team but in this case there is no doubt that one person has contributed more than anyone else to the study. He is Julian Perry Robinson who has written more of the study than anyone else and has had a great influence on the whole shape and quality of it.

Rosemary Proctor undertook the formidable task of acting as editorial assistant for the whole study and preparing an index for it.

A great debt is also owed to many people outside the institute—too many to name—for the help they have given us. This includes those who attended the early Pugwash meetings on biological warfare, those who attended meetings at SIPRI on biological and chemical warfare, those who wrote working papers for us, those who gave their time to the biological inspection experiment and many people who have visited us or helped us with advice and material at different times. It includes people from many countries, East and West, and many disciplines. It includes people with many different kinds of expertise. The amount of help they gave us—and it was far greater than we had expected at the start—was clearly an expression of their concern about the problem. We are very grateful to them all. The responsibility for what is said is, of course, ours.

September 1971
Robert Neild
Director

ATTRIBUTION

The author of Volume IV is Jozef Goldblat.

CONTENTS

Introductory Survey 17
   I. The historical background 17
   II. The inter-war period: the course of events 18
   III. The inter-war period: the main points of the CBW debate 21
        Definition of CB weapons 21
        The tear-gas issue. 22
        Prevention of CB warfare 23
        Verification proposals 25
   IV. The situation after World War II 26
   V. The recent debate. 27

Part I. Negotiations at the League of Nations and the United Nations, 1920–1969

Introductory note on the disarmament negotiating machinery 34
   I. The League of Nations 34
        The Covenant 34
        The Council 34
        The Assembly. 35
        The Permanent Advisory Commission 35
        The Temporary Mixed Commission 35
        The Preparatory Commission for the Disarmament Conference . . . 35
        The Conference for the Supervision of the International Trade in Arms and Ammunition and in Implements of War 36
        The Conference for the Reduction and Limitation of Armaments (Disarmament Conference). 37
   II. The United Nations. 38
        The Charter 38
        The Security Council. 38
        The General Assembly 38
        The Disarmament Commission 38
        The Geneva Disarmament Conference 39
        Other conferences 40
   III. Comparison of the two systems 40

Early efforts to outlaw chemical warfare 41

Chapter 1. 1920–1924 43
   I. Preliminary discussions on chemical warfare in the League of Nations 43
   II. Disclosure of CW discoveries. 45

Contents

   III. Adhesion to the Treaty of Washington. 46
   IV. Arousing public opinion about CBW 48
   V. The report on CBW. 49
      Chemical warfare 50

      Effects of irritant agents — Effects of suffocating or asphyxiating gases — Effects of toxic agents — Combined effects — After-effects of war gases — Possible effects of new discoveries — Effects of gas on animals — Effects of gas on vegetation — Effects of gas on other sources of wealth — Protection against chemical weapons — Conclusions

      Bacteriological warfare 54
      Possible effects of use and protection — Conclusion

   VI. Publicity for the report on CBW 56
   VII. Summary and comment 56

Chapter 2. 1925 58

   I. The convening of the Conference on the International Trade in Arms 58
   II. The opening phase of the conference 58
   III. The Legal Committee's consideration of CBW 60
   IV. The Military Committee's consideration of CBW 61
   V. The report of the General Rapporteurs 64
   VI. Drafting the Geneva Protocol 67
   VII. Adoption and signing of the Geneva Protocol 69
      Protocol for the Prohibition of the Use in War of Asphyxiating, Poisonous or other Gases, and of Bacteriological Methods of Warfare . . 69

VIII. Summary and comment 70

Chapter 3. 1926–1930 72

   I. The convening of the Preparatory Commission for the Disarmament Conference. 72
   II. Consideration of CBW by the Sub-Commissions of the Preparatory Commission 74
   III. The Preparatory Commission's discussion of proposals for CBW prohibition. 89
   IV. The controversy over tear gas 102
   V. Proposal for assistance to CBW-attacked countries 104
   VI. Summary and comment 104

Chapter 4. 1931 – November 1932 107

   I. The convening of the Disarmament Conference 107
   II. General debate on CBW at the Disarmament Conference . . . 108
   III. First report of the Special Committee on CBW 113
   IV. Conclusion of the first phase of the Disarmament Conference. . . 118

V.	The report of the chairman of the Special Committee 120
	Absolute or relative character of the prohibition 122
	The prohibition of preparations for chemical, incendiary and bacteriological warfare 123
	The supervision of the observance of the prohibition of preparations for chemical warfare. 125
	Sanctions in the event of the use of chemical, incendiary and bacteriological weapons. 128
	Conclusions of the report. 132
VI.	Consideration of the chairman of the Special Committee's report 133
VII.	Questionnaire submitted to the Special Committee 142
	Prohibition of the preparation of chemical, incendiary and bacterial warfare 142
	Supervision of the prohibition to make preparations for chemical, incendiary and bacterial warfare. 143
	Case of a breach of the prohibition to use chemical, incendiary and bacterial weapons against an opponent 144
VIII.	Summary and comment 144

Chapter 5. November 1932–1933 147

    I. The second report of the Special Committee on CBW 147
        Prohibition of chemical, incendiary and bacterial warfare 147

        Defensive material — Offensive material — Special case of lachrymatory substances

        Enforcement of the prohibition against the preparation of chemical warfare 155
        Cases of infringement of the prohibition to employ against an adversary chemical, incendiary and bacterial weapons 159

        Establishment of the facts — Body by which the facts of infringement are to be established — Procedure for establishing the facts of infringement — Penalties

    II. Consideration of the second report of the Special Committee by the Bureau of the Disarmament Conference 163
        Prohibition of preparations for chemical and bacterial warfare 164
        Penalties for the use of chemical or bacterial weapons 165
    III. The British draft disarmament convention 167
        Section I. Prohibition of chemical, incendiary or bacterial warfare . . 167
        Section II. Prohibition of preparations for chemical, incendiary and bacterial warfare 168
        Section III. Supervision of the observance of the prohibition of preparations for chemical, incendiary or bacterial warfare 169
        Section IV. Establishment of the fact of the use of chemical, incendiary or bacterial weapons 169

Contents

   IV. Amendments to the draft convention 170
    V. Postponement of the Disarmament Conference 172
   VI. Summary and comment 172

Chapter 6. 1935–1938 175
     I. The use of gas in Ethiopia 175
    II. Sanctions against Italy 186
   III. The use of gas in China 189
   IV. Summary and comment 191

Chapter 7. 1946–1953 193
     I. Chemical and bacteriological weapons defined as weapons of mass destruction 193
    II. Allegations of the use of bacteriological and chemical weapons in Korea and China 196
   III. Discussion in the UN Disarmament Commission of bacteriological and chemical warfare in the light of Korean and Chinese allegations 201
   IV. Discussion in the UN Security Council of bacteriological and chemical warfare in the light of Korean and Chinese allegations 207
    V. Discussion in the UN General Assembly of bacteriological and chemical warfare in the light of Korean and Chinese allegations . . 214
   VI. Summary and comment 221

Chapter 8. 1954–1967 224
     I. CBW prohibition in Germany and Austria 224
    II. CBW prohibition in the proposals for the reduction and limitation of armaments 226
   III. CBW prohibition in the programmes for general and complete disarmament 228
   IV. Allegations of use of chemical weapons in Indo-China 234
    V. UN appeal for compliance with the 1925 Geneva Protocol 238
   VI. Allegations of use of chemical weapons in the Yemen 243
  VII. Suggestion for a revision of the Geneva Protocol 247
 VIII. Summary and comment 250

Chapter 9. 1968–1969 253
     I. Proposal for separate consideration of biological methods of warfare 253
        UK working paper on microbiological warfare 254
        Discussion of the UK working paper 257
    II. United Nations decision to study the effects of CBW 260
        Secretary-General's suggestions 260
        Consideration of the proposal for a CBW study 261
   III. Report of the UN Secretary-General on CB weapons 266
        Introduction 267

Chemical and biological warfare agents. 267
　　　Production of weapons 270
　　　Conclusion 271
　　　The Secretary-General's proposals 273
　　　Consideration of the Secretary-General's report 273
　IV. WHO report on CBW hazards to health 274
　　　Qualitative considerations 274
　　　Quantitative estimates 275
　V. Summary and comment 276

Part II. Developments in 1969–1970

Unilateral renunciations of CBW 280

Prohibition of use of CB weapons 283

　　　Adherence to the Geneva Protocol 283
　　　Tear gas and herbicides 285

Prohibition of production and possession of CB weapons 290

　　　Separate or joint treatment of chemical and biological weapons . . 290
　　　Draft convention on BW 295
　　　Draft convention on CBW 298
　　　Verification 301

Summary and comment 307

Postscript 312

　　　Non-aligned states' proposals for CB disarmament 312
　　　Socialist countries' draft convention on BW 314
　　　Ratification of the Geneva Protocol 320

Appendix 1. Revised UK draft convention for the prohibition of biological methods of warfare and accompanying draft Security Council resolution, of 18 August 1970 322

Appendix 2. Revised draft convention on the prohibition of the development, production and stockpiling of chemical and bacteriological (biological) weapons and on the destruction of such weapons, submitted by Bulgaria, Byelorussian SSR, Czechoslovakia, Hungary, Mongolia, Poland, Romania, Ukrainian SSR and the USSR on 23 October 1970 326

Appendix 3. Draft convention on the prohibition of the development, production and stockpiling of bacteriological (biological) weapons and toxins and on their destruction, submitted by Bulgaria, Byelorussian SSR, Czechoslovakia, Hungary, Mongolia, Poland, Romania, Ukrainian SSR, and the USSR on 15 April 1971 331

Appendix 4. Revised draft convention on the prohibition of the development, production and stockpiling of bacteriological (biological) and toxin weapons and on their destruction, submitted by Bulgaria, Byelorussian SSR, Czechoslovakia, Hungary, Mongolia, Poland, Romania, Ukrainian SSR, and the USSR on 5 August 1971 336

Appendix 5. List of states which have signed, ratified, acceded or succeeded to the 1925 Geneva Protocol 341

References 348

Guide to CB disarmament proposals in the League of Nations and the United Nations, 1920–71 361

Addendum. Draft convention on the prohibition of the development, production and stockpiling of bacteriological (biological) and toxin weapons and on their destruction, submitted by Bulgaria, Canada, Czechoslovakia, Hungary, Italy, Mongolia, Netherlands, Poland, Romania, USSR, UK, and USA on 28 Septembr 1971 404

Country index 410

Introductory Survey

The purpose of this volume is to present a comprehensive account of the international negotiations since the 1920s to abolish chemical and biological weapons, in the hope that it may contribute to present efforts to negotiate a treaty prohibiting the production and possession of chemical and biological weapons.

The study is based on original documents, mostly from the debates in the League of Nations, the United Nations and the Geneva disarmament conference.

Part I deals chronologically with the CB disarmament[1] negotiations in the two world organizations from 1920 to 1969, while Part II gives an analytical report of the developments after 1969. Each chapter is followed by a short summary and comment.

The aim of this introductory survey is to facilitate the study of the volume by providing a perspective on the CB disarmament debate and by bringing out the main conclusions that are relevant today.

I. *The historical background*

Ancient custom condemns the use of poison or poisoned weapons in war or the use of weapons causing unnecessary suffering. During the second half of the nineteenth century this custom was codified in a number of international conventions. Thus, the Brussels Convention of 1874 on the laws and customs of war prohibited (*a*) the employment of poison or poisoned weapons, and (*b*) the employment of arms, projectiles or material calculated to cause unnecessary suffering. At about the same time the military development of chemical weapons on a significant scale began to appear as a practical reality. An attempt to forestall this development was made at the first international peace conference in the Hague in 1899, where an agreement was signed "to abstain from the use of projectiles, the object of which is the diffusion of asphyxiating or deleterious gases".

---

[1] CB disarmament means chemical and biological disarmament. CBW means chemical and biological warfare.

The old rules against the use of poison and the 1899 Hague prohibition of gases, confirmed at the 1907 Hague Conference, did not, however, succeed in preventing gas warfare during World War I. From mid-1917 onwards, especially, gas played a major role in the hostilities in the European theatres.

In the minds of the soldiers and in public opinion in general, gas appeared as a particularly repulsive and treacherous weapon, arousing feelings associated with poison. The large-scale use of gas during the war was a traumatic experience and, together with other factors, reinforced popular demands for a ban on such methods of warfare. On the other hand, chemical weapons had for the first time proved their military value in a major war; this would make it more difficult to reach an international agreement to give up their use.

The Versailles Treaty with Germany in 1919 included a special article which not only repeated the legal ban on the use of gases but specifically prohibited their manufacture in and importation to Germany. Besides the motives arising from general abhorrence of gas warfare, the efforts of the American and British chemical industries to get at the German chemical industry, the most advanced of its kind in the world, played a major role in having this article inserted in the treaty.

II. *The inter-war period: the course of events*

After the war, great hopes were set on disarmament and international cooperation. The League of Nations was established and its Covenant included an obligation for the Council to draw up an international disarmament plan. Although no such plan was ever drawn up by the Council, several commissions were set up to deal with disarmament. In fact, the first ten years after World War I witnessed a general reduction in armaments but this was carried out on national initiatives. With two exceptions—the enforced disarmament of Germany and the naval limitations agreed in Washington in 1922 and in London in 1930—no international disarmament agreements were reached during the period.

Although the prospects for disarmament were more propititious during the first half of the inter-war period, from 1920 to 1933, than during the second half, the difficulties were still very great. Germany and the Soviet Union were outcasts in the international community while the United States had retreated into isolationism. Furthermore, there was a deep split in the attitude of the leading Western powers, Britain and France, to disarmament. Britain and its Scandinavian supporters took the view that armaments were the cause of war; they therefore insisted, in principle, on proportional arms reductions. France and its Central European allies saw armaments as the result of insecurity; they therefore wanted guarantees for their national security before agreeing to arms reductions. Britain and the United States were, however, unwilling to give such security guarantees.

The problem of chemical and biological weapons was brought up at the disarmament conferences during the inter-war period. Interspersed with the conferences were several expert studies of CB weapons carried out by various bodies of the League. The main aspects of the proceedings are described in chapters 1 to 5. In brief, these were the important events concerning CB weapons:

1922. In the course of the Naval Conference in Washington a treaty relating to the use of submarines and noxious gases in warfare was signed in February 1922. By the operative clause of that treaty the parties (the United States, Britain, France, Italy and Japan) declared that the prohibition on "the use in war of asphyxiating, poisonous or other gases and all analogous liquids, materials or devices" was part of international law, and agreed to be bound thereby as between themselves. Although the treaty was ratified by the United States, among others, it did not enter into force because France opposed its provisions on submarines and therefore did not ratify it.

1924. The League of Nations had asked a number of experts to consider the effects of CB weapons. The aim was to increase public awareness of the possible dangers of CBW. The report appeared in 1924 and made the following main points: the use of poisonous gases marked the appearance of a terrible weapon; chemical weapons gave an immense superiority to any power with hostile intentions; the possibilities of camouflaging chemical preparedness were very great; and biological weapons were not particularly formidable.

1925. On the initiative of the League of Nations, an international conference to consider the supervision of the trade in arms was convened in Geneva. The conference did not achieve its main purposes but produced as a side result the Protocol for the Prohibition of the Use of Asphyxiating, Poisonous or Other Gases, and of Bacteriological Methods of Warfare. The circumstances were as follows. At the beginning of the conference the US delegation produced a draft proposal for prohibiting the export of chemical weapons. This proposal met with opposition because it was felt to be ineffective in stopping the use of chemical weapons and to be discriminatory against states which did not produce chemical weapons. The United States then suggested the adoption of a special document based on the provisions on chemical warfare contained in the Washington Treaty of 1922. After

Poland had suggested the inclusion of a specific ban on the use of bacteriological[2] methods of warfare, this became the adopted procedure and the Geneva Protocol was signed on 17 June 1925 (see page 341 for text and signatories). By June 1971 there were eighty-seven parties to the Protocol. The United States signed the Protocol in 1925 but did not ratify it.

1926–30. During these years the Preparatory Commission for the Disarmament Conference held numerous sessions. The Commission had been set up by the Council of the League in 1925 to prepare the international Conference for the Reduction and Limitation of Armaments (Disarmament Conference) which was finally convened in 1932. The Preparatory Commission arranged for several expert studies in the field of disarmament including CB disarmament. The extension of the "no first use" prohibition in the Geneva Protocol to include a prohibition on production was discussed at great length, but no agreement could be reached. Nor was there agreement on the right of reprisals or collective sanctions in the case of a violation of the Geneva Protocol.

1932–33. Of the sixty-four states of the world at that time, all except four Latin American republics participated in the Disarmament Conference in 1932–33. The conference failed for several reasons. France, the United Kingdom and the United States could not agree on collective security guarantees; Germany was no longer prepared to accept unilaterally the arms limitations imposed on it by the Versailles Treaty; Japan was not willing to agree to any restrictions in the field of armaments, being already bent on military expansion in East Asia; the Soviet Union had unsuccessfully advocated general and complete disarmament for a long time.

However, with regard to the technical aspects of CB disarmament some progress was made. A special committee of the Disarmament Conference carried out a thorough investigation of CB weapons and suggested a few positive measures for limiting the development of CB weapons. Politically more important was the acceptance of the principle of qualitative disarmament, meaning that the most offensive weapons should be abolished. In June 1932, President Hoover proposed a plan for abolishing offensive weapons, including chemical weapons. The British draft disarmament convention submitted in February 1933 contained elaborate provisions for an extensive prohibition of chemical and biological weapons. It included an absolute prohibition of the use of biological weapons even for retaliation; a ban on preparations for CBW in time of peace as well as in time of war; and a complaints procedure for investigating breaches of the prohibition.

---

[2] The terms "biological", "bacteriological" and "bacterial" (warfare, weapons, etc.) are used interchangeably in this volume as they were at the period.

While the British disarmament proposals and a less negative US attitude to collective security might have influenced the outcome of the Disarmament Conference in a positive direction a year earlier, it was too late in 1933. Germany was changing regime, and in October 1933 left the League. Japan's policy had become more and more intransigent, and it left even earlier, in March 1933. Suspicions and conflicts between the major powers were growing, and the disarmament efforts soon changed into an armaments race.

1935–36. The Italians employed chemical weapons during their invasion of Ethiopia. This was the first open breach of the Geneva Protocol. The use of chemical weapons may have shortened the war in Italy's favour. The sanctions imposed by the League of Nations against Italy had no appreciable effect.

1937. Japan began using chemical weapons in its war with China which was to continue until 1945. This instance of chemical warfare received much less international attention than the Italo-Ethiopian War, possibly because it was overshadowed by events then taking place in Europe.

1939–45. Chemical and biological weapons were not used in World War II. Several reasons have been given for this: deterrence—the two sides warned each other not to use chemical weapons at the risk of strong retaliatory action in kind; a general feeling of abhorrence on the part of governments for the use of CB weapons, reinforced by the pressure of public opinion and the constraining influence of the Geneva Protocol; and actual unpreparedness within the military forces for the use of these weapons.

III. *The inter-war period: the main points of the CBW debate*

Definition of CB weapons

During the discussions in this period there was an interesting development in the definition of CB weapons, mainly in the direction of securing the widest possible scope of agents covered by the prohibition.

The early, pre-World War I, conventions used terms such as "employment of poison or poisoned weapons" (Brussels Declaration of 1874), and "diffusion of asphyxiating or deleterious gases" (First Hague Declaration of 1899). At the same time the conventions repeated the established ban against "arms, projectiles or material calculated to cause unnecessary suffering". There is still a legal dispute about the scope of the prohibition aimed at by these two conventions, taken together; that is, whether the prohibition covered all existing chemical weapons, the development of new ones or perhaps even the development of biological weapons. This issue is discussed in detail in Volume III, *CBW and International Law.*

After the war the terminology used in the Versailles Treaty described the prohibited agents as "asphyxiating, poisonous or other gases and all analogous liquids, materials or devices". The same terminology was used in the Washington Treaty of 1922 and in the preamble to the Geneva Protocol of 1925.

Thus the term "chemical weapons" was not used in the early conventions or in the treaties concluded after the war. However, during the CBW debates in the early twenties, the classification into chemical and bacteriological weapons evolved. The terms appeared, for instance, in the 1924 expert study commissioned by the League of Nations. The obvious reason for the specific attention given at that time to bacteriological weapons was that BW for the first time appeared as a distinct, though still improbable, future possibility.

During the drafting of the Geneva Protocol at the Arms Trade Conference in 1925, Poland proposed that the prohibition on the use in war of the forbidden agents should specifically include bacteriological weapons. There was no opposition to the idea as such, but it is interesting to note that at least one important power, France, considered that the general prohibition in the Protocol on "the use in war of asphyxiating, poisonous or other gases, and of all analogous liquids, materials or devices" should have been sufficient to cover bacteriological warfare. France made no objection, however, to an explicit reference to BW which was then included in the final text of the Protocol: "the High Contracting Parties ... agree to extend this prohibition to the use of bacteriological methods of warfare".

During the debates on CBW in the second half of the 1920s another attempt was made to enlarge the scope of the prohibition by including incendiary weapons among the prohibited weapons. This was first proposed by the Kingdom of the Serbs, Croats and Slovenes in 1926 in the Preparatory Commission for the Disarmament Conference. The 1932 Conference in fact jointly considered chemical, bacteriological and incendiary weapons, and the 1933 British draft disarmament convention included special provisions with regard to incendiary weapons. Since the Disarmament Conference failed, nothing came of those efforts to extend the prohibition to incendiary weapons.

The tear-gas issue

Another question, related to the issue of the scope of the prohibition, discussed above, is whether lachrymatory agents (tear gases) and other irritant agents are included in the Geneva Protocol prohibition. This first became a controversial issue in 1930 when, in order to secure a uniform interpretation of the Geneva Protocol and to settle the doubts which had arisen on this point, the British delegation to the Preparatory Commission submitted a memorandum stating that in the British view the use in war of lachrymatory gases was prohibited by the formula "other gases" contained in the Protocol.

The French delegation supported the British in this interpretation. The only dissenting note came from the US representative who defended the use of tear gases in war on the grounds (*a*) that these agents had a humane purpose in war, and (*b*) that it would be wrong to prohibit the use in war, against an enemy, of the same agents that were used in peacetime to control civilian populations. The USA did not therefore, in 1930, agree to an interpretation according to which tear gases were prohibited under the Geneva Protocol. However, at a meeting of the Bureau of the Disarmament Conference in November 1932 the US representative took a different position, saying that there was no question of tear gas being used in war, but that the USA did not want to give up the preparation and employment of this gas for domestic police purposes. The main US concern at that time seems indeed to have been to avoid any restrictions on the peacetime use of tear gas in the United States. There is an even earlier indication, from 1922, of the USA being prepared to refrain from using tear gas in war: a military advisory committee to the US delegation to the Washington Conference proposed that gas warfare, including tear gases, should be prohibited in every form.

Prevention of CB warfare

The technical situation with regard to CB weapons was very different in the inter-war period compared with the post-World War II period. First, only chemical weapons had been developed and put to military use; biological weapons were at a much earlier stage of development and bacteriological warfare was not accorded much weight in the debate. Secondly, many of the existing chemical weapons were based on dual-purpose chemicals which had both civilian and military uses. For this reason, most experts found it pointless to prohibit the preparation in peacetime of chemical agents for use in war: it would be too easy to produce chemical weapons from civilian chemical factories. Verification of nonproduction of chemical agents for use in war was also considered largely impossible. The doubt about the possibility of stopping peacetime preparations for chemical warfare was reinforced by a belief that civilian aircraft, for instance, could easily be adapted to chemical weapons delivery.

This state of affairs led nations to rely for their security on deterrence—the threat of retaliation in kind—and sanctions. The policy of retaliation had a long historical background, culminating for many countries in the experience of World War I and its aftermath. This method of seeking security through reliance on the threat of reprisals, or retaliation, also came to permeate the Geneva Protocol as the latter developed into an agreement of "no first use", partly because of the many reservations attached to it by the parties. The reservations, which many countries made when ratifying the Protocol, usually contained the following two limitations: (*a*) the ratifying country declared that it only considered itself bound by the provisions of the Protocol with regard to other parties to the Protocol; (*b*) it declared that it would not be bound by the Protocol towards any enemy whose armed forces, or the armed forces of whose allies, failed to respect the Protocol. The reservations made by the United Kingdom, France and the Soviet Union may be cited as examples. The problem of evaluating these reservations at the present time in the light of the development of international law is discussed in Volume III.

During the CB disarmament debates which took place after the signing of the Geneva Protocol in 1925 and in preparation of the Disarmament Conference, the question of reprisals recurred frequently. In the discussion of the various proposals for going beyond the Protocol, several nations, particularly those belonging to the Little Entente, spoke in favour of granting a party which had been attacked with CB weapons a right to retaliate with the same weapons. France preferred collective reprisals in kind against an aggressor. Gradually, however, an unconditional ban on the use of biological weapons came to be favoured. This was reflected in the proposals submitted to the Disarmament Conference, that is, the right of retaliation with respect to chemical weapons but an unconditional ban on the use of biological weapons. However, at the end of this period of disarmament debates there was a minority opinion, represented by Germany, among others, in favour of not permitting any reprisals in kind—whether by the party attacked or collectively—but to rely on other collective retaliation measures. This problem was never solved at the Disarmament Conference.

The other big issue in the CBW debate was how to prevent peacetime preparations for chemical and biological warfare. As already stated, most experts were rather pessimistic as to the possibility of obtaining an effective ban on preparations, meaning research and development, production, and stockpiling. There were not only problems with regard to the civilian sector. Practically all nations insisted on the right to carry out defensive preparations in order to protect their armed forces and populations against the effects of a CBW attack. There were doubts whether a clear dividing line could be drawn between offensive and defensive preparations. Further, the majority view that reprisals in kind were to be allowed implied that some production of at least chemical agents had to be permitted. All this created seemingly insuperable obstacles to securing a ban on preparations for CBW. Even allowing for the technical problems to be solved, which they were not, there was the wider political issue of whether the major nations at the time would have been prepared to give up their CBW capabilities once and for all.

Still, the last concrete proposal for international CB disarmament in the inter-war period, the CBW chapter of the British draft convention of 1933, stated that "all preparations for chemical, incendiary or bacterial warfare shall be prohibited in time of peace as in time of war". The general prohibition covered the manufacture, import, export and possession of appliances or substances, as well as the training of armed forces in the use of the banned weapons.

Verification proposals

On a technical level, useful work was done during the inter-war period in the evaluation of different schemes for verifying prohibitions—on use, production and possession—of CB weapons. Some of them have reappeared in the recent CBW debate and may therefore still be of interest.

In relation to the prohibition of the use in war of CB weapons a rather elaborate investigating procedure was suggested at the Disarmament Conference. Under the scheme developed in the British draft convention, the Permanent Disamament Commission (PDC), which was to be set up, was responsible for establishing whether the prohibition of the use of these weapons in war had been violated. The parties might submit complaints to the PDC, which was assisted in the examination of the allegations by a special commission of investigation and which might also send commissioners to the territory in question.

In relation to the prohibition of preparations for CBW or incendiary warfare, on the other hand, the only international enforcement measures proposed in the British draft convention were that the PDC should examine the complaints put forward by any party to the treaty which alleged that the prohibition had been violated.

Another proposal, discussed at the Disarmament Conference, provided for an International Information and Documentation Service for Protection against Chemical Weapons, to be established in conjunction with a disarmament convention prohibiting preparations for CB warfare. It would have the task of collecting all the information that could be found in official trade statistics or private statistics relating to the manufacture of chemical substances in the territory of each signatory state and their import into that territory. In doubtful cases the Service might, on its own initiative or at the request of other states, ask the governments of the countries concerned to supply additional information and/or explanations. The Service would report to the PDC, which might take further measures.

At the time, however, it was not considered possible to enforce a prohibition against preparations for chemical warfare by examining the commercial statistics of the activities of chemical industries in all countries. The expert committee that studied the matter concluded that it was conceivable in theory but impossible in practice to exercise control by entrusting national or international bodies with the duty of inspecting chemical factories and of making public the character of the products manufactured, the existing stocks and the production capacity.

A proposal to establish an international cartel of chemical industries to ensure that private manufacture should not be employed for preparations for chemical and biological warfare was made during the debates in the Preparatory Commission and later brought up at the Disarmament Conference. It met objections similar to those against the proposal for the International Information and Documentation Service. There were, however, indications that the business community believed more in this idea than did the politicians.

IV. *The situation after World War II*

Chemical and biological weapons appeared in a greatly different technical and political setting after World War II compared with the inter-war period. The change was mainly due to the following factors:

The advent of nuclear weapons, the destructive power of which overshadowed everything else. The rise of nuclear weapons was probably the main reason for the relatively low attention given to CB weapons during the first years of the post-war period.

The rapid development of nerve gas weapons during the 1940s and 1950s. With the discovery of the V-agents, weapons were created which were many times more deadly than earlier chemical weapons.

The developments in microbiology making biological warfare appear, particularly during the 1960s, a much more threatening possibility than it had seemed before the war.

The United Nations first became involved with the question of CBW in 1947 during discussions on how to define mass destruction weapons. On the initiative of the United States, weapons of mass destruction were then defined to include, *inter alia,* "lethal chemical and biological weapons". The actual resolution was worked out in the Commission for Conventional Armaments and passed by the Security Council in August 1948.

In 1948 the Secretary-General made an appeal for an expert study of the problems involved in the control of CB weapons, but the proposal was not followed up. A few years later, in 1952, there was a debate in the United Nations on the allegations that the United States was practising biological and chemical warfare in China and Korea. The allegations and debate are summarized in Chapter 7 of this volume and in Appendix 4 of Volume V on *The Prevention of CBW*.

Generally, however, there was not much interest in CB weapons in the disarmament community during the 1940s and 1950s. From 1959 several plans for general and complete disarmament (GCD) were submitted by the great powers. All these plans included chemical and biological weapons, usually to be abolished at the second stage of the plan.

The one special case of CB disarmament during the period concerned Germany. All armaments production in Germany was forbidden in 1945. In 1954, when the Western Allies permitted limited arms production in West Germany in connection with the forming of the Western European Union, special restrictions were introduced for CB weapons (in addition to atomic weapons). The production of such weapons was entirely prohibited in West Germany, and a control machinery, still in force, was established to enforce the prohibition. The experience of this case is evaluated in Volume V, Appendix 3.

While the Geneva Protocol had been ratified or acceded to by over forty countries before the war, there were in the early post-war years still no prospects of a ratification by the United States, the most important non-party to the Protocol. In 1947 the Protocol was even removed from the Senate calendar, where it had been since before the war. In 1952 the United States openly declared that it regarded the Protocol as obsolete. The Soviet Union and its allies, on the other hand, stood forward during the period as firm supporters of the Geneva Protocol, a position they had taken since the 1920s.

V. *The recent debate*

During the second half of the 1960s the situation changed, and there was renewed interest in CB disarmament which was even much stronger than in the 1924–1933 period. Several factors contributed to this:

The use of chemical weapons (irritant agents and anti-plant agents) in Viet-Nam, and the general abhorrence of that war.

The concern expressed by scientists about military misuse of scientific developments, in particular the advances in microbiology.

The search for partial disarmament measures as a way of relieving international tension.

In 1966 Hungary raised the CBW issue in the United Nations, mainly with reference to the US use of chemical warfare agents in Viet-Nam. Supported by the other Socialist countries, Hungary requested a resolution which would demand strict compliance by all states with the Geneva Protocol, condemn any use of CB weapons and declare that the use of those weapons for the purpose of destroying human beings and the means of their existence constituted an international crime. The United States strongly opposed this proposal and, together with other Western powers, secured unanimous agreement on a revised draft resolution which called for strict observance by all states of the principles and objectives of the Geneva Protocol and invited all states to accede to the Protocol. The USA maintained, however, that the Protocol did not apply to tear gases and herbicides. The resolution was adopted by the UN General Assembly in December 1966.

In 1967 Malta raised the CBW issue in the United Nations, saying that the Geneva Protocol was obsolete, and asked for an extensive revision of the Protocol. Many countries, including the USSR and its allies as well as Sweden, were critical of the Maltese move, because they thought it might lead to a weakening of the existing legal prohibition of CBW.

In July 1968, at the Geneva Disarmament Conference, the United Kingdom suggested a treaty prohibiting the production and possession of biological warfare agents. This was the first time a split in the treatment of chemical and biological agents for disarmament purposes had ever been suggested. The proposal acutely reflected the fact that at least some of the major powers did not think the time was ripe for prohibiting the production of chemical weapons.

The Socialist countries and a number of non-aligned countries were very critical of the British proposal which eventually was supported by the United States. The Soviet Union and its allies countered the British proposal, which had been presented in the form of a draft treaty in the spring of 1969, by proposing, in September 1969, a treaty prohibiting the development, production and possession of both chemical and biological weapons. The two draft treaties, which differed considerably not only with regard to their scope but also with regard to the way in which the prohibition should apply to agents, appliances and weapons and with regard to verification machinery, are analysed in Part II of this volume.

The UN Secretary-General's report on the effects of CB weapons was released in July 1969. It stated that if CB weapons were ever to be used on a large scale in war, no one could predict how enduring the effects would be and how they would affect the structure of society and human environment; the dangers would apply as much to the aggressor country as to the country attacked and protective measures would not be of much use; the risk of proliferation of CB weapons was considerable. In the foreword to the report, the Secretary-General urged that all countries should accede to the Geneva Protocol and confirm that the Protocol applied to the use in war of all existing and possible future CBW agents, including harassing agents; further, an agreement should be reached to stop the development, production and stockpiling of all CBW agents and to eliminate existing stocks.

In addition to the UN report, a report on the health effects of CB weapons was prepared by a group of World Health Organization consultants.

The controversy over whether tear gases, other irritant agents and anti-plant agents were covered by the Geneva Protocol flared up again in the United Nations in the autumn of 1969. On the initiative of Sweden a draft resolution was submitted according to which the Assembly would declare that the Geneva Protocol embodied the generally recognized rules of international law prohibiting the use in international armed conflicts of all biological and chemical methods of warfare; further it would be declared contrary to those rules to use any chemical agents of warfare which had direct toxic effects on man, animals or plants, or any biological agents of warfare which were intended to cause disease or death in man, animals or plants and which depended for their effects on their ability to multiply in the person, animal or plant attacked.

The United States strongly opposed the draft resolution and again asserted that tear gases, other irritant agents and anti-plant agents were not covered by the Geneva Protocol. In the course of the debate the participants used many arguments from the tear-gas debate in the early 1930s. Finally, however, the resolution was adopted with 80 votes in favour, 3 against (the United States, Australia, Portugal) and 36 abstentions. A few months after the tear-gas debate in the United Nations, the British Government took the view that the use of CS (a tear gas used in Viet-Nam) and other such gases not significantly harmful to man in other than wholly exceptional circumstances was not prohibited under the Geneva Protocol. The British statement, made in February 1970, represented a change in the position which the United Kingdom had upheld since 1930.

The United States Government announced a new CBW policy in a statement by President Nixon in November 1969. The three important decisions were: (*a*) unilateral renunciation of biological warfare including disposal of existing stocks of biological weapons; in February 1970 it was made clear that the renunciation also embraced the so-called toxins; (*b*) renunciation of first use of both lethal and incapacitating chemical weapons; and (*c*) submission of the Geneva Protocol to the Senate for ratification.

During 1970 and the first months of 1971 the CB disarmament debate was intensive but no rapprochement could be reached between the position of the USA and the UK in favour of biological disarmament only, and the position of the Socialist countries, supported by the non-aligned nations, in favour of joint biological and chemical disarmament. Various compromise schemes were suggested, such as a Moroccan proposal for a treaty which would commit the parties to both biological and chemical disarmament, the biological disarmament to become effective at once and the chemical disarmament to become effective within a specified period, during which the parties would be committed to negotiate provisions for verification.

During these years the problem of verifying a prohibition of production and possession of chemical and biological weapons has occupied an increasingly large place in the debate. This is a feature common to all the post-war disarmament debates. The usual pattern can also be observed: the United States and its allies insist on effective international verification measures and the Soviet Union and its allies insist on national means of verification. But this has not always been so with respect to CB disarmament. In 1928 the Soviet Union, in its draft Convention on the Reduction of Armaments, proposed the establishment of a Permanent International Control Commission which, *inter alia,* would supervise the prohibition of preparations for chemical warfare. Four years later, during the CBW debates at the Disarmament Conference, the United States took the view that verification of the prohibition on chemical warfare preparations could be exercised only by governments and not by an international body.

Of the verification ideas now being discussed, one of the more useful concepts seems to be that of verification by challenge and/or invitation. The idea is that the burden of proof is put on the country against which allegations are made. The country concerned can then free itself of the allegations by allowing an international inspection on its territory. This idea first came up in the debate over a nuclear test ban in the 1960s, but would seem to apply to the CBW field as well.

Many of the verification measures which have recently been proposed are similar to those discussed forty years ago. They include:

A complaints procedure whereby complaints could be lodged with the UN Secretary-General and the Security Council (the British draft) or the Security Council only (the Socialist countries' draft). A scheme for verification by challenge and/or invitation would be part of this procedure.

The establishment of an international centre for documentation and information on CBW.

A procedure for securing an open exchange of information among scientists.

The establishment of special governmental agencies to keep check on the manufacture, export and import of agents that could be used for the production of CB weapons.

A detailed examination of the flow of relevant material in each state in order to detect conversion of agents to warfare purposes; the checking would be carried out through *ad hoc* investigations by experts provided by the United Nations.

With respect to verification the main issue is that the United States and the United Kingdom consider that verification of non-production of chemical warfare agents, while absolutely essential in their view, is impossible using any method so far suggested. The most common reason put forward is the close dependence of military chemical production on the civilian industry. The Soviet Union and its allies contend that national means of verification, together with the possibility of making complaints to the Security Council, are sufficient for supervising a production ban on chemical agents.

At the end of March 1971, the contents of the CB disarmament debate changed when the Soviet Union and its allies came over to the view of the United States and the United Kingdom and proposed a treaty for biological disarmament only.[3] Despite the dismay of the non-aligned countries, such a treaty, albeit with a provision for further negotiations concerning chemical disarmament, seems in the summer of 1971 to be the likely outcome of the CB disarmament negotiations. The points at issue and the general prospects for CB disarmament are discussed at greater length in Volume V on *The Prevention of CBW*.

---

[3] The proposal is described in the postscript to this volume.

Part I. Negotiations at the League of Nations and the United Nations, 1920–1969

Introductory note on the disarmament negotiating machinery

I. *The League of Nations*

The League of Nations was created after World War I, with the stated objective of promoting international cooperation and achieving international peace and security.

The Covenant

Under the Covenant of the League of Nations (Article 8) the members of the League recognized that "the maintenance of peace requires the reduction of national armaments to the lowest point consistent with national safety and the enforcement by common action of international obligations". They undertook to interchange "full and frank information as to the scale of their armaments, their military, naval and air programmes and the condition of such of their industries as are adaptable to warlike purposes".

The Council

According to the provisions of the Covenant, the Council of the League was to be composed of the five great powers (UK, USA, France, Italy and Japan) as permanent members, and four others elected periodically by the Assembly. Since the USA did not accede to the League, the number of permanent members was four until September 1926, when Germany was appointed a permanent member. That number reached six in 1934, when the USSR joined the League. Japan and Germany having withdrawn from the League during 1935, there were again only four permanent members.

The number of non-permanent members also varied; it was raised successively to eleven.

The Council had the duty of formulating plans for the reduction of armaments. It was also to advise how "the evil effects" attendant upon the manufacture by private enterprise of munitions and implements of war can be prevented, "due regard being had to the necessities of those Members of the League which are not able to manufacture the munitions and implements of war necessary for their safety".

There was no clear division of power between the Council and the Assembly.

In many cases the former acted on suggestions originating in other organs of the League.

The Assembly

The Assembly, consisting of representatives of the members of the League, had the character of a general conference of states. It was empowered "to deal at its meetings with any matter within the sphere of action of the League or affecting the peace of the world". It initiated projects and formulated general policies in the field of disarmament. The principal instrument of the Assembly in this work was a committee on the reduction of armaments (Third Committee), to which each delegation could nominate a delegate and technical advisers.

The Permanent Advisory Commission

In 1920, in accordance with Article 9 of the Covenant, the Council of the League established a Permanent Advisory Commission for Military, Naval and Air Questions, composed of experts from the general staffs of the states represented on the Council. The Commission was instructed to consider plans for the reduction of armaments. Subsequently its functions became limited to giving technical advice to the various bodies set up *ad hoc* at different times to formulate disarmament proposals.

The Temporary Mixed Commission

In 1921, pursuant to an Assembly resolution, the Council set up a new group to deal with disarmament questions—a Temporary Mixed Commission. It consisted of six persons of recognized competence in political, social and economic matters, six members of the Permanent Advisory Commission, four members of the Provisional Economic and Financial Committee of the League and six members of the governing body of the International Labour Organization. It was thus predominantly a civilian body and remained in being only until 1924.

The Preparatory Commission for the Disarmament Conference

In 1925, the Council, upon the request of the Assembly, constituted a Preparatory Commission for the Disarmament Conference.

The Preparatory Commission consisted solely of government representatives—members of the Council and countries not represented on the latter but deemed to occupy a special position in relation to disarmament. The basis of the composition of the Preparatory Commission was later modified so as to permit retiring members of the Council to keep their seats on the Commission. The USSR and the USA, non-members of the League, were invited to send representatives to the Commission. The membership thus varied from twenty-one to thirty-two. A Joint Commission of members of the various technical organizations of the League and the Labour Office was set up by the Council to assist the Preparatory Commission.

The Preparatory Commission created two Sub-Commissions. Sub-Commission A was composed of a military expert, a naval expert and an air expert for each of the countries represented on the Commission; its terms of reference included questions of a technical nature relating to disarmament. Sub-Commission B consisted of a representative of each delegation; its task was to deal with all non-military questions related mainly to the economic and financial aspects of disarmament.

Committees were also created to deal with special technical problems. The Preparatory Commission held six sessions and dissolved in 1930, after preparing a draft convention on the reduction and limitation of armaments and a final report.

The non-membership of certain important states in the League made it necessary to conduct some of the work in the field of disarmament at specially organized conferences.

The Conference for the Supervision of the International Trade in Arms and Ammunition and in Implements of War

The Conference was convened on 4 May 1925. It set up a General Committee which was entrusted with the first reading of the draft convention on the regulation of trade in arms; a Military, Naval and Air Technical Committee for the examination of questions relating to armaments; a Legal Committee; committees dealing with export licenses, statistics and transit as well as with zones subjected to a special régime of supervision.

The Bureau of the Conference was composed of the president and the vice-president of the Conference, the chairmen of the committees and representatives of eight states including the USA.

The Conference ended on 17 June 1925 with the signing of a convention for the supervision of the international trade in arms and ammunition and in implements of war, which never entered into force. Another document signed on the same day—the Protocol for the Prohibition of the Use in War of Asphyxiating, Poisonous or Other Gases, and of Bacteriological Methods of Warfare—was ratified by a number of countries and remains in force.

The Conference for the Reduction and Limitation of Armaments (Disarmament Conference)

The Conference met on 2 February 1932. It was attended by sixty states, including eight states not members of the League. Actually, sixty-four states were invited by the Council in 1931. Iraq, after becoming a member of the League, was invited in November 1932. The four states which had been invited but which were never present at the conference were: Ecuador, Nicaragua, Paraguay and Salvador. (For the list of states represented at the conference see page 107.)

The Conference was not *stricto sensu* a League conference. All technical facilities were provided by the League, but the delegations were accredited to the Conference.

Two principal organs were set up: the General Commission and the Bureau. The General Commission was composed of representatives of all delegations, a chairman, a vice-chairman, and a rapporteur. The Bureau consisted of the president of the Conference, the honorary president, the fourteen vice-presidents of the Conference, the vice-chairman of the General Commission and the presidents of the four Commissions on Land Armaments, Naval Armaments, Air Armaments and National Defence Expenditure, respectively.

While the General Commission was responsible for major decisions of principle, the Bureau was charged with the preparation of proposals for submission to the Conference and coordinated the work of the various organs of the Conference.

Altogether, apart from the General Commission and the Bureau, there were about forty commissions and committees, most of which were assigned tasks of technical study and report. One of them was a special committee set up to consider the problems of chemical, bacteriological and incendiary weapons. It consisted of Australia, Brazil, Denmark, France, Germany, Italy, Japan, the Netherlands, Poland, Spain, Switzerland, the UK, the USA, and the USSR.

On 16 March 1933, the United Kingdom delegation submitted a draft convention which was adopted on 27 March as a basis of discussion, and accepted unanimously on 8 June, after a first reading debate, as the basis of the future convention. It was understood that, before the second reading, the governments would endeavour by negotiations to reduce existing political difficulties. But the agreement was not attained.

The Conference was never closed; its further convocation was merely postponed by a decision of the Council of the League of 22 January 1936. The Preliminary Report of the president of the Conference was issued in 1936.

II. *The United Nations*

The United Nations came into being in 1945 "to save succeeding generations from the scourge of war".

The Charter

Under the United Nations Charter the UN members have undertaken "to maintain international peace and security" (Article 1).

To promote this purpose, which is among the primary purposes of the organization, "with the least diversion for armaments of the world's human and economic resources", special responsibilities have been conferred on the Security Council in connection with disarmament and the regulation of armaments.

The Security Council

According to the Charter, the Security Council is composed of five permanent members (China, France, USSR, UK and USA) and non-permanent members elected by the General Assembly for a term of two years. In 1965, the number of the latter was increased from the original six to ten.

The Council is responsible for formulating, with the assistance of the Military Staff Committee (consisting of the chiefs of staff of the permanent members of the Council or their representatives), plans for the establishment of a system for the regulation of armaments. The Committee has played no part in the discussions of such plans.

The lack of agreement between the permanent members in the Security Council has made it impossible for that organ to carry out those responsibilities.

The General Assembly

The General Assembly of the members of the United Nations is entrusted with considering the principles governing disarmament and the regulation of armaments, and is empowered to make recommendations with regard to such principles to the members of the organization or to the Security Council or to both.

The Disarmament Commission

The first General Assembly resolution, adopted in 1946, established an Atomic Energy Commission with the task of making proposals for the elimination from national armaments of atomic weapons and of all other major weapons adaptable to mass destruction. The Commission was composed of one representative from each of the states represented on the Security Council, and Canada when that state was not a member of the Council.

Another group—the Commission for Conventional Armaments—was established in 1947, with the same composition as the Council, to submit proposals for the general regulation of armaments and armed forces and for safeguards in connection with such regulation and reduction.

Both commissions were dissolved in 1952 to be replaced by a single Disarmament Commission with the same membership. The Disarmament Commission was later expanded to include all the members of the United Nations.

In 1954 the Disarmament Commission created a Sub-Committee consisting of Canada, France, the Soviet Union, the United Kingdom and the United States, to search for a comprehensive plan of disarmament. It met in private until 1957.

During subsequent years, the major powers found it more convenient to set up negotiating bodies, committees and conferences, linked to but not constituting an integral part of the United Nations.

The Geneva Disarmament Conference

In 1960 a Ten-Nation Committee on Disarmament (TNCD) was formed with the participation of Bulgaria, Czechoslovakia, Poland, Romania and the USSR on the one side, and Canada, France, Italy, the United Kingdom and the United States on the other. It was replaced in 1961 by the Eighteen-Nation Committee on Disarmament (ENDC), which added to the original ten members eight non-aligned countries: Brazil, Burma, Ethiopia, India, Mexico, Nigeria, Sweden and the United Arab Republic.

The ENDC was enlarged in 1969 by the inclusion of Argentina, Hungary, Japan, Mongolia, Morocco, the Netherlands, Pakistan and Yugoslavia. The new name of the committee is "The Committee on Disarmament". It holds conferences at Geneva and, as a rule, meets for a spring and a summer session. France is not participating.

The Conference of the Committee on Disarmament (CCD) discusses both partial measures, including the prohibition of chemical and biological warfare, and a programme for general disarmament, and is now the main forum for disarmament negotiations (apart from the Strategic Arms Limitations Talks between the USA and the USSR). It submits annual reports to the UN General Assembly which considers these reports, more particularly in the First Political Committee consisting of all the members of the Assembly, and formulates recommendations in the form of resolutions.

Introductory note

Other conferences

There were also conferences dealing with specific disarmament subjects, such as the conference of experts from four Western and four Eastern European countries to study the possibility of detecting violations of an agreement on suspension of nuclear weapons tests; the tripartite conference (USSR, USA and UK) on the discontinuance of nuclear tests; the conference of experts from five Western and five Eastern European countries to study measures for the prevention of surprise attacks.

In August 1968, a Conference of Non-Nuclear-Weapon States was convened to examine problems related to the non-proliferation of nuclear weapons. It adopted resolutions and a declaration which were transmitted to the UN General Assembly.

A proposal was also discussed in the United Nations to hold a world disarmament conference to which all countries would be invited. The intention was to facilitate the participation of the People's Republic of China. The idea has not been realized.

III. *Comparison of the two systems*

In the League of Nations system there was far greater emphasis upon disarmament, as an independent approach to peace, than in the United Nations. The Covenant of the League sought to establish a legal basis for disarmament, while the UN Charter attached primary importance to collective security. The League's disarmament machinery was more elaborate and more technically oriented than that of the United Nations.

The reason for the difference between the two approaches may lie in the fact that the Covenant was written at the end of World War I which, as many then believed, was caused by the arms race prior to that war, while a few decades later the prevalent feeling was that World War II might have been avoided if the big powers had maintained an adequate military potential as well as readiness to use it. Besides, unlike the Covenant, the Charter was drafted when the war was in full progress and when planning a system of disarmament might have seemed ill-timed. The early involvement of the United Nations in the matter of disarmament was prompted chiefly by the use of the atomic bomb in 1945.

Early efforts to outlaw chemical warfare

The latter half of the nineteenth century and the early part of the twentieth century saw determined attempts by the leading states to mitigate the evils of war by agreeing to codes of rules regulating its conduct. International documents were adopted prohibiting the use of certain types of weapons—some explosives and incendiaries, as well as poison and gases.

Thus, in the Declaration of St. Petersburg, signed on 29 November – 11 December 1868, the contracting parties considered that the employment of arms which uselessly aggravate the sufferings of disabled men, or render their death inevitable, would be contrary to the laws of humanity, and undertook: "to renounce, in case of war among themselves, the employment by their military or naval troops of any projectile of weight below 400 grammes, which is either explosive or charged with fulminating or inflammable substances".

The International Declaration, adopted at the Brussels Conference on 27 August 1874, recalled that the laws of war do not recognize in belligerents an unlimited power in the adoption of means of injuring the enemy, and especially prohibited the employment of "poison or poisoned weapons".

These efforts continued at the International Peace Conferences conducted later, at the Hague, Netherlands. At the first of those conferences, in 1899 (acts signed on 29 July), the powers agreed to abstain from the use of "projectiles the sole object of which is the diffusion of asphyxiating or deleterious gases". At the second, in 1907 (acts signed on 18 October), the prohibition "to employ poison or poisoned weapons" was reiterated and included in the "Regulations respecting the Laws and Customs of War on Land".

In spite of those regulations, poison gas was widely used by both sides in World War I and caused heavy casualties.

In an appeal to the belligerents of 6 February 1918, the International Committee of the Red Cross said:

We wish to-day to take a stand against a barbaric innovation. ... This innovation is the use of asphyxiating and poisonous gas, which will it seems increase to an extent so far undreamed of.

... We now hear of new volatile poisons, large-scale production of which is all the easier as the raw material is ready to hand. We are shown missiles loaded with these poisonous gases spreading death—and a horrible death it is—not only among the fighting forces, but behind the lines among an inoffensive population over a wide area in which all living things will be destroyed. We protest with all the force at our command against such warfare which can only be called criminal.

The Commission on the Responsibility of the Authors of the War and on Enforcement of Penalties, constituted by the Allied Powers at the preliminary peace conference in Paris, 1919, included in the list of war crimes: "Use of deleterious and asphyxiating gases".

As a result, the Treaty of Peace with Germany, concluded at Versailles on 28 June 1919, stated that: "The use of asphyxiating, poisonous or other gases and all analogous liquids, materials or devices being prohibited, their manufacture and importation are strictly forbidden in Germany" and that "The same applies to materials specially intended for the manufacture, storage and use of the said products or devices". (Article 171.)

A similar provision was contained in other peace treaties, which concluded World War I,[1] and in the Treaty of Berlin, 1921, between the USA and Germany.

---

[1] Art. 135, Treaty of St. Germain, 1919, with Austria; Art. 82, Treaty of Neuilly, 1919, with Bulgaria; Art. 119, Treaty of Trianon, 1920, with Hungary; Art. 176, Treaty of Sèvres, 1920, with Turkey (this treaty was replaced by the Treaty of Lausanne, 1923, which did not contain such a provision).

Chapter 1. 1920–1924

The Covenant of the League of Nations, which entered into force on 10 January 1920, defined the obligations of its members with regard to disarmament. The activities of the organization in this field were aimed at the reduction and limitation of all types of national armaments then in existence. Chemical weapons constituted only one of the items on the disarmament agenda. It came up, however, early in the debate in view of the shock that the employment of gas in the war had produced in world public opinion.

I. *Preliminary discussions on chemical warfare in the League of Nations*

On 17 May 1920 during the fifth session of the Council of the League, the British representative raised the problem of the use of poisonous gas in warfare. The British government felt it desirable that the subject be studied internationally "with a view to some agreement being reached".

The Permanent Advisory Commission for Military, Naval and Air Questions, constituted by the Council on 19 May 1920 (pursuant to Article 9 of the Covenant of the League of Nations) and composed of representatives of the general staffs of the states members of the Council, was entrusted with the study of the problem. [1]

The following questionnaire was considered by the Commission at its first session, on 3–5 August 1920:

1. Does the experience of the last war, and the increase in the employment of gases, authorize us to go back upon the opinion, so far universally admitted, that the use of this weapon is fundamentally cruel?

2. In cases where its employment is permitted, should a limitation of the quantity to be used be imposed in the same way as the proposed limitation for the armies, armaments and navies authorized for each country?

3. In cases where its employment is prohibited, are there measures in existence to render this prohibition effective? Would it be possible, for

example, to prohibit experiments in the laboratories, or to take measures whereby, in case of war, commercial factories could be prevented from immediately adapting their manufactures to military purposes?

4. In a general way, what opinion would the Permanent Commission be able to give the Council concerning international regulations on this matter?

On 20 October 1920, at its second session, the Permanent Advisory Commission arrived at the following conclusions:

1. The employment of gases is a fundamentally cruel method of carrying on war, though not more so than certain other methods commonly employed, provided that they are only employed against combatants. Their employment against noncombatants, however, must be regarded as barbarous and inexcusable.

2. It would be useless to seek to restrict the use of gases in wartime by prohibiting or limiting their manufacture in peacetime.

3. The prohibition of laboratory experiments is out of the question.

As to possible international regulations with regard to asphyxiating gases, the Commission considered itself not competent to supply an answer. [2]

The opinion of the Permanent Advisory Commission was included in the report to the Council of 22 October 1920. In discussing the report the British representative on the Council said that if, as the Commission held, it was really impossible to prevent laboratory research and to make it impossible for states to accumulate stocks of the material necessary for the production of gas, any theoretical condemnation of the use of poison gas would only put those powers who respected it at the mercy of a less honest power prepared to make use of gas in case of war.

The Council, however, condemned the use of gas. It decided to refer back to the Permanent Commission the consideration of the methods which might ensure an effective control of the production of gas, and to ask the governments for their views as to the penalties which should be imposed upon states making use of poison gas (in light of Article 16 of the Covenant dealing with sanctions). [3]

On 12 December 1920, at its eleventh session, the Council approved the forms of a questionnaire worked out by the Permanent Commission, to be forwarded to states for the exchange of information regarding armaments. The questionnaire contained, among other things, a chart showing armoured cars and tanks, and machines for projecting inflammable, asphyxiating and tear gas, etc. [4]

The International Committee of the Red Cross appealed to the members of the League to reach agreement on the absolute prohibition of the use of gas as a weapon, however delivered, whether by drift, missiles or otherwise.

II. *Disclosure of CW discoveries*

During the disarmament debate in the Third Committee of the League of Nations Assembly, dealing with the question of the reduction of armaments, Lord Robert Cecil of South Africa stated on 28 September 1921 that the use of poison gas would be greatly diminished if each nation were convinced that this weapon could be turned against itself. The best method for preventing the use of poison gas was to appeal to the scientists of all countries to make all discoveries public. In that way no nation would consider itself in a privileged situation with regard to the others.

The British representative thought that such action would not be very effective, since governments would carry on their research in secret in the laboratories. Nor did the French representative think that perfect publicity could be attained; the suppression of the use of poison gases could only be achieved by means of an agreement between governments.

A unanimous resolution was eventually adopted stating that the Temporary Mixed Commission should be asked to examine—in consultation with the Permanent Advisory Commission—whether it was feasible to address an appeal to the scientific men of the world to publish their discoveries in poison gas, and similar subjects, so as to minimize the likelihood of their being used in any future war. The Temporary Mixed Commission, including persons competent in political, social and economic matters, had been established by the Council on 25 February 1921 to prepare proposals for the reduction of armaments. [5]

The report of the Third Committee contained the following motivation of the adopted resolution:

> It is common knowledge that inventions have been made or perfected since the war whereby wholesale destruction of the civil population would be possible by the dropping of poison bombs and the like from the air, nor is there any reason to suppose that the limits of invention in these fiendish devices have been reached. It is obvious that if it were known that every nation was armed with such weapons they would all hesitate to make use of them for fear of the consequences to themselves of their employment by their antagonist; but if some nation believed itself to be in possession of a secret invention unknown to its rivals, the temptation to use it would be overwhelming. Much therefore, could be done to render impracticable the employment of these weapons if there were no secrecy about them. . . .[6]

The Second League of Nations Assembly approved the report of the Third Committee on 1 October 1921. [7]

The Temporary Mixed Commission considered the question at its third, fourth and fifth sessions and concluded that an appeal of the nature suggested by the Second Assembly was not a practical measure.

The main reasons which led the Commission to that conclusion were as follows:

1. Publication of discoveries would probably have results contrary to the general trend of legislation in the world which was to regulate and control the distribution of lethal weapons;

2. Any invention with regard to the use of gas to be employed in war must be tested on a large scale. This can only be done by scientists who work under government orders and with government subsidies. Such scientists would, of course, carry out their investigations under an absolute pledge of secrecy to the governments which they serve.

Under such circumstances only the less dangerous inventions would be published and this would be worse than no publication at all, since it would tend to produce a feeling of false security.

3. Even if the really important inventions were published by the scientists of some countries, there could be no method of ensuring that the same would be done by all. This, in the event of war, would place those countries who responded to the appeal at a disadvantage as compared with those who did not.

The Temporary Mixed Commission felt that if such an appeal were made it would not achieve the object aimed at by the Assembly, i.e., to minimize the likelihood of poison gas being used in any future war.

Lord Cecil then suggested that people should realize what the new methods of warfare would mean in the future, not only to the armies in the field, but to the civilian population at home. He thought it would be useful to collect, partly from existing publications and partly by enquiries from experts, the facts necessary for an authoritative exposition of the subject. The Commission decided to appoint a subcommittee to investigate the matter.

III. *Adhesion to the Treaty of Washington*

On 6 February 1922, as a result of the Conference on the Limitation of Armaments, held in Washington, a treaty was signed between the USA, the UK, France, Italy and Japan relating to the use of submarines and noxious gases. Article 5 of the treaty read:

The use in war of asphyxiating, poisonous or other gases, and all analogous liquids, materials or devices, having been justly condemned by the general opinion of the civilized world and a prohibition of such use having been declared in the Treaties to which a majority of the civilized Powers are parties,

The Signatory Powers, to the end that this prohibition shall be universally accepted as a part of international law binding alike the conscience and practice of nations, declare their assent to such prohibition, agree to be bound thereby as between themselves and invite all other civilized nations to adhere thereto.

Formidable obstacles had been standing in the way of reaching the above accord. The extent of opposition to prohibiting the use of gas in war was reflected in the report of the subcommittee of experts on poison gas, set up at the Conference, which argued that such a prohibition would be unrealistic because many high explosives produce deadly gases; research could not be restricted or supervised, nor could the manufacture of gases be controlled; the use of chemicals by an unscrupulous enemy would give him a critical advantage of surprise, if the attacked nation was unprepared. The only limitation considered practicable was to prohibit the use of gases against cities and other large bodies of noncombatants.

If the Conference, nevertheless, succeeded in proclaiming the prohibition, it was due, in the first instance, to the diplomatic efforts of the US delegation.

The League's Temporary Mixed Commission examined a proposal that the members of the League should be urged to adhere to the Treaty of Washington concerning asphyxiating and poisonous gases, but decided that no action should be taken until information was received that the treaty had been forwarded to the non-signatory states for adhesion, after ratification[1] by the signatory powers. [8]

The matter was later discussed by the Third Committee of the League Assembly in September 1922.

The representative of Colombia thought that it would be advisable to recommend that the Assembly adopt a convention condemning the use of poisonous gases. The text would be identical to that of Article 5 of the Washington Treaty.

The representative of Australia proposed that the Council should recommend to the members of the League and other nations to adhere to the Treaty of Washington.

The representative of Norway, however, was opposed to the idea, maintaining that "man was incapable of humanizing war and could only endeavour to abolish it".

The delegate of Colombia having agreed to withdraw his proposal in favour of that of the delegate of Australia, the Third Committee proposed that the Assembly request the Council to recommend the members of the League and other nations to adhere to the Treaty of Washington. [9]

The Third Assembly approved the Third Committee's proposal in re-

---

[1] The treaty never came into force because of French objections to its provisions on submarines.

solution VII adopted on 27 September 1922. [10] The Council decided to put the question on the agenda of the envisaged international conference for the limitation of naval armaments.

IV. *Arousing public opinion about CBW*

In a report to the Assembly submitted on 22 September 1922, the Third Committee said that the Temporary Mixed Commission was well advised in appointing a special subcommittee to study the development of chemical warfare. (See page 46.)

The subcommittee was composed of four members. It adopted a programme of work which defined the task entrusted to it in the following way:

The aim was to show to the public opinion of the world the effects which would be produced by the most powerful means of destruction placed at the service of modern warfare by modern science.

It would be borne in mind that henceforward an armed nation, utilizing the whole of its human and material resources, would attempt to strike not only at the combatants on the enemy's front, but at the whole enemy nation—its population, its riches and its resources of every kind. War of this sort, which carried destruction beyond the fighting lines and which rendered opposing nations vulnerable to the extreme limits of their territories, had been made possible by the increasing range of modern guns, by the far-reaching activity of air forces and by conveying and disseminating in other ways the means of destruction.

Without discussing the legitimacy of such practices, the committee would merely seek to discover what was possible in warfare, whether permitted or not by the laws of war, in order that the public might have an accurate conception of the dangers which it had to fear.

In these circumstances, it was desirable to obtain from the most qualified experts as detailed and complete a statement as possible of the effect which would be produced on human life, animal life, vegetable life, and on the wealth and resources of all kinds of a country attacked at any point within its territory by:

1. chemical warfare by means of the most powerful explosives, chemical products and gases, as already practiced and as further developed since the last war;

2. bacteriological warfare by means of microbes or any other agents, if, in defiance of all human laws, its effectiveness should induce nations to adopt it.

The programme was submitted to a certain number of experts, chosen after consultation with the Health Committee of the League. [11]

On 3 July 1923, the Council, on the suggestion of the Temporary Mixed Commission, invited the governments represented at the 1922 Washington Conference (see above) to consider the possibility of making available to the Commission the report on chemical warfare, which had been drawn up by their experts at that Conference. [12]

In September 1923, the Third Committee recommended that the Fourth Assembly adopt a resolution expressing an interest in a report on the probable effects of chemical discoveries on future wars and formulating a request that the Council and the Temporary Mixed Commission ensure by all possible means the fullest publicity to such a report.

The Third Committee also discussed a proposal, put forward by the representative of Hungary, that efforts should be devoted to investigating and publishing information concerning the research undertaken for discovering means of safeguarding humanity from the disastrous consequences of chemical and bacteriological warfare. The use of such remedies would have to be made accessible to inhabitants of all countries of the world, in the same manner as remedies against contagious diseases.

The Hungarian proposal was rejected. The Committee did not feel that it was in a position to widen the mandate of the special subcommittee. [13]

The Third Committee's recommendations were adopted by the Fourth Assembly on 29 September 1923. [14] On 10 December 1923 the Council endorsed the Assembly's resolution on the matter. [15]

V. *The report on CBW*

The subcommittee appointed to consider the question of chemical and bacteriological warfare appealed to chemists, physiologists and bacteriologists in various countries. The following experts replied: Professor André Mayer of the Collège de France; Professor Angelo Angeli of the Royal Institution of Higher Studies at Florence; Professor Pfeiffer of Breslau; Professor J. Bordet of the Pasteur Institute in Brussels; Professor W. B. Cannon of the Harvard School of Medicine; Professor Th. Madsen of Copenhagen; Senator Paterno of Rome University; and M. J. Enrique Zanetti of Columbia University.

There were also the report of the subcommittee on poison gas of the Washington Conference and the memorandum on its findings.

The papers received dealt with the known effects of chemical warfare and the possible effects of bacteriological warfare, and are summarized here.

Chemical warfare

What are commonly called "asphyxiating gases" include in point of fact not only gases but also solid or liquid substances which are pulverized in the air or scattered on the ground by various technical methods (expansion after compression, dissemination by explosion, evaporation by heat, etc.). These substances include a wide range of compounds which are harmful to the human body or to animals. Some of these compounds were invented and manufactured for military purposes but the majority of the injurious substances are ordinary substances which are used daily for industrial purposes. For example, the gas wave which was released during the war and which took a French division by surprise and decimated it was composed of chlorine, a substance in daily use for whitening and disinfection, and also used as an "intermediate substance" in a large number of chemical preparations. Moreover, there is very little difference between the manufacture of pharmaceutical products and that of injurious substances used in war.

Chemical weapons are capable of producing the most varied physiological effects. Their power, their efficiency and their diversity are as unlimited as those of pharmacology or any branch of chemistry. They are at the disposal of any great industrial power which has chemical works. Nothing is easier than to divert from their peaceful uses substances which are regularly manufactured, or rapidly to adapt industrial machinery to the manufacture of harmful substances (Professor André Mayer).

Many high explosives produce toxic gases that frequently cause death as do those termed chemical warfare gases; limitations on the use of the latter would probably result in misunderstandings upon the outbreak of war (Report of the Sub-Committee on Poison Gas, Washington Conference).

The various agents used by chemical warfare could be classified according to their effect on the human body: irritant (lachrymatory, sneeze-producing and blistering) agents; suffocating or asphyxiating agents; and toxic agents.

Effects of irritant agents

Lachrymatory agents act on the mucous membrane of the eyelids, thereby producing a copious flow of tears accompanied by pain and for all practical purposes blinding the victim for as long a period as he remains in the atmosphere impregnated with this agent.

Professor Enrique Zanetti contended that the blinding effect of these gases was purely temporary, being caused only by irritation of the membrane of the eyelids and not by any deep-seated effect on the eyeball or optic nerve; the effect usually passed in a few hours or a few days at the most, and although the victim was as completely put out of action as if his eyes had been gouged out, there was no record of permanently serious effects being produced thereby.

Professor André Mayer, however, pointed out that if taken in large quantities all lachrymatory gases were fatal.

Sneeze-producing agents induce intense irritation of the nasal passages with consequent violent sneezing and intolerable headaches. The chief object of the employment of these substances is to prevent the soldiers who have been gassed by them from keeping on their masks, the uncontrollable sneezing resulting in the loosening of the mask and exposing the victim to the action of other toxic products which may be fired concurrently or immediately after the sneeze-producing gas.

Blistering agents were the most important of all gases used in World War I. The so-called mustard gas, also called "Yperite", caused serious lesions to the skin and mucous membrane. Whenever the skin is exposed even to the vapour exhaled from the slow evaporation of this gas, large blisters appear within a period of two to eight hours. The severity of this blistering depends on the length of exposure. The action of this gas produces necrosis of the mucous membrane and leaves a raw surface extremely susceptible to infection. For that reason, in cases where the lungs were affected by vapours of this gas, results were frequently fatal. Moreover, soil which is saturated with "Yperite" contaminates by contact persons who pass over it. Any articles which have been impregnated with the gas remain dangerous for a number of days.

Effects of suffocating or asphyxiating gases

These gases cause fatal damage to the lungs. When inhaled, they cause haemorrhage in the air cavities of the lungs. Of all gases in this category, carbon oxychloride, also known as phosgene, was the one most effectively employed in World War I. Other agents directly affect the blood, e.g., carbon monoxide, which usually causes death by syncope.

Effects of toxic agents

The toxic agents of the nervous system, such as derivatives of prussic acid, kill by instantaneously suppressing the functions of the nervous system. These gases, however, as known so far, produce the paralyzing effect only when they are used in fairly high concentration.

Combined effects

It should not be supposed that each of the substances which was used in warfare possessed only one of the enumerated properties. Most of them combined several. The alteration in the strength used, which may be obtained by changing the method of filling shells or the concentration of fire, completely transforms the injurious effects. Combined effects consisting of various destructive actions may be obtained either by releasing several substances together or by using one substance having several properties (Professor André Mayer).

After-effects of war gases

There is a general belief that although chemical agents cause fewer deaths than shell or bullets, the lesions caused by them leave traces which permanently affect the victims, particularly their lungs. This impression is not borne out by statistics of the American Army or by those of the British and French. The percentage of cases of tuberculosis was not any greater among the wounded from gas than among the wounded from any other cause (Professor Enrique Zanetti).

Possible effects of new discoveries

The above-mentioned gases comprise the groups which were employed during the war. It is conceivable, said Professor Enrique Zanetti, that other gases may be discovered that would interfere with other functions of the body. However, Senator Paterno thought that there was no ground to believe that new substances of greater military value than any yet known could be discovered and manufactured on a large scale, and with regard to asphyxiating gases he concluded that "we must neither hope nor fear" that the progress of chemistry will lead to any greater success in the discovery of these gases than in discovering explosives.

Effects of gas on animals

The effects of gases are essentially the same for human beings as for animals. Among animals there seem to be certain differences in sensibility but these are only of degree and comparatively small (Professor Enrique Zanetti).

Effects of gas on vegetation

Vegetation exposed to the action of such gases as chlorine and mustard gas shows some withering of the green leaves, but in no case has destruction been reported; the condition of vegetation on and near chemical warfare experimental fields has not shown any serious effects from repeated exposure to toxic gases in the concentrations likely to be produced on battlefields (Professor Enrique Zanetti).

Effects of gas on other sources of wealth

The effects would be indirect through the paralyzing action as, for example, the shutting-down of factories through the gassing of the surroundings, so as to render them unapproachable to workmen. However, the dropping of a few airplane bombs filled with a high-power lachrymatory gas would effectively shut down a factory for as long as a month without causing any considerable destruction of life or property such as would result from long-range shelling or bombing with high explosives. In the case of mine pits and galleries, a thorough drenching with a persistent gas, such as mustard gas, or even a simple lachrymatory gas, would render them unapproachable, perhaps for months. In the course of time the gas would completely disappear.

No agent is known to produce a chemical destruction of sources of wealth except through its action on the human element connected with their exploitation (Professor Enrique Zanetti).

Protection against chemical weapons

Some protection is obtained by the use of insulating and filtering apparatus, if the concentration is not too great.

Against the skin lesions produced by blistering gases, no satisfactory means of protection is known.

Conclusions

Though the experience of the war has shown that no fortifications or armour can resist the force of modern explosives, men themselves could at least find safe shelter from them in trenches, caves or dug-outs sunk deep underground. But poisonous gases can go everywhere, both in the open and into dug-outs, and if the concentration exceeds a certain limit even masks become useless; men are thus without any means of defence, and, even in those places which were formerly regarded as safest, they cannot escape death (Professor Angelo Angeli).

The chemical weapon gives an immense superiority to any power with hostile intentions. An injurious substance studied in secret (and this study may be carried on anywhere), manufactured in large quantities (and this manufacture can be carried out in any chemical works), and launched unexpectedly against an unprepared population, can completely destroy every shadow of resistance. The nations of the world must realize how terrible is the threat which thus hangs over them (Professor André Mayer).

The possibilities of camouflaging chemical preparedness are very great. Mask protection research can be quietly carried on under the guise of research for development of masks to be used in the chemical industry and the name of pharmacological research could hide a multitude of sins. Chemical warfare on protected troops has not introduced any such horrors as generally believed, neither is it likely to introduce them if protection keeps pace with new developments (Professor Enrique Zanetti).

The use of poisonous gases marked the appearance of a new and terrible weapon, and it is difficult to foresee what forms this weapon might take as time goes on (Professor Angelo Angeli).

Bacteriological warfare

It is a crime against humanity to think even of attempting to use for the destruction of life and health the achievements of modern medical science whose aim is the safeguarding of human life and the prevention of epidemics. The spread of epidemics cannot be limited to the enemy forces but the whole population, including women and children, would fall victim to the disease; indeed, noncombatants would in most cases suffer most from its destructive effects (Professor Pfeiffer).

Professor J. Bordet was not inclined to believe that it would be possible in practice to produce epidemics of cholera, typhoid, or even plague in enemy countries. The diseases in question were well-known and were easily diagnosed; attempts to produce them would be discovered with little difficulty. There was, however, a germ which caused an insidious disease that was often very hard to diagnose and which might possibly be employed successfully to harm an adversary, i.e., the micrococcus of Malta fever. Individuals were very readily infected with its cultures which produced a lingering disease, rarely fatal, but very depressing.

Possible effects of use and protection

Professors Pfeiffer, Bordet and Madsen were of the opinion that bacteriological warfare would have little effect on the actual issue of a contest in view of the protective methods available for circumscribing its effects.

The pollution of drinking water by cultures of typhus or cholera germs would be combated by filtering, by treating the waters of rivers with chlorine, and by preventive vaccination.

The propagation of plague by pest-infested rats would be as dangerous for the nation employing this method as for its adversary.

As regards the poisoning of weapons, the germs which could be employed would not preserve their dangerous properties if they were prepared a long time beforehand and allowed to dry on metallic surfaces. Nor if placed in a projectile would these germs resist the shock of discharge, the rise of temperature and the violence of explosion which destroys all life. The only method presenting a certain danger would be that of dropping glass globes filled with germs from airplanes.

The use of infectious germs at the front might react on the forces of the nation resorting to such methods. The germs might attack those who employed them (Professor J. Bordet).

The majority of the experts expressed the view that bacteriology could not produce infective substances capable of destroying a country's livestock and crops. Professor Cannon did not concur in this latter opinion; he admitted the possibility of airplanes disseminating over wide areas parasites capable of ravaging the crops. Professor J. Bordet saw the possibility of producing contagious diseases among domestic animals. Cultures of glanders had been discovered in the German Embassy at Bucharest, with instructions for producing glanders in the Romanian cavalry. It might not seem very difficult to propagate rinderpest among the enemy's herds. Of cattle diseases, this was the one which would be most suitable for the purpose; but means existed to combat it.

Conclusion

The scientists did not consider the bacteriological weapon as particularly formidable. [16]

The main opinions of the consulted experts were incorporated in the Temporary Mixed Commission's report on chemical and bacteriological warfare of 30 July 1924. The Commission stated:

With regard to the possible use of the chemical arm against civilians:

It may be said that such a development of warfare would be too horrible for use and that the conscience of mankind would revolt at it. It may be so, but in view of the fact that in modern wars such as the last one, the whole population of a country is more or less directly engaged, it may well be that an unscrupulous belligerent may not see much difference between the use of poison gas against troops in the field and its use against the centres from which those troops draw the sinews of war.

Noting therefore, on the one hand the ever increasing and varying machinery of science as applied to warfare, and on the other, the vital danger to which a nation would expose itself if it were lulled into security by over-confidence in international treaties and conventions, suddenly to find itself defenceless against a new arm, it is, in the opinion of the Commission, essential that all nations should realize to the full the terrible nature of the danger which threatens them.

With regard to bacteriological warfare:

... Although the conclusions drawn may be comparatively reassuring for the present, they nevertheless direct attention to the possibilities which the development of bacteriological science may offer in the future. ... [17]

VI. *Publicity for the report on CBW*

On 25 September 1924, the Third Committee of the League Assembly examined the report of the Temporary Mixed Commission and expressed the opinion that the essential object of the work which the Commission had carried out was to direct attention to the serious nature of the methods of destruction which modern science placed at the disposal of combatants, thereby demonstrating the necessity of taking steps to prevent war itself. [18]

Upon recommendation of the Third Committee, on 27 September 1924, the Fifth Assembly adopted a resolution by which it requested the Council to publish the report of the Temporary Mixed Commission on the probable effects on warfare of chemical discoveries and, if advisable, to encourage the work of making information on this subject generally accessible to the public; noted the facility and rapidity with which factories producing chemical substances required in peacetime can be transformed into factories for chemical warfare; and recommended that the attention of public opinion throughout the world be drawn to the necessity of endeavouring, in the first place, to remove the causes of war by the pacific settlement of disputes and by the solution of the problem of security, in order that nations may no longer be tempted to utilize their economic, industrial or scientific power as weapons of war. [19]

The League Council, at its meeting on 30 September 1924, considered the Fifth Assembly's resolution on chemical warfare the best method of drawing the attention of the public to this question. The Council trusted that the delegates of the governments which had unanimously voted in favour of the resolution would do everything in their power to give it the greatest possible publicity in their countries. [20]

VII. *Summary and comment*

The important rules of conduct in war established in the latter half of the nineteenth and the early part of this century, in particular those forbidding the use of gases, were disregarded in World War I. Gas was used extensively and caused heavy casualties.

Germany, as well as Austria, Bulgaria, Hungary and Turkey, were prohibited under the peace treaties from manufacturing or importing asphyxiating, poisonous or other gases and all analogous liquids, materials or devices. These were the first international instruments containing a prohibition to produce gas for warlike purposes. Their importance however was limited: they were imposed on defeated countries.

The search for measures freely agreed to, and of universal application, so as to eliminate effectively the threat of chemical warfare, characterized the efforts of the League of Nations from the early years of its existence, and was a constant item on the agenda of the League's disarmament debate.

An idea was put forward in 1921 to abolish secrecy about chemical weapons, to appeal to scientists to make public their discoveries, and thus "render impracticable the employment of those weapons". The reasoning was that if every nation were armed with chemical weapons, the states would be deterred from using them for fear of the consequences to themselves. After consideration by a special League body, the proposition was found unrealistic.

Another suggestion was for all the League members to adhere to the Washington Treaty prohibiting the use of gas, which had been signed by the great powers in 1922. The treaty, however, was not ratified by all its signatories and never entered into force.

A proposal for an authoritative report on the effects which would be produced by the use of chemical and bacteriological weapons was accepted. The aim was to arouse public opinion and make it aware of possible dangers. The report, a predecessor of a similar document issued in 1969, was based upon contributions by chemists and bacteriologists from various countries and appeared in 1924.

The main points made were that the use of poisonous gases marked the appearance of a terrible weapon; chemical weapons gave an immense superiority to any power with hostile intentions; the possibilities of camouflaging chemical preparedness were very great. Lachrymatory agents were classified as chemical warfare agents, a point of relevance in later discussions.

Bacteriological weapons were not considered particularly formidable, but attention was drawn to the possibilities which the development of bacteriological science might offer in the future.

The report paved the way to the Geneva Protocol of 1925.

Chapter 2. 1925

I. *The convening of the Conference on the International Trade in Arms*

On 27 September 1924, the Fifth Assembly of the League of Nations requested the Council to submit to the governments of state members and non-members of the League the draft convention relating to the control of the international trade in arms, munitions and implements of war, drawn up by the Temporary Mixed Commission, and to ask these governments whether they were prepared to take part in a conference to be convened for the purpose of discussing the draft. [1]

On 30 September 1924, the Council decided to act in accordance with the Assembly's resolution. [2]

Following an exchange of correspondence between the Secretary-General and the governments and as a result of subsequent resolutions of the Council, the Conference for the Supervision of the International Trade in Arms and Ammunition and in Implements of War convened at Geneva on 4 May 1925.

In addition to the limitation of trade in all arms, the Conference considered proposals for the prohibition of the use of chemical and bacteriological warfare. The discussion culminated in the signing of a protocol which provided for such prohibition.

II. *The opening phase of the Conference*

On 5 May 1925, at the plenary meeting of the Conference, the United States representative stated that one of the most important suggestions, which his delegation would present in detail, concerned measures to deal with the traffic in poisonous gases, "with the hope of reducing the barbarity of modern warfare". [3]

On 7 May 1925, when the General Committee of the Conference, which was entrusted with the first reading of the text, was discussing the definition of categories of arms to be covered by the convention (Chapter I, Article 1), the representative of the United States reiterated the desire of his government that some provision be included in the convention relating to the prohibition of the exportation of asphyxiating, poisonous and deleterious gases. He read out a draft which ran as follows:

> The use in war of asphyxiating, poisonous or other gases and all analogous liquids, materials or devices has been justly condemned by the general opinion of the civilized world, and a prohibition of such use has been declared in treaties to which a majority of the civilized Powers are parties. The High Contracting Parties, therefore, agree absolutely to prohibit the export from their territories of any such asphyxiating, poisonous or other gases, and all analogous liquids, intended or designed for use in connection with operations of war.

Such a provision, said the US representative, would be in the interests of humanity and peace. He admitted that there were difficulties in drawing a dividing line between gases used in warfare and those used for legitimate industrial purposes, as well as in resolving the problem of equal treatment of the producing and the non-producing countries.

The representative of Brazil supported the United States proposal in principle; he pointed out, however, that if the producing states remained free to manufacture gases and to employ them, the result would be to create an inadmissable inequality between the producing and the non-producing states. He recalled that at the Santiago Pan-American Conference[1] the delegates of Colombia and Uruguay had pointed out that these new methods of warfare were an excellent means of defence for weak countries in view of the facility with which they could be procured, and their terrible effects.

The representative of Poland found the United States suggestion very useful and felt that a similar action would be indispensable with regard to bacteriological warfare.

The representative of Hungary, while accepting the views of the USA

---

[1] The Fifth International Conference of American States, held at Santiago, Chile, 25 March–3 May 1923, adopted a resolution on the reduction and limitation of military and naval expenditures. In the Fifth Agreement the Conference resolved:
"(c) To recommend that the Governments reiterate the prohibition of the use of asphyxiating or poisonous gases, and all analogous liquids, materials or devices, such as are indicated in the Treaty of Washington, dated February 6, 1922."
Similar instruments, signed by American states, include:
1. Convention on the Limitation of Armaments of Central American States, signed at Washington, 7 February 1923, which in Article 5 read:
"The Contracting Parties consider that the use in warfare of asphyxiating gases, poisons, or similar substances as well as analogous liquids, materials or devices, is contrary to humanitarian principles and to international law, and obligate themselves by the present Convention not to use said substances in time of war."
2. Resolution, adopted by the Inter-American Peace Conference held at Buenos Aires in December 1936, which proscribed the use of chemical elements of warfare; and recommended to Governments that in the pacts they conclude for the limitation of armaments they should exclude by statute such methods of war as use of poison gas, poisoning of water, etc.

and Poland, thought that the means of defence against chemical warfare should be excluded from the prohibition.

On 8 May 1925, the following alternative text regarding chemical warfare was submitted by the United States:

To the end of lessening the horrors of war and of ameliorating the sufferings of humanity incident thereto, the High Contracting Parties agree to control the traffic in poisonous gases by prohibiting the exportation of all asphyxiating, toxic or deleterious gases and all analogous liquids, materials and devices manufactured and intended for use in warfare under adequate penalties, applicable in all places where such High Contracting Parties exercise jurisdiction or control.

The Polish proposal concerning bacteriological warfare was worded in the following way:

With reference to the suggestions made by the United States delegation regarding the materials used for chemical warfare, and inasmuch as the materials used for bacteriological warfare constitute an arm that is discreditable to modern civilisation, the Polish delegation proposes that any decisions taken by the Conference concerning the materials used for chemical warfare should apply equally to the materials used for bacteriological warfare.

The Hungarian proposal was formulated as an addition to the United States text:

... it being understood that such import and export prohibition shall not apply to methods of defence against asphyxiating, poisonous, or other similar gases employed as a means of warfare.

The texts were referred to the Military and Legal Committees of the Conference. [4]

III. *The Legal Committee's consideration of CBW*

The Legal Committee, at its meeting of 11 May 1925, set up a subcommittee to examine the question of chemical and bacteriological warfare. [5]

On 19 May 1925 the subcommittee reported to the Legal Committee that it might carry out its task in either of two distinct ways: by inserting a special provision in the convention for the supervision of the international trade in arms, or by submitting a text under which all states would prohibit the use of gas.

The representative of Belgium considered the second method preferable, as it went much farther than the mere prohibition of exportation. This was also the opinion of the representatives of the British Empire and the Irish Free State. When the Hungarian representative raised the question of placing no restrictions on defensive weapons, the Romanian delegate stated that the effect of the Hungarian amendment would be to re-introduce gas warfare, which had been prohibited under the terms of treaties and decisions of international conferences. [6]

On 20 May 1925 the Legal Committe forwarded to the General Committee the following alternative suggestions:

1. Prohibition by means of an article in the Convention of the exportation of asphyxiating, poisonous or other similar gases, and all analogous liquids, materials or devices;
2. A declaration, either in the Final Act or in a separate document, laying down that the use of the said gases in time of war was contrary to international law;
3. A statement, in an appropriate article of the Convention, that the use of gases in war was prohibited by international law;
4. To allow, in regard to the exportation of the means of defence referred to in the Hungarian amendment, an exception to be made to the conditions laid down in the Draft Convention. [7]

The General Committee decided to submit these four suggestions to the Military, Naval and Air Committee, asking it to examine in the first place suggestions 1 and 4.

The United States delegation then introduced, in an official statement of 23 May 1925, a simplified text reading

The High Contracting Parties agree to prohibit the export from their territory of all asphyxiating, toxic or deleterious gases, and all analogous liquids, exclusively designed or intended for use in connection with operations of war,

which, in its view, fell entirely within the scope of the Draft Convention. [8]

IV. *The Military Committee's consideration of CBW*

The Military Committee discussed chemical and bacteriological warfare on 25, 26 and 27 May 1925.

The representative of the Kingdom of the Serbs, Croats and Slovenes said that in case of a prohibition of export, non-producing countries would be placed in a very dangerous position of inferiority, having been deprived of means of defence. A prohibition of export would have to be preceded by the establishment of security guarantees. Unless all gas-producing countries subscribed to a solemn undertaking not to employ gas in warfare, the non-producing countries would find themselves obliged, and would also find means, to manufacture gas for themselves.

The Hungarian representative explained that his proposal to exclude from the prohibition the means of defence against chemical warfare should not be understood as allowing the use of "defensive" gases, but as covering only gas masks, i.e., personal means of defence.

The Romanian delegate, however, insisted that not only shells, bullets, bombs and gases, but also gas masks might be employed both for defensive and offensive purposes. Moreover, to prepare a reserve of gas masks, experiments with gas were necessary in order to test their effectiveness; thus a state which was free to manufacture masks would also have to manufacture all kinds of gas in order to test them. This would be equivalent to authorizing the manufacture of gases. He thus alluded to the restrictions imposed upon Hungary in the field of armaments under the Treaty of Trianon.[2]

The delegate of France also saw a danger in admitting the possibility of chemical warfare by regulating the methods of defence. The adoption of the Hungarian proposal would, in his view, weaken the moral and effective scope of the desired prohibition of such warfare.

As a result of these interventions the representative of Hungary withdrew his proposal.

The representative of the British Empire was convinced that unless the use of gas in warfare was entirely prohibited, the United States proposal to ban the export of gas would not go far to prevent its use in future wars. The US proposal would constitute a menace to non-producing states, for it would induce an unscrupulous producing state to use gas in an attack upon a non-producing state. Under such circumstances, it would prove necessary for all countries, for the sake of their defence, to prepare against the use of gas; but such preparation was impossible unless a certain quantity of gas was available for the purpose of testing the various appliances used in anti-gas defensive measures. To prohibit export of gas, therefore, would inevitably mean that every country would be obliged to produce the gas it needed for its experiments, and thus all countries would gradually become gas-producers.

The representative of Italy described the grave position in which countries which did not produce toxic gases would find themselves if the trade in gas were prohibited without a previous prohibition of chemical warfare. He quoted the conclusion reached by Dr Eliot, President of Harvard Uni-

---

[2] The Treaty of Trianon of 4 June 1920, Part V, Article 119, contained the following provision:
"The use of flame throwers, asphyxiating, poisonous or other gases, and all similar liquids, materials or devices being prohibited, their manufacture and importation are strictly forbidden in Hungary. Material specially intended for the manufacture, storage or use of the said products or devices is equally forbidden.
"The manufacture and importation into Hungary or armoured cars, tanks or any similar machines suitable for use in war are equally forbidden."

versity, at a conference on "Science and War" which had taken place in the United States:

> ... No considerable war can hereafter be fought, or even undertaken, except by a nation which has highly developed manufacturing and chemical industries. Now, that is on the whole rather an encouraging consideration, because it throws out, so to speak, from nations capable of aggressive war a very large proportion of the nations of the world. It reduces to a comparatively small number the nations capable of carrying on war. Isn't this a great gain for the world as a whole? It is not a gain, of course, for the backward, non-manufacturing nations, because it is to them a plain exhibit of a condition fundamentally mortifying....

Thereupon, the Italian delegate contended that if states undertook not to have recourse to chemical warfare, prohibition of trade in toxic gases would logically be the necessary corollary to that. The prohibition of traffic in toxic products was in reality of far less importance than the wider undertaking, especially since, from the practical point of view, it would be extremely difficult to prevent the traffic in such products. In any event, provisions adopted in regard to the use and the trade in toxic gases would have to be universal if they were to be effective. The Swedish delegate shared this view. He said that abolition of trade in asphyxiating gases could only be contemplated if the manufacture and use of the products in question for military purposes were completely abolished. He also emphasized that any convention for the prohibition of chemical warfare could never properly come into force unless all producing countries, including the great countries not represented at the Conference, adhered to it. (The USSR was not represented at the Conference.)

The Brazilian representative, on the other hand, pointed out that there existed appliances of a definite nature to be used in chemical warfare. There were shells manufactured specifically for the purpose and containing well-known products; there were airplane bombs of the same kind, etc. It would therefore be possible to prohibit the export of such appliances without preventing the trade in substances which could also be used in time of peace. This was the most urgent measure to be taken. It should be completed by the prohibition to manufacture anything connected with chemical warfare. The two measures ought, in turn, to be followed by the adoption of a universal agreement entirely outlawing chemical warfare.

The majority of the Committee thus considered that the prohibition of the exportation and importation of chemical and bacteriological arms would not have the effect of putting a stop to the future use of chemical and bacteriological methods of warfare because it would in no way stop the producing states from making use of those weapons; besides, such prohibition would be very difficult to apply as it was practically impossible to discriminate between chemical products used for industrial, pharmaceutical or other purposes and those which might be used in chemical warfare. There was also a feeling that the Conference, which had been convened to deal with international trade in arms and munitions, did not possess the necessary powers to take a decision in regard to the prohibition of chemical warfare.

The Polish delegation suggested that the question of the prohibition of chemical and bacteriological warfare should be placed on the agenda of the envisaged conference on the manufacture of arms and implements of war, while the British and Italian delegations thought that a special conference should be convened for the purpose.

In an effort to meet the points raised by other delegations, the United States delegation submitted yet another draft in which the prohibition of the traffic in means of chemical warfare was coupled with a universal prohibition of the use of poison gases. But the Committee adopted a resolution based on a joint proposal put forward by the British, Italian and Polish delegations. It read:

> The Military, Naval and Air Technical Committee:
> Unanimously recognizing that chemical and bacteriological warfare has been justly condemned by the opinion of the civilized world, and to the end that this prohibition shall be universally accepted as a part of the international law binding the conscience and practice of nations, considers that every possible effort should be made to secure as soon as possible a universal Convention forbidding this warfare;
> Considering that the prohibition of the export of chemical and bacteriological arms is, in most cases, practically impossible, and would, moreover, be of no effect until all nations undertook to abstain from chemical and bacteriological warfare of all kinds;
> Proposes that this larger political issue, namely, the prohibition of chemical and bacteriological warfare should be considered by a special Conference which should be convoked at an early date and at which all States would be represented. [9]

V. *The report of the General Rapporteurs*

The reports of the Legal and Military Committees were included in the report of the General Rapporteurs.

The General Rapporteurs stated that there was reason to hope—in view of the purely humanitarian aims underlying the proposals submitted by the United States and Poland—that a conference convened with a view to eliminating from future wars chemical and bacteriological arms would be likely to secure direct cooperation of all nations throughout the world. They supported the opinion of the Military Committee and suggested that a resolution be included in the Final Act of the Conference, calling for a universal convention which would deal with the problem of the prohibition of chemical and bacteriological warfare in general. [10]

On 5 June 1925, during the discussion of the report of the General Rapporteurs in the General Committee, the representative of Italy (who was also rapporteur of the Military Committee), in an attempt to substantiate the thesis that the prohibition of trade could not be effective, quoted the opinions of scientists (included in the Temporary Mixed Commission's Report of 30 July 1924) that the materials required for chemical warfare were in everyday use in peacetime industries.

The Swiss representative said that it was hardly possible for the Conference, after having had the question of chemical warfare submitted to it, to break up without at least referring to and affirming once again the existing international engagements in this sphere; evidence must be furnished of the earnest desire that chemical and bacteriological warfare should be formally condemned. To this end he proposed a text, to be inserted in the Final Act of the Conference, which would recognize that the use of poisonous and similar gases was already forbidden by the law and conscience of mankind and that a codification of this principle in a universal convention would constitute an additional guarantee; definite rules as to the application of the principle would also have to be laid down.

In supporting the Swiss position the representative of Japan stressed that unless the use of asphyxiating gases as a military device was first condemned, it would not be possible effectively to prohibit their exportation. A similar view was voiced by the representative of the Netherlands.

The delegate of China suggested that the League's Committee on the Codification of International Law should be asked to investigate the legal provisions which could be introduced into international law to prohibit the use of gases and bacteriological means of warfare.

The delegate of Colombia expressed his dislike for recommendations and declarations of the kind suggested by the Swiss representative for inclusion in the Final Act; such a procedure, as a rule, merely cloaked the weakness of a conference. An important advance had already been taken towards the abolition of asphyxiating gases, and it was inconceivable merely to adopt a resolution concluding with so feeble a *voeu* as the one to embody the principle referred to in international law. That would be a retrograde step as compared with Article 5 of the Treaty of Washington.

The United States representative then recommended that a resolution based on Article 5 of the Treaty of Washington of 6 February 1922 (see page 46) be adopted with a view to its immediate signature by the delegations which might be prepared to sign; it could be open for signature for a reasonable time by other states, whether represented at the Conference or not. He added that, if it should be felt for any reason that the action suggested could not be taken, the President of the United States would be glad to extend an invitation for a special conference at Washington with a view to framing a convention on chemical warfare.

The representative of the British Empire approved the United States alternative proposal concerning a special conference.

The representative of France, reiterating his country's desire to limit the horrors of war and to prevent the advancement of science from adding further to them, quoted the following sentence from the military regulations of France on the conduct of the larger units: "Faithful to the international undertakings which France has signed, the French Government will, on the outbreak of war, and in agreement with the allies, endeavour to obtain from enemy governments an undertaking that they will not employ gas as a weapon of war." As to the United States proposals, he preferred the adoption of a text—in the form of a protocol—which would be open for signature by states. He thought it was the most direct and the most expeditious procedure to follow, but would not be opposed to summoning a special conference, should there be objections to the first solution. At the same time he warned that international morality was not sufficient to ensure respect for engagements not to have recourse to some particular form of warfare, unless the parties felt that they had behind them sanctions which the signatory powers had subscribed to in advance. The only way of giving effect to the noble aspirations which had been expressed with regard to chemical and bacteriological weapons, was to let the state which violated its undertakings know that it would find arrayed against it the forces of the whole civilized world.

The delegate of Norway consistently upheld the position that it was impossible, once war broke out, to prevent the use of the most horrible methods; any formula which would be adopted, would merely be another stone on the road paved with good intentions.

The General Committee adopted the United States proposal to establish a protocol embodying Article 5 of the Treaty of Washington, or some similar formula to be worked out by the Drafting Committee. [11]

VI. Drafting the Geneva Protocol

The Drafting Committee drew up the following text:

At the time of signing the Convention for the control of the international trade in arms, ammunition and implements of war of today's date, the undersigned Plenipotentiaries, in the name of their respective Governments:

Considering that the use in war of asphyxiating, poisonous or other gases, and of all analogous liquids, materials or devices, has been justly condemned by public opinion, and that the prohibition of such use has been declared in treaties to which the majority of Powers of the world are parties;

And to the end that this prohibition shall be universally accepted as a part of international law, binding alike the conscience and the practice of nations;

Declare that the High Contracting Parties, so far as they are not already parties to treaties prohibiting such use, accept this prohibition and agree to be bound thereby as between themselves.

The High Contracting Powers will exert every effort to induce other States to adhere to the present Protocol. Such adhesion will be notified to the Government of the French Republic, and by the latter to all signatory and adhering States, and will take effect on the date of the notification by the French Government.

The present Protocol, of which the French and English texts shall both be authentic, shall be ratified as soon as possible. It shall bear today's date and shall be open for signature until September 30th, 1926.

Each Power shall address its ratification to the Government of the French Republic, which shall at once notify the deposit of such ratification to each of the signatory and adhering Powers.

The instruments of ratification and adhesion shall remain deposited in the archives of the French Government.

The present Protocol will come into force for each signatory Power as from the date of deposit of its ratification and, from that moment, each Power will be bound as regards other Powers which have already deposited their ratifications.

On 8 June 1925, the General Committee discussed the Drafting Committee's text.

The delegate of Poland drew attention to the omission of the prohibition of bacteriological warfare. He said that the results of such warfare would exceed even the horror of the devastation caused by the use of chemical methods. As far as production was concerned, the bacteriological weapon had several advantages over the chemical one; it could be manufactured more easily, more cheaply and with absolute secrecy. Unlike poison gas, the action of which was generally of short duration and restricted to a limited area, cultures of microbes, once secretly let loose in any place, might, due to their speedy multiplication and their ever-increasing virulence, easily occasion epidemics affecting great masses of men, animals and even plants. The Polish representative mentioned some of the most deadly weapons in warfare against the human race, including cholera, typhoid fever, plague, tetanus, glanders and botulism. Animals might be infected, for instance, by cultures of the germ of glanders, cattle plague, etc. Bacteriological warfare could be waged against the cultivation of plants, vineyards, orchards and fields. The consequences of bacteriological warfare would thus be felt equally by the armed forces of the belligerents and the whole civilian population, even against the desire of the belligerents, who would be unable to restrict the action of such weapons.

He reminded the Committee that the discussions which had taken place at the Conference with regard to chemical warfare referred also to bacteriological warfare, and moved that the Protocol should be completed accordingly.

The US representative accepted the amendment proposed by Poland. The French representative thought that the formula used by the Drafting Committee was wide enough to cover bacteriological warfare, but had no objection to making an explicit reference to it.

The Polish proposal was adopted.

The representative of the Irish Free State suggested that paragraph 4 of the Protocol should be so amended as to omit the words: "so far as they are not already parties to treaties prohibiting such use". Since one of the treaties referred to was the Treaty of Washington of 1922, which had not come into force, all the powers should sign the Protocol.

The Irish suggestion, along with some Japanese drafting proposals to bring the text more in line with the formulations of the Washington Treaty, was referred back to the Drafting Committee which was also entrusted with the final wording of the Polish amendment. [12]

On 9 June 1925, the Drafting Committee reported its opinion with regard to the observations made by the Irish representative, stating that the text of the sentence in question should remain in the draft; there could be no doubt that all the powers represented at the Conference, whether bound by previous treaties or not, would have to sign the Protocol. Most of the other amendments were embodied in the new text of the Protocol. [13]

The Irish delegation withdrew its suggestion but declared that states which were already parties to similar treaties would not be excused thereby from signing the Protocol.

VII. *Adoption and signing of the Geneva Protocol*

On 10 June 1925 the text of the Protocol, as eventually proposed by the Drafting Committee, and as subsequently amended following the Turkish suggestion to delete the first part of the first sentence of the preamble, was unanimously adopted by the General Committee. [14]

On 16 June 1925 the Conference decided to ask the Secretary-General of the League of Nations to draw the attention of the Committee of Jurists for the Codification of International Law to the Protocol as well as to the clause in the Washington Treaty relating to the prohibition of chemical warfare. [15]

On 17 June 1925, the following text of the Protocol was approved by the Conference.

**Protocol for the Prohibition
of the Use in War of Asphyxiating, Poisonous or Other Gases,
and of Bacteriological Methods of Warfare**

The Undersigned Plenipotentiaries, in the name of their respective Governments:

Whereas the use in war of asphyxiating, poisonous or other gases, and of all analogous liquids, materials or devices, has been justly condemned by the general opinion of the civilized world; and

Whereas the prohibition of such use has been declared in Treaties to which the majority of Powers of the world are Parties; and

To the end that this prohibition shall be universally accepted as a part of International Law, binding alike the conscience and the practice of nations;

Declare:

That the High Contracting Parties, so far as they are not already Parties to Treaties prohibiting such use, accept this prohibition, agree to extend this prohibition to the use of bacteriological methods of warfare and agree to be bound as between themselves according to the terms of this declaration.

The High Contracting Parties will exert every effort to induce other States to accede to the present Protocol. Such accession will be notified to the Government of the French Republic, and by the latter to all signatory and acceding Powers, and will take effect on the date of the notification by the Government of the French Republic.

The present Protocol, of which the French and English texts are both authentic, shall be ratified as soon as possible. It shall bear to-day's date.

The ratifications of the present Protocol shall be addressed to the Government of the French Republic, which will at once notify the deposit of such ratification to each of the signatory and acceding Powers.

The instruments of ratification of and accession to the present Protocol will remain deposited in the archives of the Government of the French Republic.

The present Protocol will come into force for each signatory Power as from the date of deposit of its ratification, and, from that moment, each Power will be bound as regards other Powers which have already deposited their ratifications.

In witness whereof the Plenipotentiaries have signed the present Protocol.

Done in Geneva in a single copy, this seventeenth day of June, One Thousand Nine Hundred and Twenty-Five.

(For a list of signatures, ratifications, accessions and successions to the Protocol, see page 342.)

Three more documents were signed on the same day: Convention for the Supervision of the International Trade in Arms and Ammunition and in Implements of War; Declaration regarding the Territory of Ifni (the inclusion of that territory in the special zones with a regime of supervision established in Chapter III of the Convention); and Protocol of Signature which recognized the right of states enumerated in the preamble of the Convention to sign all or any of the above-mentioned instruments at any date prior to 30 September 1926. [16]

Unlike the Protocol prohibiting the use of gases and bacteriological weapons, the Convention for the supervision of trade in arms never entered into force.

VIII. *Summary and comment*

In 1925, CB weapons were discussed at a conference in Geneva devoted to regulating trade in arms.

The proceedings of the conference were marked by a conflict of interests between the weapon-producing countries and the non-producers. The latter considered regulations concerning trade in arms to be a restriction putting them in a position of inequality, unless restrictions were imposed upon the manufacturers of weapons. The controversy grew even sharper with regard to chemical and bacteriological weapons, when the question of a total ban on trade in these weapons was raised. The non-producers felt that such a measure would not only prove unjust, leaving the have-nots without means of defence, but would be unrealistic as long as chemical and bacteriological arms remained legal means of warfare. Hence they insisted on the prohibition of production, or at least of the use of those weapons, and expressed the desire to include the ban in a binding international agreement. Another important demand was that security guarantees be given by the producing countries to the non-producers.

There were similar differences between the haves and have-nots some forty years later, in the course of negotiations on the non-proliferation of nuclear weapons. The principle involved was, roughly speaking, the same. In order to stop effectively trade in chemical and bacteriological weapons i.e., their dissemination—argued the non-producers at the 1925 Conference—the manufacturing powers would have to renounce their production and use. To check horizontal proliferation of nuclear weapons, i.e., their acquisition by new states—argued the non-nuclear-weapon participants in the non-proliferation debate—vertical proliferation of such weapons, i.e., their production, accumulation and use, would also have to be banned. The former succeeded in outlawing the use of chemical and bacteriological weapons. The latter received, in conjunction with the Non-Proliferation Treaty, some security guarantees.

The Geneva Protocol signed on 17 June 1925 reproduced the terms of the Washington Treaty of 1922, and included the prohibition of BW—an important addition for which Poland was responsible. It thus prohibited the use in war of asphyxiating, poisonous or other gases, and of bacteriological methods of warfare.

No observations were made, either in the course of the discussion, or at the signing of the Protocol, by any country, with regard to the scope of the prohibition.

Many states which later ratified or acceded to the Protocol qualified their adherence to it by two-fold reservations: that the Protocol was binding on the state making the reservation only in its relations with the parties to the Protocol; that it would cease to be binding on the state making the reservation in its action against an enemy state if the latter's armed forces or allies failed to respect the prohibitions laid down in the Protocol.

While China, France, the United Kingdom, and the USSR became parties to the Geneva Protocol within a few years after its signing, the United States government, which initiated and pushed through the Protocol, did not ratify it. In the 1920s, it was prevented from doing so by the concerted opposition of the American chemical industry and the military, in particular the Chemical Warfare Service.

The Geneva Protocol was before the US Senate until 1947, when it was withdrawn by the US President along with a number of other treaties. It was re-submitted for approval in August 1970.

Chapter 3. 1926–1930

I. *The convening of the Preparatory Commission for the Disarmament Conference*

By virtue of the League of Nations Council resolution of 12 December 1925, a commission was set up entrusted with the preparation for the Conference for the Reduction and Limitation of Armaments (often referred to as the Disarmament Conference). [1]

The Preparatory Commission held six sessions from 1926 to 1930. At its first session (18–26 May 1926), the representative of Belgium said that it would be a waste of time to discuss the limitation of the number of guns and rifles while ignoring modern means of warfare, such as chemical means. Speaking about the so-called industrial mobilization, he pointed out that a dyeworks could be converted into a formidable instrument of war within a period varying, according to circumstances, from forty days to forty hours. However, signing conventions prohibiting the use of gases in war was no solution. There was not a single country which, after signing such conventions, would not find it necessary to produce gases as a defensive weapon, or in order to prepare a counter-offensive and—as was known—defensive and counter-offensive measures inevitably provoked an offensive. The offensive, in turn, meant destruction of cities and the degradation of mankind. According to a US source there were gases which could permeate a radius of 50 miles around New York, destroying all life. The Belgian delegate expressed the belief that there should be general supervision with regard to poisonous gases. Preparation of these instruments of death in laboratories could not be allowed.

The British representative pointed out that, although one might not be able to limit the amount of gas produced, one could limit the instruments enabling its use, thus limiting the "utilizability" of gas.

He conceded that forbidding chemical warfare altogether would be useful, but doubted whether that would, by itself, provide sufficient security. There was scarcely a country, certainly not one of the larger countries, which was not carrying on experiments in connection with chemical warfare, perhaps not for the purpose of using chemical weapons aggressively but in order to be prepared should they be used aggressively against itself. Thus, neither the Treaty of Washington which condemned the use of poisonous gases, nor the Geneva Protocol of 1925, had lessened the activities of various nations in preparing for such warfare in the future. As to the Belgian suggestion concerning supervision, he wondered what control would do to check the possibility of manufacturing poisonous gases.

The Italian delegate said that the possible disadvantages and dangers of applying international control outweighed any advantages it might offer. To be useful, control would have to cover the whole economic life and, more especially, the industrial life of the country. If a country was suspected of bad faith, every corner of the producing factories would have to be examined, a special service would have to be organized to search for hidden material, etc.; all such searches would probably reveal nothing. Besides, a country could quite easily render any control illusory and ineffective. Control would certainly lead to friction and generate bad feeling; an atmosphere of misunderstanding and resentment would be created. Countries must be trusted when they undertook to observe the pacts they had signed.

The representative of France felt that the League of Nations and the states represented in it had already done their duty in trying to avert the peril of chemical and bacteriological warfare by signing the Geneva Protocol of 1925.

A similar opinion was expressed by the representative of the Kingdom of the Serbs, Croats and Slovenes. He said that the Geneva Protocol constituted a solemn undertaking of states to renounce chemical warfare. It had been unanimously approved by all the delegations attending the Conference on the Supervision of the International Trade in Arms, including the German delegation—Germany being a great power which had not adhered to the Washington Treaty, and also one of the largest chemical powers in the world. He wondered whether the Commission was entitled to question the value of the Protocol which had solved the thorny question of chemical warfare finally and categorically. This was the best solution and the only effective one; all others were half-measures. It was desirable that the big military and chemical powers should be the first to ratify the Protocol. (France was the only country which by then had ratified it.) Others would no doubt follow suit. Otherwise, by casting doubt upon the value of the Protocol and the good faith of its signatories, the Commission might alienate public opinion, and the spectre of chemical warfare would again raise its head. This would compel all states to proceed with or institute chemical armaments, a step which would inevitably lead to a fresh armaments race. He suggested that it was necessary to include among chemical weapons incendiary material which, when thrown from airplanes, might be almost as dangerous to large centres of population as poison gas. [2]

There was, however, no opposition to examining certain technical problems relating to chemical warfare.

II. *Consideration of CBW by the Sub-Commissions of the Preparatory Commission*

At the British delegate's proposal, the Preparatory Commission decided to refer to the competent subcommissions—Sub-Commission A and Sub-Commission B—a series of questions "without prejudice to any convention or rule of international law on the subject".

Sub-Commission A was composed of a military expert, a naval expert and an air expert for each of the countries represented on the Preparatory Commission. Its terms of reference included items of a technical nature relating to disarmament.

Sub-Commission B was composed of a representative of each delegation; its task was to deal with all non-military questions, related mainly to the economic and financial aspects of disarmament, submitted to it by the Preparatory Commission. It was authorized to obtain opinions on these questions from any organizations or persons it thought fit to consult, and in particular from the Joint Commission.

The Joint Commission was constituted by the League of Nations Council resolution of 12 December 1925 to assist the Preparatory Commission for the Disarmament Conference. Composed of members of the various technical organizations of the League and the Labour Office, it was entitled to call in the assistance of such experts as it considered desirable.

The questions submitted by the Preparatory Commission were as follows:

To Sub-Commissions A and B

1. (*a*) Can factories normally and legitimately employed for chemical purposes, including dyeworks, be quickly adapted to manufacture poison gases?

(*b*) If the answer to the above is in the affirmative, how long would it take to effect the change?

(*c*) Can any proposals be made to prevent or hinder chemical factories from being used for the production of poisonous gases?

To Sub-Commission A

2. (*a*) What are the means which would probably be employed for spreading gas and what would be the apparatus required?

(*b*) How long would it take to manufacture this apparatus, and how long would it take to superimpose this apparatus on the normal equipment of an airplane?

(*c*) Would the length of time referred to immediately above vary in the case of military or civilian aircraft?

3. (a) What is the information in existence as to the effect of the distribution of poisonous gas over closely populated districts?

(b) Have any experiments been carried out on this subject?

(c) Apart from the difficulty of equipping the entire population of a city with gas masks, are there any gases known against which a gas mask affords no protection?

Sub-Commission A was also invited to consider what effective sanctions could be proposed for the enforcement of the international undertaking not to employ poison gas or bacteria in warfare. In addition, Sub-Commission A was requested to investigate the consequences from the military point of view of inserting in the convention relative to disarmament, or in that regarding the prohibition of certain forms of warfare, provisions similar to those contained in the Statute of the International Labour Office (Articles 411 to 420 of the Treaty of Versailles).[1]

Sub-Commission B was requested to ascertain the consequences of such insertion from the economic point of view. [3]

Sub-Commission A submitted its report to the Preparatory Commission in December 1926. [4]

Sub-Commission B, in exercise of the power conferred upon it by the Preparatory Commission entrusted the Joint Commission with the study of the questionnaire on chemical warfare. The Joint Commission thought it necessary to consult the opinion of specially qualified experts in the chemical industry. For this purpose it appointed a committee consisting of four of its own members and the following experts: Professor Livio Cambi, Director of the Institute of Industrial Chemistry at the University of Milan, Italy; M. Joseph Frossard, Director-General of the Kuhlmann Institutes in Paris, France; Professor Just, industrial chemist and former university professor, Germany; Dr A. T. de Mouilpied, industrial chemist, Great Britain; Dr E. Zanetti, Professor of chemistry at Columbia University, USA.

The report of the committee of experts was included in the report of the Joint Commission, which in turn was embodied in the report of Sub-Commission B. The latter, dated 30 November 1926, was submitted to the Preparatory Commission for the Disarmament Conference. [5]

---

[1] Articles 411–420 of the Treaty of Versailles of 28 June 1919 are contained in Part XIII dealing with the Organization of Labour. They provide for the right of any member to file a complaint with the International Labour Office if it is not satisfied that any other member is securing the effective observance of a convention which both have ratified; communication of the complaint to the government in question; appointment and composition of the Commission of Enquiry; possible recommendations of the Commission, including measures of an economic character against a defaulting government; publication of the report; possible reference to the Permanent Court of International Justice; conditions for the discontinuance of measures taken against the defaulting government.

In forwarding its report, Sub-Commission B, which had not considered the substance of the questions dealt with in the Joint Commission's report, pointed out that the opinions included therein were those of experts, and were not the offical views of the governments of the countries of which the experts happened to be nationals, and still less of the governments which had no nationals on the Joint Commission.

The replies of the Sub-Commissions to the questions submitted by the Preparatory Commission are summarized here.

Question 1(a): Can factories normally and legitimately employed for chemical purposes, including dyeworks, be quickly adapted to manufacture poison gases?

Sub-Commission A replied:

Generally speaking, chemical factories, especially dyeworks and factories connected therewith, can be very quickly adapted to the manufacture of poisonous gases. In the dye industry many of the intermediates themselves are poisonous chemicals which are capable of immediate use in chemical warfare, while others are intermediates for the manufacture of chemical warfare agents. As a general rule, chemical warfare agents are similar in composition to commercial chemicals and are made by similar processes. The raw materials for the chemical warfare agents are commercial products, and the commercial uses of the more important of these raw materials are well developed. It is apparent, therefore, that chemical factories normally and legitimately employed for chemical purposes, including dyeworks, could be quickly, often immediately, adapted to the manufacture of poisonous gases.

Sub-Commission B replied:

Chemical factories can be adapted to manufacture poison gases; the rapidity of such adaptation varies according to circumstances, as described in the reply to question 1 (*b*).

Question 1 (b): If the answer to the above [question 1 (*a*)] is in the affirmative how long would it take to effect the change?

Sub-Commission A replied:

The time required for adaptation depends on the state of the chemical industry or factory and the nature of the gas to be made. Chlorine, bromine and phosgene, which are articles of commerce, can be used as poisonous gases; hence the plants which produce them for industrial purposes are immediately available for war use. Only a short time would be needed to convert works which do not carry the poisonous chemical to its final stage. Their adaptation would be a comparatively simple matter. This applies to works manufacturing yperite, chloracetophenone, methyl chloroformate, diphosgene, bromacetone, chloracetone and similar products.

Azo plants used in dye-making can be converted without delay to manufacturing chemical warfare arsenicals, as these processes need skilled personnel rather than complicated apparatus.

The intermediates necessary for the manufacture of certain gases are used in peacetime production, for example in dye-making. With a well-developed chemical industry and carefully prepared plans, peacetime production could be largely turned over to war production within three months.

Although a country with a large chemical industry would be able to begin quantity production of chemical warfare agents in a very short time, the determining factor probably would be the production of special containers for war use of chemical warfare agents rather than the conversion or employment of chemical plant for the production of such agents. However, for emergency use simple types of containers can easily be improvised. Another important factor in the matter of quantity production of chemical warfare agents is the availability of a sufficient supply of raw materials.

Sub-Commission B replied:

No time is required for adaptation in the case of poison gases which are produced in large quantities in industry.

For poison gases which are manufactured in industry, but in quantities insufficient for war requirements, two cases have to be considered:

The period required for adaptation is almost nil if the production can be intensified by modifying such factors as the degree of utilization of existing installations, recourse to reserve units, the technical, physical and chemical conditions of the reactions, etc.

If increasing production necessitates increasing the number of existing installations, the period of adaptation would be from a few weeks to a few months, varying according to the scientific, technical and material resources of the country and its industrial organization.

In the case of poison gases not yet manufactured in industry, the period required for the adaptation of factories is difficult to determine, but depends primarily on the resources of the country concerned. That period might be considerably extended if it is necessary to establish entirely new factories with special material in a country industrially ill-equipped. On the other hand, it would be much shorter if the equipment of the existing factories could be adapted rapidly to manufacture new products, as is often the case, and if the existing chemical industry is sufficiently important. Countries which have, in addition to large factories, a strong industrial organization, and which have standardized the plant employed by the chemical industry, more particularly in the metallurgical, ceramic and other industries, enjoy considerable advantages. Such countries are particularly well-equipped for obtaining rapidly the material necessary for new manufactures.

Question 1 (c): Can any proposals be made to prevent or hinder chemical factories from being used for the production of poisonous gases?

Sub-Commission A replied:

In practice, it is impossible to prevent or hinder the manufacture of poisonous gases in peacetime. The industry is in the hands of private enterprise; the industrial interests can put up an effective opposition to prohibition of the manufacture of poisonous chemicals which are needed for commercial purposes and which may be used for chemical warfare purposes, and to any obligation to use other methods of manufacturing synthetic products. It will in any case always be impossible to gain any knowledge of discoveries and to prevent the study of poisonous substances in the laboratories.

Until a universally effective and absolute prohibition as to the employment of means of chemical warfare is brought into existence, it seems impracticable to submit any proposals tending to prevent or hinder chemical factories from being used for the production of poisonous gases.

Sub-Commission B replied:

It is absolutely impossible to suppress the manufacture of poisonous gases, such gases being the current products of industry or intermediate agents indispensable for obtaining other products.

It may perhaps be possible to institute between the industries in different countries agreements which would be sanctioned by the states concerned and would provide more particularly for the rationing of manufacture. Such agreements might cover both the nature of the products and the quantities manufactured. They could allay much of the anxiety which would undoubtedly be caused by a state whose neighbours see its industrial chemical power increasing to disquieting proportions. They would make it possible, at the same time, to exercise stricter supervision as regards prohibition to manufacture certain products which appear to be of use only for military purposes.

It would be expedient to abolish all subsidies to official laboratories and private institutions whose object is to promote research in the matter of poisonous gases for purely military purposes.

Question 2 (a): What are the means which would probably be employed for spreading gas and what would be the apparatus required?

Sub-Commission A replied:

Gas can be used in projectiles (i.e., artillery shell or trench mortar bombs), in airplane bombs, as sprays from aircraft or motor vehicles, in gas cylinders, in thermogenerators and in simple containers opened by an explosive charge. If use by aircraft is considered, bombs of steel, cast iron, porcelain or glass may be used, as well as spraying devices; the latter may be ordinary iron or steel containers from which the gas in liquid form can be ejected by pressure or under the action of gravity on opening a tap or valve.

All aircraft fitted with smoke projectors can very easily use that apparatus for spreading poisonous gas.

Question 2 (b): How long would it take to manufacture this apparatus, and how long would it take to superimpose this apparatus on the normal equipment of an airplane?

Sub-Commission A replied:

Certain types of airplane bombs can be made very quickly, as the design can be simple; spraying devices can be improvised from ordinary trade articles with great rapidity. The apparatus can be added to the normal equipment of an airplane practically at a few hours notice. Properly designed and constructed equipment may, however, take several weeks to make and install. Artillery shell and similar projectiles for gas would take several weeks to make. Gas cylinders, on the other hand, are used in industry for the transport of chlorine, and supplies are immediately available.

Question 2 (c): Would the length of time referred to immediately above vary in the case of military or civilian aircraft?

Sub-Commission A replied:

Depending upon the type of aircraft and the character of the apparatus, there appears to be no reason why the devices mentioned cannot be attached with equal readiness to civilian as well as military aircraft.

The delegations of Bulgaria, Czechoslovakia, France, Poland, Romania and the Kingdom of the Serbs, Croats and Slovenes added to the above statement that in certain cases no time would be required; this would be the case of civilian aircraft fitted with smoke projectors for the purpose of commercial publicity.

Question 3 (a): What is the information in existence as to the effect of the distribution of poisonous gas over closely populated districts?

Sub-Commission A replied:

The information as to the effect of the distribution of poisonous gas over closely populated districts is not sufficiently exact to take as a basis for a final conclusion.

The effect upon civilians inhabiting such districts is not comparable to the experience of the use of gas in shell and similar projectiles under war conditions.

Question 3 (b): Have any experiments been carried out on this subject?

Sub-Commission A replied:

To the knowledge of the Sub-Commission, no experiments as to the effect of the distribution of poisonous gas over closely populated districts have been carried out.

The delegations of Belgium, Finland, France, Poland, Romania and the Kingdom of the Serbs, Croats and Slovenes added that trials carried out in certain countries for special purposes, such as covering wooded or cultivated areas with insecticide, seem to have been successful. They afford a valuable indication of the necessary conditions for sprinkling with poisonous gas a town or roads, cross-roads or open spaces over which enemy troops have to pass.

Question 3 (c): Apart from the difficulty of equipping the entire population of a city with gas masks, are there any gases known against which a gas mask affords no protection?

Sub-Commission A replied:

The best gas masks will protect the eyes and lungs against all known gases likely to be used as offensive agents. Typical service gas masks do not protect against carbon monoxide, but the employment of this gas in war is improbable owing to technical difficulties; however, special masks can be produced to deal with it, if required. Blistering agents which attack the skin, such as mustard gas, require protective clothing in addition to gas masks.

While it is improbable that any new gas will be developed against which existing gas masks afford no protection, the possibility of such surprise development must be borne in mind. This might result in serious casualties before adequate protection could be provided.

Sanctions for the enforcement of the international undertaking not to employ poisonous gas or bacteria in warfare

Sub-Commission A:

A. The delegations of Belgium, Bulgaria, Czechoslovakia, Finland, France, Poland, Romania and the Kingdom of the Serbs, Croats and Slovenes submitted the following reply:

In view of the fact that a country in possession of a chemical industry will always be able, within a period of time varying between a few hours and a few weeks, to make use of that industry in order to manufacture poisonous substances and use them in warfare, the only effective sanction, from the technical point of view, which can prevent a state from violating its undertakings in connection with the prohibition of chemical warfare consists in the possibility of immediate reprisals by the same chemical means.

The fear of such reprisals would probably be sufficient to prevent any state from resorting to chemical warfare. The more forcible the reprisals envisaged, the more effective would be their preventive force.

All states in possession of a chemical industry should therefore undertake:

to put at the disposal of any state which is attacked by gas the raw materials, chemical products and means of operation necessary for reprisals;

to engage themselves in joint reprisals, so far as distance permits, by the use of chemical means against the state which has committed an act of aggression by the use of gas.

This undertaking would not justify any special preparation for chemical warfare in peacetime, since reprisals can always be carried out easily by means of the aircraft available for use, without measures of mobilization, and since a chemical industry can be converted very quickly, the moment it becomes necessary, to the manufacture of the product required for reprisals.

B. The delegations of the Argentine, the British Empire, Chile, Germany, Italy, Japan, the Netherlands, Spain, Sweden and the United States of America stated that they were unable to associate themselves with the proposal for sanctions. In their view, Sub-Commission A was not competent to express any opinion upon such a proposal because the considerations underlying it were essentially political.

C. The delegation of the Netherlands made, in addition to B above, the following declaration:

The delegation cannot under any circumstances agree to the employment of chemical warfare as a sanction, for the following reasons:

In the majority of cases it will be difficult to determine definitely whether any country has really used poisonous gas as a weapon of war.

Reprisals involving the use of such means as have been condemned by the civilized world, and especially their use by countries outside the conflict, are not acceptable in principle.

The starting of a general chemical war for reasons of sanctions would be deplorable for mankind as a whole, for in this way use would be made of a means of warfare which has been generally condemned by the whole of humanity.

D. The delegation of Germany submitted, in addition to B above, the following statement:

At the time of the Conference on the Supervision of the International Trade in Arms, Germany stated that it was ready to agree without any reservation to any international rules for the purpose of abolishing chemical warfare. It is precisely from the point of view of that statement that the measures suggested as sanctions might give rise to objections, for, if the contemplated sanctions are allowed, the states applying them would have legal authority to use gas as a means of warfare. If the desire is to abolish chemical warfare, any measures, even in the form of sanctions, which are calculated to make chemical warfare general and to make it a form of warfare recognized by international law, must be avoided.

In a special declaration the delegations of Belgium, Czechoslovakia, Finland, France, Poland, Romania and the Kingdom of the Serbs, Croats and Slovenes made it clear that they rejected any interpretation of the replies given by the Sub-Commission which would tend to diminish the gravity of the menace existing in the possibility of chemical warfare.

In particular, they drew the attention of the Preparatory Commission to the following points:

1. Although it is true that in some cases it might take three months to convert peacetime products of the various chemical industries into poisonous gas for use in war, it must be borne in mind that the manufacture of many poisonous products with a view to their use in war would be very rapid and very easy for a country with a developed chemical industry, particularly for the manufacture of dyestuffs and pharmaceutical products, provided that the country also has the necessary raw materials and trained technical personnel well supplied with laboratories and adequate means for research work. A chemical industry, and especially the scientists and technical experts essential to it, cannot be improvised. As a result, a country with a highly developed chemical industry is at any time prepared for chemical warfare. On the other hand, those countries whose chemical industries are less well-developed or non-existent would require more or less considerable space of time in order to obtain the products necessary for chemical warfare.

2. It is possible that the gas masks of some countries stop all known gases. It is also probable that, as soon as a poisonous gas is identified, a suitable protecting mask can be devised fairly quickly, granted that well-equipped laboratories and trained experts are available.
It must, however, be noted:

that the manufacture of new masks is always a relatively slow matter;

that the masks known and used in the First World War did not give complete protection against arsines;

that it is always comparatively difficult to give protection against vesicants;

that a heavy bomb filled with poison, falling inside a building, would be capable of producing such a concentration of poisonous gas that everyone would succumb to it, even if protected by a mask;

that poisonous gases are potentially unlimited in number, that the use of a new poisonous gas would always be certain of producing a very successful surprise effect, and that it is this circumstance which distinguishes gas warfare from the use of explosives where the element of surprise appears to have been eliminated.

3. Although no direct experiments have been made in regard to the bombarding of densely populated districts with poisonous products, and although experience from the First World War on this matter is inconclusive, it should be observed:

that, according to the information available to Sub-Commission A, experiments have been made upon animals, and poisonous gases have been spread over forests for the purpose of destroying insects;

that a bombardment by poisonous substances would probably produce much more serious effects upon a town than a bombardment by explosives, particularly if the poison used were persistent in its effects and sufficiently dense to penetrate into shelters, cellars, etc.; if it were contained in heavy bombs capable of poisoning the atmosphere to a very high degree; and if incendiary bombs were at the same time used to hamper the organization of aid;

that it was ascertained during the war that the effects of yperite, for example, can be felt for a fortnight, and necessitate total evacuation of certain areas.

The delegations (Belgium, Czechoslovakia, Finland, France, Poland, Romania and the Kingdom of the Serbs, Croats and Slovenes) further noted that no effective technical sanction had been proposed to Sub-Commission A to prevent the use of chemical weapons in war, except the possibility of immediate reprisals. The delegations which considered that the question of reprisals was exclusively political suggested no other measure capable of having the same effect.

Under these circumstances:

In view of the fact that in most cases the preventive supervision of chemical preparations would be ineffective, because these preparations could be undertaken only at the moment they were required; and

In view of the fact that, for the same reasons, the limitation or suppression of chemical armaments would be either impossible or ineffective;

The mentioned delegations, while realizing the practical difficulties of organizing collective reprisals and the political and moral problems which such organization might raise, desired to place on record that, in the absence of these reprisals, the *preventive* effect of which they thought might be decisive, there were no technical means of preventing chemical warfare.

Consequences from the military point of view of inserting in the convention provisions similar to those contained in the Statute of the International Labour Office

Sub-Commission A replied:

The delegations of the Argentine and Germany expressed no opinion on the question.

A. The delegations of Chile, Italy, Japan and the USA stated that the insertion of such provisions would result in the establishment of a procedure by which the Council—or any other body suggested in order to include non-members of the League of Nations—if called upon to deal with a complaint regarding military measures taken by a state adhering to the convention, might order an enquiry with a view to ascertaining whether the complaint was well-founded. The delegations were of the opinion that such enquiries would in general prove fruitless and illusory. The suspected state, learning of the enquiry and its object long before it was actually carried out, could take steps to render it abortive; the enquiry would lack the factor essential for success, namely, unexpectedness. Even if it could be given the character of unexpectedness—though this hardly seemed possible—the enquiry, to be successful, would have to be carried out under strict and definite rules which would involve serious commitments in regard to military secrecy on the part of the state subjected to the enquiry. If the procedure in question were to be admissible, all the states adhering to the convention would have to accept the serious consequences which were inherent in the control entailed by enquiries of the kind proposed. The delegations could not agree to such control. They quoted the view, with which they agreed, of the Permanent Advisory Commission which in September 1921, referring to the expediency of exercising control over the declarations of countries members of the League of Nations in the matter of the undertakings entered into under Article 8 of the Covenant,[2] unanimously expressed the following opinion:

> Either control would be exercised independently of the power to be controlled, which does not appear to be consistent with its right of sovereignty;
> Or the nature and the time of such control would be decided upon by the government of the power to be controlled, and in these circumstances the suspicion attaching to the information provided by it would necessarily extend to a control which was also limited by it.
> In a word, the undertakings contained in Article 8 are based on a belief in the pledged word, and the Permanent Advisory Commission does not consider that it is either opportune or conducive to great efficiency to substitute mistrust for this belief.

B. The delegations of Belgium, the British Empire, Czechoslovakia, Finland, France, the Netherlands, Poland, Romania, the Kingdom of the Serbs, Croats and Slovenes, Spain and Sweden submitted a text in which they stated:

1. In cases in which it is possible to institute an enquiry on the sole basis of public documents or documents communicated by the government without proceeding to direct enquiries on the spot, it would be necessary to set up a special commission, both competent and impartial, which would be instructed to draw up an objective report after examining the complaint and the documents submitted.

2. In cases of preparations for aggression and in all cases in which the time required for the employment of the preceding method would be incompatible with the nature of the infraction, examination of the complaint may necessitate a direct enquiry carried out on the spot, as soon as possible. This is necessary to allow the discovery of a sudden increase in armaments, to hinder the accumulation of secret armaments, and to verify the existence of the characteristic features of the first preparations for an aggression, such as a considerable strengthening of units and peacetime effectives of a state, measures for partial or general mobilization—particularly the in-

---

[2] Article 8 of the Covenant of the League of Nations provided for the reduction of national armaments and the enforcement by common action of international obligations; formulation by the Council of plans for such reduction, subject to revision at least every ten years; obligation of the governments not to exceed the limits of armaments fixed in these plans without the concurrence of the Council; exchange of information between the members of the League as to the scale of their armaments, their military, naval and air programmes and the condition of such of their industries as were adaptable to warlike purposes.

dividual calling-up of reservists—requisitions, the manufacture of war material, etc.

(The delegations of the Netherlands and Sweden did not associate themselves with the preceding paragraph.)

From the military point of view, the logical consequences of the provisions contemplated would be the adoption of precise measures of execution corresponding to the special object of the enquiry and conferring on the latter technical guarantees of effectiveness, in the absence of which the enquiries would be liable not only to be needlessly vexatious, but also to give false guarantees of security and thus increase the dangers of aggression.

The decision to proceed to such enquiries would involve the following consequences:

(a) As regards the procedure prior to the enquiry, it would be essential to fix very short time-limits, so as to allow for the fact that infractions in the matter of armaments might in a very short time have much more serious consequences than those arising in the case of infractions of labour legislation.

(b) One of the conditions which is most necessary for the effectiveness of an enquiry is its element of surprise. It should be capable of being carried out by observing definite and strict rules.

3. The following military consequences would also result from the foregoing:

(a) possibility of secret mobilization plans being disclosed;

(b) possibility of war inventions and military secrets being disclosed;

(c) grave risks of friction between states;

(d) possibility of unjustifiable requests for enquiries made with the sole object of ascertaining secrets relating to the national defence of certain states.

Sub-Commission A did not consider whether there were any measures which might obviate these serious difficulties.

(The delegations of Czechoslovakia, Finland, France, Poland, Romania, the Kingdom of the Serbs, Croats and Slovenes and Spain did not associate themselves with paragraphs (c) and (d) above.)

4. The technical consequences of inserting, in a convention relative to disarmament, provisions for enquiries sufficiently accurate and rapid to be effective from the military point of view, would be:

To enable states exposed to risks of aggression to have military preparations contrary to the provisions of the convention detected in time, established and arrested;

To add to the military, naval and air conditions for the security of those states, and to enable them also to calculate the reduction of armaments to which they would agree.

(The delegations of the British Empire and Sweden did not associate themselves with the preceding paragraph.)

Consequences from the economic point of view of inserting in the convention provisions similar to those contained in the Statute of the International Labour Office

Sub-Commission B replied:

A convention on the prohibition of chemical warfare can not be concluded between the states, unless there are agreements between the industries concerned. It will be impossible to deal successfully with the various technical, industrial and commercial aspects of the question without the assistance of qualified representatives of the industries.

Two types of violation of a convention will have to be considered.

If the infraction is due to competition between the industries, it is necessary that it be settled directly, between them, by application of the regulations to be laid down in the agreements concluded between the industrial groups of different countries. In this case there would be no need for state intervention.

If the violation of the convention results from clandestine manufacture for export for military purposes or from the exercise of pressure by one of the states on its own industry, the application of the procedure of objections, investigations and sanctions would be called for.

The fact that the convention has been established after the conclusion of agreements between chemical industries would not create any special economic difficulties for investigation on the spot. It would be advisable not to send too many commissions, but to have recourse, if possible, to a single investigator of assured competence.

Economic sanctions would eventually be to the detriment of the states recognized as culpable and would tend to strengthen the feeling of general security.

The Joint Commission included the following proposal:

With a view to suppressing chemical and bacteriological warfare, each state undertakes to take the necessary measures to establish as a crime in common law and to punish with suitable penalties any exercises or training by military persons or civilians in the use of poisons and bacteria and, in particular, any exercises or training by air squadrons, ...

although—it admitted—the political and military nature of the proposal placed it outside the sphere of the Commission's work. [4 and 5]

The following important points were made in the course of the discussion at the third session of Sub-Commission A, which was held from 27 September to 5 November 1926, and which led to the conclusions of that body. The representative of the Kingdom of the Serbs, Croats and Slovenes reiterated the view that the best way of dealing with chemical warfare was to ratify the Geneva Protocol. But sanctions were needed, for a resort to chemical warfare might take place notwithstanding the fact that it had been prohibited in an international agreement. The chances were that no country would decide to use that method of attack, knowing that there were effective and rapid sanctions which could be applied, and that others would help the attacked country. That would be especially true if the resources concentrated against the attacker were such as to exceed his own.

The United States representative said that quite extensive experiments had been carried out in his country, one of the objects of which was to determine what might be the effects of gassing thickly populated areas; the experiments had been made on animals. On the basis of data available to it, the US delegation was of the opinion that the effects would not be more serious than in the case of employment of the same amount of high explosive shell in the same district, at the same time and under the same conditions. As a matter of fact, the resulting deaths from the use of gas would probably be fewer.

The French delegate pointed out that in perfectly legitimate chemical research work it was possible to hit upon a substance which might be poisonous; nothing could prevent a country from keeping these discoveries secret or from making schemes for the manufacture of poisonous substances, probably with the same raw materials, personnel and equipment. He did not agree that the effect of gas bombing would be no greater than that caused by high explosives, especially in the case of persistent gases and also burning gases. He recalled, in this context, the considerable losses suffered by troops marching through gorges which previously had been bombarded with mustard-gas shell.

The representative of the British Empire warned that the experiment of spreading insecticide over the tops of trees could not be compared with the spreading of poison gas in wartime.

The German delegate said that attempts had been made in his country to destroy dangerous forest insects with arsenic powder sprinkled by airplanes on the leaves which were subsequently to be eaten by caterpillars. The experiments were inconclusive. [6]

III. The Preparatory Commission's discussion of proposals for CBW prohibition

On 25 April 1927, at the third session of the Preparatory Commission, the delegations of Belgium, Czechoslovakia, Poland, Romania, and the Kingdom of the Serbs, Croats and Slovenes submitted the following proposal:

> The High Contracting Parties undertake to abstain from the use in war of asphyxiating, poisonous or similar gases, and of all analogous liquids, substances or processes.
> They also undertake to abstain from the use of all bacteriological methods of warfare.
> They undertake, moreover, not to permit the importation, the exportation or the manufacture on their territory of substances utilizable for chemical or bacteriological warfare, when they are imported, exported or manufactured with a view to such use.

The representative of Germany, in welcoming the proposal as aiming at suppressing chemical warfare, observed that it would not be enough to make the use of gases in war illegal. The use of such substances in wartime required their preparation in peacetime. If in time of peace the states agreed not to prepare for such use, not to train military airmen in bombing, not to prepare for infecting high roads and whole districts, not to train specialists to use poisonous substances in war—if, in other words, they abstained from developing the necessary conditions for employing the chemical weapon—only then would it be effectively abolished in war. It was not the importance of the chemical industry in a country which was decisive. The use of chemical weapons depended rather on exercising and training technical personnel, on mechanical equipment and technical means, the existence of which was the very basis and *sine qua non* for the use of chemical and bacteriological methods of war.

The French representative insisted that the Preparatory Commission should take into consideration the opinion formulated by a group of eight countries in Sub-Commission A, namely, that the only effective sanction which could prevent a state from violating its undertakings in connection with the prohibition of chemical warfare consisted in the possibility of immediate reprisals by the same chemical means.

At the suggestion of the German delegation, the following sentence was added to the above text, as paragraph three:

> They also undertake to abstain from any preparation in peacetime of the use of the methods of warfare stated in the two preceding paragraphs.

The proposal thus amended was included under Chapter IV of the Draft

Convention for the Limitation and Reduction of Armaments, which resulted from the first reading. [7]

Chapter IV of the Draft Convention was discussed in the Preparatory Commission from 20 to 24 April 1929. [8] The Commission had before it the proposal quoted above, the reports of Sub-Commissions A and B, and observations of the German representative to the following effect:

> The use of the chemical weapon is at present entirely prohibited by the protocol on the use of poison gases (as chemical substances could only be employed along with other weapons, it would be more accurate to speak of the chemical weapon and not of chemical warfare). It would, nevertheless, be desirable to renew the prohibition in the disarmament convention. Otherwise, it might happen that it would be permissible for certain states signatories of this convention to employ chemical methods of warfare (i.e. states which had not ratified the gas protocol), while other states would be forbidden to do so (those which had ratified the protocol). Further, the prohibition to employ the chemical arm would have to be supplemented by a prohibition to prepare for its use.
>
> It would be desirable to consider whether the prohibition to employ chemical methods of warfare could not be made more effective by a general restriction in the use of the most important weapon by which chemical substances can be employed in war, i.e. the air weapon. The possibility might be considered of absolutely forbidding the dropping from the air of substances used in war, and the preparation of this form of warfare. This prohibition would apply not only to the dropping of gas bombs, but also to the dropping of high explosive and incendiary bombs and to all other possible forms of chemical warfare waged from the air. It would thus have the merit of contributing towards mitigating the sufferings of the civilian population in time of war. In addition, this prohibition would be directed against military weapons which can be used particularly for purposes of aggression. [9]

The representative of Colombia considered that "for the weaker nations which did not possess hundreds and thousands of guns, formidable dreadnoughts, armoured tanks, torpedo-boats, submarines, Berthas, and what not, it would be a fool's game to assist in abolishing a method of offence which might be at their disposal and be of assistance to them in defending all that they held most dear". He said further that inventing and employing, in the most scientific manner possible, the most destructive and abominable implements of war would one day drive war itself definitely from the face of the earth.

The Belgian representative, however, recalled the old principle of international law forbidding the use of poison in war and quoted relevant provisions of international agreements to that effect. He stated that although the prohibition of chemical warfare had been embodied in the Geneva Protocol of 1925, its duration would not necessarily be the same as that of the draft convention under discussion. Besides, some countries had adhered to the Protocol with reservations, and others without reservations.[3] The Commission ought to try to secure uniformity; the best way would be for a definite undertaking to be given by all countries and to be included in the convention.

Referring to the wording of the proposal contained in Chapter IV of the draft convention, the representative of the British Empire wondered whether it was possible to prohibit, as completely as was laid down therein, the preparation of any substance which might be used in time of war for lethal purposes, but which was a legitimate substance for use in commerce in time of peace. The Japanese delegate could not see any practicable way of preventing the preparation of poisonous gases.

The US representative stressed that preparation of defence against chemical warfare should not be prohibited.

The German representative explained that the intention was not to prohibit industries from making gases which could be used for chemical warfare, nor was it intended, at least for the time being, to prohibit the manufacture of gas masks. The meaning of the paragraph he had proposed was that countries should not stock materials for chemical warfare with the intention of using them in war; that bombs must not be prepared with chemical gases; and that soldiers must not be specially trained for chemical warfare.

The representative of the USSR indicated that, with the exception of the last two paragraphs of the proposed text, Chapter IV was merely a repetition of the Geneva Protocol of 1925. He suggested that, since the Protocol was not likely to be improved upon, there should be a resolution urging all governments to accelerate the ratification of the Protocol. An additional document, which would include new provisions not contained in the Geneva Protocol, could then be adopted.

An agreement or protocol in regard to chemical warfare could be reached independently of any reduction or limitation of armaments. If, on the other hand, a special chapter on chemical warfare were to be included in the Draft Convention for the Reduction and Limitation of Armaments, ratification of the Geneva Protocol of 1925 would certainly be further delayed by those governments which had not been in a hurry to ratify it, and they would have an excuse for doing so, while waiting for the adoption of the draft convention.

The Soviet delegate also insisted that preparation for the use of chemical and bacteriological methods of warfare should be prohibited not only

---

[3] By spring 1929, thirteen states had ratified or acceded to the Geneva Protocol of 1925.

in time of peace but also in time of war. His proposal presented to the Commission read as follows:

I. Whereas the fundamental points in Chapter IV of the 1927 Draft reproduce the Protocol on chemical and bacteriological warfare signed at Geneva on 17 June 1925, the Preparatory Disarmament Commission decides, with the object of bringing the above-mentioned Protocol into force as soon as possible, to omit Chapter IV of the 1927 Draft and to adopt the following decision:

The Preparatory Commission for the Disarmament Conference appeals to all States which have signed the Protocol of June 17, 1925, but which have not ratified it to ratify it with as little delay as possible.

II. The Preparatory Commission similarly approaches all States which have signed the above-mentioned Protocol and proposes that they should sign a supplementary Protocol annexed thereto and consisting of the following articles:

Article I. All methods of and appliances for chemical aggression (all asphyxiating gases used for warlike purposes, as well as all appliances for their discharge, such as gas-projectors, pulverisers, balloons, flame-throwers and other devices) and for bacteriological warfare, whether in service with troops or in reserve or in process of manufacture, shall be destroyed within three months of the date of the entry into force of the present Convention.

Article 2. The industrial undertakings engaged in the production of the means of chemical aggression or bacteriological warfare indicated in Article I shall discontinue production from the date of the entry into force of the present Protocol.

Article 3. In enterprises capable of being utilized for the manufacture of means of chemical and bacteriological warfare, a permanent labour control shall be organized by the workers' committees of the factories or by other organs of the trade unions operating in the respective enterprises with a view to limiting the possibility of breaches of the corresponding articles of the present Protocol. [10]

The text of the protocol, intended to supplement the Geneva Protocol of 1925 prohibiting the use of chemical and bacteriological warfare, was based on the Draft Convention on the Reduction of Armaments, submitted by the delegation of the USSR on 23 March 1928. [11]

The first two articles of the supplementary protocol reproduced almost word for word Articles 31 and 32 under Chapter IV of the Soviet draft convention, dealing with chemical methods of warfare. Article 3 of the supplementary protocol was based on Article 44 under Chapter VII of the draft convention.

Chapter VII dealt with control and provided among other things:

The establishment of a Permanent International Commission of Control consisting of an equal number of representatives of the legislative bodies and of the trade unions and other workers' organizations of all states participating in the convention. (Among the duties of the Commission would be to select the places, the procedure and the technical conditions for the destruction of the material, and the preparation of all the necessary supplementary technical agreements);

The establishment of a Permanent International Committee of Experts acting under the orders of the International Commission of Control;

Investigations to be carried out on the spot by the Permanent International Commission in the event of reasonable suspicion of a breach of the convention and of the subsequent supplementary agreements on the reduction and limitation of armaments; and the appointment by the Commission for this purpose of special commissions of enquiry;

The organization, in enterprises for the production of war material or in enterprises capable of being utilized for the manufacture of armaments, of a permanent labour control by the workers' committees of the factories or by other organs of the trade unions operating in the respective enterprises, with a view to limiting the possibility of breaches of the corresponding articles of the convention. A similar control would be set up in the various branches of the chemical industry, a list of which was to be drawn up by the Permanent International Commission of Control;

An undertaking by the contracting parties to furnish the Commission of Control, within the time limits fixed by it, with full information as to the situation of their armed forces.

The Japanese delegation felt that it would suffice to stipulate in the draft convention that its acceptance involved, *ipso facto,* the acceptance of the Geneva Protocol of 1925 by states which had not as yet acceded thereto. [12]

The delegation of Italy suggested that the provisions of the Geneva Protocol be inserted in the draft convention, or that the text of Chapter IV be replaced by an article worded as follows:

The High Contracting Parties which have not yet ratified or signed the Protocol for the Prohibition of the Use in War of Asphyxiating, Poisonous or Other Gases and of Bacteriological Methods or Warfare, signed at Geneva on June 17th, 1925, undertake to ratify it or accede to it as soon as possible. [13]

The first alternative implied the same effect as the Japanese proposal. The second alternative did not go that far and would make it possible for a government after ratifying the convention not to be bound by the Geneva Protocol.

Following the Spanish delegate's remark that it would be difficult to adhere to an absolute prohibition vis-à-vis governments which had themselves made use of the methods of war in question, the representative of Belgium suggested that a distinction be made between the undertaking to abstain from the use of asphyxiating, poisonous or similar gases, which normally could be observed only subject to reciprocity, and the undertaking to abstain from the use of bacteriological methods which in all cases constituted a crime against international law. Bacteriological warfare was necessarily directed against the entire population, and no civilized government would like to be guilty of such a crime even against a criminal government which had itself resorted to those methods.

Accordingly, and in view of the serious objections raised against the prohibition of preparation in peacetime for the use of methods of chemical warfare, as well as against restrictions on the importation, exportation and manufacture of the substances in question, the representative of Belgium proposed that the relevant paragraphs be deleted and that Chapter IV be drafted as follows:

> The High Contracting Parties undertake, subject to reciprocity, to abstain from the use in war of asphyxiating, poisonous or similar gases, and of all analogous liquids, substances or processes.
> They undertake unreservedly to abstain from the use of all bacteriological methods of warfare. [14]

The French delegate then submitted the following draft of paragraphs 3 and 4 of Chapter IV:

> Paragraph 3. The High Contracting Parties also undertake to abstain from any preparation in peacetime with a view to the use in war of the methods stated in the two preceding paragraphs, and undertake as soon as the Convention is put into force to take effectual steps to prevent private persons from making preparations in their territory for the use of such methods in war.
> Paragraph 4. The High Contracting Parties undertake, moreover, to take effectual steps to prevent the manufacture in their territory, the importation or the exportation of substances utilizable for chemical or bacteriological warfare, so far as these have no normal utility in peacetime. If such substances have a normal utility in peacetime, the High Contracting Parties undertake to restrict the importation, exportation or manufacture of those substances to commercial requirements. [10]

The Romanian and Serb-Croat-Slovene delegations proposed to add the following text as paragraphs 5 and 6.

> Paragraph 5. The High Contracting Parties undertake to place at the disposal of any state which is the victim of aggression by means of poisonous or bacteriological substances such raw materials, products and appliances as may be necessary to meet this aggression.
> Paragraph 6. The High Contracting Parties further undertake to participate themselves, as far as distance will allow, in collective reprisals by employing the chemical and bacteriological means at their disposal against the state which has been guilty of an aggression by such means. [10]

The last paragraph, to which the Polish delegation also subscribed, differed from the Belgian proposal in that according to the latter bacteriological warfare was prohibited in all circumstances and not only against the enemy who himself respected such prohibition.

The German delegation's proposal read:

The High Contracting Parties mutually undertake not to launch weapons of offence of any kind from the air by means of aircraft, nor to employ unpiloted aircraft controlled by wireless or otherwise carrying explosive or incendiary gaseous substances.

They further undertake to make no preparations of any kind for the use of the weapons of offence referred to in the previous paragraph.

The German representative also proposed to substitute for the term "chemical warfare" the term "chemical arms". [10]

The delegate of Persia sent in a communication which contained the following proposal:

The Persian Delegation holds that it will be preferable either to adhere to the amendment put forward by the Italian Delegate or to consider in Chapter IV of the Draft Convention what urgent steps could be taken by the Council of the League of Nations to compel parties to renounce chemical warfare and observe the provisions of the 1925 Protocol. [10]

In the course of the discussion of the French draft (see page 94), the French representative explained that his proposal was not meant to prevent the manufacture of gas masks.

In reply to questions asked by the Greek delegate, he stated that if it were eventually found that supervision was not practicable, the provisions would have to be revised; engagements of the kind proposed would be meaningless in the absence of supervision.

The representative of the Netherlands wondered whether paragraph 3 of the French proposal would prohibit the preparation of defence against a chemical attack, not merely defence in the strict sense of the term—by the use of masks, for example—but also defence by a counter-attack. The question would become even more important if the Commission adopted the Belgian proposal which made the prohibition of chemical warfare subject to reciprocity. He also thought that the new criterion established by paragraph 4 of the French proposal, namely, that: "If such substances have a normal utility in peacetime, the High Contracting Parties undertake to restrict the importation, exportation, or manufacture of those substances to commercial requirements", was not of a practical nature. Although a government could, within certain limits, verify the requirements of its own country, it would find it very difficult to determine the commercial requirements of the world.

The Soviet delegate observed that a system of supervision with regard to the reduction of armaments could not be employed to supervise the manufacture of materials intended for chemical warfare. The Soviet delegation, in its proposal, provided therefore for special supervision.

The representative of the British Empire found the wording of the French proposal somewhat loose. How would it be possible, he asked, to ensure that the limit of commercial requirements had not been exceeded? If a works manufacturing chemical products carried out a foreign order, how could it be known that only a part of this order was intended for strictly commercial purposes?

The United States representative, referring to the special situation of his country, pointed out that the United States federal laws would not enable his government to subscribe to a text providing for "... steps to prevent private persons from making preparations ...". Such measures could only be taken by the individual states, each within its own jurisdiction. The importation and exportation of the substances in question could be regulated by the federal government, but their manufacture could be supervised only by the states. The federal government could not intervene.

The Italian delegate stated that if the exercise of supervision were implied, his delegation could not accept the proposal. The Japanese and Chilean representatives took the same position.

The delegate of Belgium said that apart from the delicate question of supervision, the French proposal raised a new problem in that it applied the prohibition to private persons. It would then have to be proved that the manufacture, importation or exportation of gases by a private person was intended for a purpose other than ordinary industrial purpose, or else the manufacturer would have to prove the contrary. Moreover, paragraph 3 of the French text appeared to prevent any means of studying methods of self-defence and to prohibit analytical laboratories or gas chambers which soldiers could be trained to enter with masks.

The French representative replied that he would withdraw his amendment provided that paragraphs 3 and 4 of the original text were deleted.
[15]

In submitting the Romanian and Serb-Croat-Slovene proposal regarding sanctions, (see page 94) the Romanian representative referred to the relevant part of the report which Sub-Commission A presented in 1926. He argued that, faced with the prospect of sanctions, any government would hesitate before resorting to chemical and bacteriological warfare. On the one hand, it would have to consider possible reprisals and, on the other, its responsibility to its own nationals. Persia supported this proposal.

The representative of the Netherlands disagreed saying that countries which stood outside a conflict could not intervene with methods which had been condemned by the whole civilized world. The argument gained even more strength when not only chemical but also bacteriological arms were considered. The adoption of the proposal advanced would come as a shock to world public opinion. It was impossible to contemplate an undertaking by any country to supply any other country attacked by the bacteriological arm with the necessary raw materials for retaliatory action. There were also political arguments against the adoption of the proposal. If, for example, country A were attacked by country B, the members of the League were obliged, under Article 16 of the Covenant, to assist country A. If, however, the latter, in defending itself, employed the chemical or the bacteriological arm, then—according to the proposal before the Commission—the members of the League would be obliged to assist country B. But it would be absurd for them to be helping both sides at once.

The representative of the USSR agreed with the above remarks and said that he would oppose any scheme of punitive action.

The representative of France saw no difficulty in laying down the principle that action should be taken against a country which broke its solemn undertaking not to employ the chemical arm. A case such as that described by the delegate of the Netherlands ought not to arise. The country in the right would know that it was in its interest not to use illicit methods of warfare, if it did not wish to lose the promised help. If, therefore, help was given to a country attacked, and if that country was attacked with the chemical arm, would not all those giving assistance be entitled to use that arm?

The Romanian delegate, supported by the French delegate, suggested that the opinions on the question of sanctions should be brought to the attention of the Disarmament Conference. In the meantime, governments might study the problem and give a definite reply at the Conference. [15]

The representative of the USSR said that the Soviet proposal was covered by the French proposal which had met with opposition. He thus saw no point in having his text discussed at this stage. [10]

During the consideration of the Belgian proposal (see page 94), the German delegate said he would prefer to omit the word "reciprocity" because he thought it was a matter of course, and also because he felt that the moral importance of the clause would be weakened by the introduction of such a term.

The representative of Persia stated that he would not wish to weaken the scope and moral effect of the Geneva Protocol by introducing an element of reciprocity; on the other hand, he wished the text to include guarantees for insufficiently protected countries, such as Persia, which had no chemical industries.

The representative of the Netherlands believed that it would be better to have a text saying exactly what was meant rather than to leave room for doubt. Otherwise, no one would be sure of the obligations incurred.

The Japanese delegate said that ratifications could be accompanied by a reservation with regard to reciprocity, made by any country desiring to make it.

The representative of Greece recalled that in whatever form an international law might be promulgated, the law would imply, *nolens volens,* the principle of reciprocity. Thus it was obvious that whenever any one state was sufficiently unscrupulous to violate the clause concerning chemical and bacteriological warfare, other states, and particularly the state victim of such violation, would be absolved and might employ against the aggressor the same methods as he himself had used. As to the distinction, drawn by the representative of Belgium, between chemical and bacteriological warfare on the point of reciprocity, the Greek delegate said that when it was a question of employing methods of injuring the enemy as an organized combatant force, he could see no difference between the various means which might be used, and the distinction between combatants and non-combatants was becoming more and more difficult to establish in actual practice. Further, he expressed doubts whether, in a convention aiming at the limitation and reduction of armaments, it was desirable to have a special chapter with provisions prohibiting chemical and bacteriological warfare. The proper place for such a prohibition would be in a convention on the laws and usages of war, and, in any case, this had been embodied in the Geneva Protocol of 1925. There was no point in going back upon a text which had already been ratified by a number of countries; it was too late to try to improve it. Chapter IV of the convention should, therefore, be omitted entirely. He proposed another method of establishing a close connection between the draft convention and the prohibition of chemical and bacteriological warfare, namely by adding at the end of the convention an article to be worded as follows:

> The ratification of the present Convention implies, *ipso facto,* the accession without any reservation of each of the High Contracting Parties to the Protocol concerning the prohibition of the use of asphyxiating, poisonous, or other gases and of bacteriological methods of warfare, signed at Geneva on June 17th, 1925.

Thus, not only would the acceptance of the Geneva Protocol be obligatory for all who definitely accepted the Convention on the Reduction and

Limitation of Armaments, but also a confusion of texts would be avoided. At the same time it would not preclude an amplification, if necessary, of the Geneva Protocol by more complete and detailed provisions.

The delegate of Turkey accepted the above formula with the exception of the words "without any reservation". [16]

The discussion narrowed down to the issue whether it was desirable to retain the clauses relating to chemical warfare in a chapter of the convention, or whether, seeing that the prohibition of the use of certain kinds of weapons had no direct relation to the reduction of armaments—which was the object of the convention—these provisions should rather be placed at the end of the convention in a separate article providing that the ratification of the convention would automatically involve accession to the Geneva Protocol of 1925, or be inserted in some other document.

The Soviet representative thought that the best course would be to address an invitation to countries to ratify the Geneva Protocol, and to leave out Chapter IV. In his view, the Commission would only discredit its work if the Protocol, as it had already been accepted by the majority of states, were included in the convention, in its existing form.

The Spanish representative felt that if, after a lengthy discussion, the Commission took the course of purely and simply abolishing Chapter IV, such action might be misunderstood by public opinion.

The French delegate considered that a gap would be left in the convention if Chapter IV were deleted, for the convention would limit land, sea and air armaments and would be silent as to chemical armaments. The latter were prohibited only in a Protocol which had not been ratified by all countries. The convention would thus in fact seem to authorize chemical armaments.

The representative of the Kingdom of the Serbs, Croats and Slovenes contended that in order to strengthen the Geneva Protocol it was essential that the Commission should consider inserting in the Convention a provision prohibiting resort to chemical warfare and should, if necessary, elaborate it in closer detail.

The Polish delegate went even further in stating that the Geneva Protocol was obsolete. He referred to the Briand–Kellogg Pact of 1928, which contained a general prohibition of resort to war,[4] and said that its provisions

---

[4] The Pact of Paris (also known as the Briand–Kellogg Pact) of 27 August 1928, was drawn up outside the League of Nations. It read as follows:

"The High Contracting Parties solemnly declare in the names of their respective peoples that they condemn recourse to war for the solution of international controversies, and renounce it as an instrument of national policy in their relations with one another." (Article I.)

"The High Contracting Parties agree that the settlement or solution of all disputes

should not be overlooked by the Commission. The Geneva Protocol was nothing more than a declaration to the effect that chemical and bacteriological warfare must be prohibited. Nothing else could be done in 1925, but it should not be asserted that progress was impossible. In spite of the pessimistic view held by Sub-Commissions A and B, he thought it was possible to find more effective means than those recommended in 1925.

The Chinese delegate was prepared to accept the Belgian proposal as being more effective than the inadequate Geneva Protocol of 1925.

The United States representative said that if the provisions of the first two paragraphs of Chapter IV were observed in good faith, any supplementary provisions were entirely superfluous. On the other hand, if the provisions of the first two paragraphs were not observed in good faith, there would be no use in a supplementary provision.

The Chilean delegate made a suggestion which, he hoped, might to a certain extent satisfy those who desired to refer to the Geneva Protocol or have the Protocol reproduced. Chapter IV would, according to this suggestion, begin with the following words:

> The High Contracting Parties once more condemn recourse to chemical warfare, which is prohibited under numerous treaties and under the Geneva Protocol of 1925, and undertake, subject to reciprocity...

The representative of the British Empire preferred the Belgian proposal, because he thought there was an objection in principle to laying down in one convention that it involved acceptance of a totally different convention. [16]

On 23 April 1929, by eleven votes to ten the Preparatory Commission adopted the proposal of Belgium which retained the first two paragraphs of Chapter IV, introducing the principle of reciprocity in the first paragraph and a declaration in absolute terms in the second paragraph.

The text adopted was therefore as follows:

Chapter IV—Chemical Warfare

The High Contracting Parties undertake, subject to reciprocity, to abstain from the use in war of asphyxiating, poisonous or similar gases and of all analogous liquids, substances or processes.

They undertake unreservedly to abstain from the use of all bacteriological methods of warfare. [16]

---

or conflicts of whatever nature or of whatever origin they may be, which may arise among them, shall never be sought except by pacific means." (Article II.)

The signatories were: Australia, Belgium, Canada, Czechoslovakia, Eire, France, Germany, Great Britain, India, Italy, Japan, New Zealand, Poland, South Africa and the United States.

The Pact went further than the Covenant of the League of Nations in that it contained a general prohibition of resort to war. It came into force on 25 July 1929.

In view of this decision taken by the Commission, the Turkish delegation made a declaration the substance of which was:

The condition of reciprocity was implied in the spirit of the Protocol of 1925 concerning the prohibition to make use in war of asphyxiating, poisonous or similar gases and bacteriological means;

By its introduction into the Draft Convention under consideration the reservation in the Belgian amendment might be construed as an interpretation in regard to the right of reciprocity of states which had ratified the Protocol of Geneva, but which had not thought it necessary to formulate a reservation already implied in their undertakings;

It was not for the Preparatory Disarmament Commission to adopt a provision such as to constitute an interpretation of an international act concluded independently;

The Belgian amendment, adopted by a majority of the Commission, could not be regarded in any case as having an exclusive bearing on the Geneva Protocol and thus determining its absolute and conditional character;

The right of states which had ratified the Protocol of 1925 without reservation to invoke the tacit condition of reciprocity was not affected.

The Commission was asked to take note of the declaration and insert it in the minutes. [17]

Upon the Polish delegation's proposal the Commission took the following decision:

The Preparatory Commission for the Disarmament Conference reserves the right to submit to the Conference proposals concerning the chemical and bacteriological arm, in order to supplement and extend the provisions of the 1925 Protocol. [16]

Another resolution, separate from the draft convention, was adopted on the same day, in accordance with the proposal submitted by the Soviet delegation:

The Preparatory Commission for the Disarmament Conference recommends all states signatories to the Protocol of June 17th, 1925, which have not yet done so to ratify it as soon as possible. [16][5]

The proposal, submitted by the Romanian and Serb-Croat-Slovene delegations, which provided for a system of mutual assistance and sanctions against a state which might violate the prohibition to make use of asphyxiating or bacteriological means of warfare, was not put to the vote and the delegations which had submitted it reserved their right to revert to it at the Disarmament Conference.

---

[5] On 12 June 1929 the Council of the League of Nations authorized the Secretary-General to call the attention of the governments to the recommendation of the Preparatory Commission concerning the ratification of the 1925 Protocol.

The German proposal, concerning the prohibition of launching gaseous substances from the air (see pages 90 and 95), was discussed in the **Preparatory Commission on 24 April 1929**, in connection with air warfare.

The German representative said that the prohibition of using chemical arms would be incomplete if the dropping of bombs were not generally prohibited. He was supported by the Soviet delegate who quoted Article 7 of Section II of the Soviet draft convention, reading:

All implements of war directed primarily against the civil population which does not directly take part in the armed conflict (military aircraft and chemical weapons) must be destroyed...

The representative of the Netherlands was also in favour of adopting the German proposal.

However, in view of the opposition of France, the British Empire and the United States, the German proposal was rejected. [17]

IV. *The controversy over tear gas*

On 2 December 1930, the British delegation submitted a memorandum concerning the question whether the use of tear gas was or was not to be regarded as contrary to the 1925 Geneva Protocol and the provisions of Chapter IV of the Draft Convention for the Limitation and Reduction of Armaments.

The memorandum recalled that in the Geneva Protocol there was a discrepancy between the French and English versions. The French word "*similaires*" (with reference to gases) appeared in the English text as "other". Basing itself on the English text, the British government had taken the view that the use in war of "other" gases, including lachrymatory gases, was prohibited. It also considered that the intention was to incorporate the same prohibition in the draft convention under discussion.

It was highly desirable—stressed the memorandum—that a uniform construction should prevail as to whether or not the use of lachrymatory gases was considered to be contrary to the Geneva Protocol of 1925 and/or to Chapter IV of the draft convention.

Replying to the British memorandum, the French delegation issued a special note:

I. All the texts at present in force or proposed in regard to the prohibition of the use *in war* of asphyxiating, poisonous or similar gases are identical. In the French delegation's opinion, they apply to all gases employed with a view to toxic action on the human organism, whether the effects of such action are a more or less temporary irritation of certain mucous membranes or whether they cause serious or even fatal lesions.

II. The French military regulations, which refer to the undertaking not to use gas for warfare (*gaz de combat*) subject to reciprocity, classify such gases as suffocating, blistering, irritant and poisonous gases in general, and define irritant gases as those causing tears, sneezing, etc.

III. The French Government therefore considers that the use of lachrymatory gases is covered by the prohibition arising out of the Geneva Protocol of 1925 or Chapter IV of the Draft Convention.

The fact that, for the maintenance of internal order, the police, when dealing with offenders against the law, sometimes use various appliances discharging irritant gases cannot, in the French delegation's opinion, be adduced in a discussion on this point, since the Protocol or Convention in question relates only to the use of poisonous or similar gases in *war*.

The interpretation given in the British memorandum was explicitly accepted, in addition to France, by the delegations of Romania, Yugoslavia, Czechoslovakia, Japan, Spain, China, Italy, Canada, Turkey, and the Irish Free State. The Soviet representative also interpreted the relevant paragraph as meaning that the use of all gases, including irritant gases, was prohibited, but objected to the French statement insofar as—in his opinion—it legalized the use of these gases by police forces.

The only opposition came from the US representative who said that a statement as to the poisonous or non-poisonous, or the lethal or non-lethal, qualities of smokes and gases, which might come under the term of the text of Chapter IV, would require a study by technical experts with specialized knowledge of the subject, supported by expert specialized medical knowledge as to the properties and probable physical and pathological effects of these various agents in normal and abnormal concentrations. The problem was essentially one of doing away with agents which caused unnecessary suffering. On the other hand, one should not be led to bring into disrepute the employment of agents which not only were free from the reproach of causing unnecessary suffering, but which achieved definite military or civil purposes by means in themselves more humane than those in use before their adoption. There would be considerable hesitation—continued the US delegate—on the part of many governments to bind themselves to refrain from the use in war, against an enemy, of agents which they had adopted for peacetime use against their own population on the ground that those agents, while causing temporary inconvenience, caused no real suffering or permanent disability.

He then suggested, and the Preparatory Commission agreed, to appeal to all governments which intended to send representatives to the Disarmament

Conference to devote very careful study to the question—the importance of which the Commission recognized—so that the problem might be settled in all its aspects by the Conference. [18]

V. *Proposal for assistance to CBW-attacked countries*

On 5 December 1930, during the last reading in the Preparatory Commission of Part V—formerly Chapter IV—of the draft convention, the Polish representative made the following declaration:

> Though recognizing the moral value of international instruments forbidding the use in warfare of chemical and bacteriological methods of war, we nevertheless feel that it is necessary to make provision, in addition to these instruments, for practical preventive and executory measures. These measures should be such as to render chemical or bacteriological attack, if not impossible, at any rate difficult, and should limit the chances of success and efficacy of such an attack. They should also constitute a fresh guarantee that no violation of the undertakings solemnly signed could be committed without involving very unpleasant consequences for the guilty state.
>
> In this connexion, therefore, it would be desirable to consider the possibility of concluding a Convention for affording international aid on as liberal a scale as possible to any country chemically or bacteriologically attacked. As such aid would be essentially of a humanitarian nature (sanitary, scientific, etc.) it should meet with general approval.
>
> This problem might be studied in due course by the League of Nations.

The delegations of Finland, Romania and Yugoslavia associated themselves with this statement. [19]

The report of 9 December 1930 of the Preparatory Commission for the Disarmament Conference contained the text adopted by the Commission as well as some of the suggestions and opinions put forward by various delegations. [20]

VI. *Summary and comment*

A commission convened by the League of Nations to prepare for a disarmament conference met from 1926 to 1930. With regard to CBW, the consensus was that steps should be taken to speed up the process of ratification of the 1925 Geneva Protocol. An appeal to this effect was initiated by the USSR and made by the commission.

At the same time, there was a feeling that the Protocol was insufficient, that a mere declaration of determination not to have recourse to the prohibited means of warfare might not stand the strain of actual hostilities, and that the envisaged convention on the reduction and limitation of armaments should include more stringent measures.

Such measures, in the view of a good number of delegations to the Preparatory Commission, would have to provide an assurance that no preparations for chemical and bacteriological war were carried out, that in case of violation sanctions would be applied against the violator and that the country attacked would be given prompt and effective assistance.

The Soviet proposal for a supervised discontinuance of production of CB weapons and of appliances for their discharge, as well as for the destruction of stocks, was not found acceptable.

The expert bodies of the Preparatory Commission expressed the view that preparations for chemical war could not be detected and prevented: chemical factories can be quickly adapted to manufacturing poison gases; some types of these gases are current commercial products of industry; apparatus required for spreading the chemicals can be easily produced or improvised; enquiries into complaints concerning violation of obligations are likely to be ineffective.

The USA, Japan and Italy were strongly opposed to international control.

The discussion of sanctions for the enforcement of the prohibition of CBW revealed unbridgeable differences of opinion between those who believed in the preventive or deterrent effect of collective reprisals, and those who placed their faith in the morally binding force of international commitments.

A suggestion by Yugoslavia to include incendiary material in the category of chemical weapons was echoed by Germany, and the USSR proposed to abolish flame-throwers along with other "methods of and appliances for chemical aggression".

Another important suggestion, made by Germany, to prohibit bombing from the air as a means of launching poisonous gases, was formally rejected.

For the first time, the applicability of the Geneva Protocol to tear gases was debated when Britain had put forward a memorandum affirming—after some indications that this view was not shared by all countries—that the use of tear gases in war was prohibited by the Protocol. France supported Britain's position. The US delegation alone spoke against this interpretation, arguing that the agents in question caused no real suffering or permanent disability and permitted achievement of military or civil purposes by more humane means. The matter was referred to the Disarmament Conference.

The text eventually adopted for inclusion in the draft convention for the reduction and limitation of armaments was confined to the prohibition of use of CB weapons but a distinction was made between chemical and bacteriological warfare as to the nature of the prohibition. While the former was forbidden subject to reciprocity, the latter was forbidden unreservedly. It was argued that biological warfare constituted a crime in all cases, even when waged against an enemy who had first resorted to such methods.

The outcome of the Preparatory Commission's efforts was not of great consequence. Nevertheless, the examination of the manifold aspects of chemical and bacteriological warfare prepared the ground for a more thorough discussion at the Disarmament Conference.

Chapter 4. 1931–November 1932

I. *The convening of the Disarmament Conference*

On 22 May 1931, the Council of the League of Nations decided to convene at Geneva the Conference for the Reduction and Limitation of Armaments (Disarmament Conference). [1]

The Conference opened on 2 February 1932. The following states were invited:

Abyssinia
Afghanistan (became member of League on 27 September 1934)
Union of South Africa
Albania
United States of America (non-member of League)
Argentine Republic
Australia
Austria
Belgium
Bolivia
Brazil (non-member of League)
United Kingdom of Great Britain and Northern Ireland
Bulgaria
Canada
Chile
China
Colombia
Costa Rica (non-member of League)
Cuba
Czechoslovakia
Denmark
Dominican Republic (observer until 7 July 1932, afterwards represented in usual manner)
Egypt (non-member of League)
Ecuador (not represented at Conference; became member of League on 28 September 1934)
Estonia
Finland
France

Germany (left Conference on 23 July 1932; returned to Conference on 14 December 1932; left Conference on 14 October 1933. Notified withdrawal from League on 21 October 1933; the withdrawal took effect in 1935)
Greece
Guatemala
Haiti
Honduras
Hungary
India
Iraq (became member of League on 3 October 1932. Invited to Conference on 2 November 1932; accepted invitation on 20 November 1932)
Irish Free State
Italy
Japan (notified withdrawal from League on 27 March 1933; the withdrawal took effect in 1935)
Latvia
Liberia
Lithuania
Luxemburg
Mexico (became member of League on 12 September 1931)
Netherlands
Nicaragua (not represented at Conference)
Norway
New Zealand
Panama

107

Paraguay (not represented at Conference)
Peru
Persia
Poland
Portugal
Romania
Salvador (not represented at Conference)
Saudi Arabia (non-member of League)
Siam
Union of Soviet Socialist Republics (became member of League on 18 September 1934)
Spain
Sweden
Switzerland
Turkey (became member of League on 18 July 1932)
Uruguay
Venezuela
Yugoslavia

By the time the Conference met, thirty-three states had either ratified or acceded to the Protocol for the Prohibition of the Use in War of Asphyxiating, Poisonous or Other Gases and of Bacteriological Methods of Warfare, of 17 June 1925.

The basis for the work of the Conference in dealing with the problem of chemical and bacteriological warfare was Part V (which consisted of Article 39) of the draft convention framed by the Preparatory Commission.

II. *General debate on CBW at the Disarmament Conference*

In the course of the general discussion at the plenary meetings of the Conference, delegations submitted proposals orally and in the form of special memoranda.

France proposed that the use by airplanes and by land or naval artillery of projectiles which were specifically incendiary or which contained poison gases or bacteria should be forbidden, whatever the objective. An international police force to prevent war, and coercionary forces to repress war and to bring immediate assistance to any state victim of aggression, would have to be set up.

The United Kingdom attached great importance to the maintenance of the provisions of the draft convention, relating to gas and bacteriological warfare. [2]

The United States advocated the total abolition of lethal gases and bacteriological warfare. [3]

Italy expressed readiness to agree to an organic plan of qualitative limitation of armaments, which would comprise the abolition of aggressive chemical and bacteriological weapons of every kind, as well as revision of the laws of war with a view to ensuring more complete and effective protection of the civilian population. [4]

The Soviet Union reverted to the proposal, which it had presented to the

Preparatory Commission, regarding the destruction of all means and apparatus for chemical, incendiary and bacteriological warfare, and re-submitted its draft convention which in Chapter IV provided for such measures. (See page 92.)

Sweden recommended that study should be made of the possibility of restricting preparations for chemical and bacteriological warfare. [5]

Denmark said that if the use of chemical and bacteriological weapons were to be prohibited, the preparation, manufacture or training in the use of such weapons must be absolutely abolished and an effective system of supervision established. It should be possible to group chemical factories in an international cartel of manufacturers of chemical products, supervised by the organ of control suggested in connection with control of armaments. The problem of aviation was intimately bound up with that of chemical warfare and incendiary bombs. The prohibition of one of these arms would involve the prohibition of the other. [6]

Czechoslovakia wished that a more complete system of prohibition against chemical and bacteriological warfare be established and was prepared to extend that prohibition to other forms of offensive warfare, more especially the aerial bombardment of the civilian population and of cities, particularly the capitals, of countries at war. The measure should be supplemented by a system of penalties. [7]

Turkey declared that the use of asphyxiating gases and bacteriological weapons being prohibited, all peacetime preparations for their use should be likewise prohibited; also all material designed for the projection of the above-mentioned weapons should be forbidden. [8]

Norway welcomed the suggestions aimed at prohibiting arms of a particularly offensive character, including arms which constituted a special danger to the civilian population. But it considered it not enough to prohibit the use of certain arms; also their manufacture in peacetime and any act which constituted preparation for their use, including the training of the necessary personnel, should be prohibited. [9]

The Netherlands expected from the Conference a categorical statement concerning absolute prohibition of the use of poisonous and harmful gases and of any preparation for chemical warfare. [10]

Switzerland specified the following methods of warfare to be prohibited: all forms, without exception, of chemical and bacteriological warfare; bombardment from the air; incendiary bombs. [11]

Yugoslavia found it essential to prohibit the preparation and use of chemicals and bacteria in war, even for purposes of legitimate defence. Such forms of warfare, besides being cruel and inhuman in the extreme, were intended for the purpose of extermination, inasmuch as they chiefly threatened the civilian population. Attacking women and children, they destroyed all hope of any possible revival of the belligerent nations. But the prohibition of these forms of war would be useless if no provision was made for sanctions. If, after careful and conscientious consideration, it was found impossible to devise sanctions, Yugoslavia would not be prepared to agree to a purely academic prohibition of chemical and bacteriological warfare without the most explicit reservations. [12]

Austria supported measures aimed at the abolition of aggressive arms of all kinds—the abolition of chemical and bacteriological warfare and of bombardment from aircraft and of all preparations made in peacetime for those purposes. [13]

Germany suggested that the prohibition of the military utilization of asphyxiating, poisonous or similar gases and all similar liquids, matters or processess as well as of all means of bacteriological warfare should be extended to the preparation of the utilization of these weapons. [14]

Haiti proposed that all states should undertake to abolish forthwith chemical and bacterial arms, renounce the bombardment from the air of civilian population and the use of offensive armaments. [15]

India was ready to cooperate in devising means for protecting the civilian population against ruthless methods of warfare. It supported the total abolition of lethal gas and bacteriological warfare, and the use of poison in general. [16]

The general discussion, which ended on 24 February 1932, confirmed the existing consensus that the use of chemical and bacteriological weapons should be condemned and prohibited. There was also a wide measure of agreement that the prohibition should not be subject to reciprocity. A number of delegations moved that there should be a prohibition of the preparation of those weapons and of training in their use.

The debate continued in the General Commission of the Conference, which was established on 8 February 1932, and entrusted with the preliminary study and coordination of plans and proposals placed before the Conference.

On 11 April 1932, the US representative submitted a draft resolution by which the Commission would express the belief that the abolition of aggressive weapons would constitute a first and essential requisite, not only for the reduction of armaments, but for the establishment of security. The Commission would resolve that "the following weapons are of a peculiarly aggressive value against land defences—tanks, heavy mobile guns and gases—and as such should be abolished" and that "an undertaking by the states not to avail themselves of the afore-mentioned weapons in the event of war is equally essential".

The US proposal was considered insufficient by the French delegation, inasmuch as it mentioned only land armaments; it was also considered inadequate in regard to supervision and sanctions. [17]

The Italian delegation, with a view to the practical application of the principles it had formulated at the plenary session of the Conference, submitted a memorandum suggesting, among others, the following measures:

> The High Contracting Parties agree to abolish the use in time of war of chemical weapons of all kinds and particularly to prohibit all asphyxiating, toxic, lachrymatory or similar gases, all liquids or other substances or devices producing results similar to the above-mentioned gases and bacteriological methods of all kinds.
>
> Accordingly, the High Contracting Parties undertake:
>
> (1) To destroy, within a period of (x) months as from the entry into force of the Convention, all quantities of chemical and bacteriological substances of the kinds mentioned above constituting reserve depots or material for experiment, as well as the plant serving for their manufacture and all appliances serving for their utilization. Nevertheless, plant capable of direct employment by the chemical and pharmaceutical industry for non-military purposes may be retained on condition that it is strictly utilized for the needs of peaceable industries.
>
> (2) To destroy, within a period of (x) months as from the entry into force of the Convention, all artillery or hand ammunition and projectiles of all kinds loaded with chemical and bacteriological substances of the above-mentioned categories and intended for discharge by aircraft.
>
> (3) Not to manufacture in future chemical and bacteriological substances of the above-mentioned kinds specifically intended to harm the belligerents or the civil population, with the exception, however, of chemical or bacteriological substances capable of being utilized for peaceable industrial and scientific purposes and for such purposes only. They also undertake not to manufacture appliances for the utilization of the said substances.
>
> (4) Not to maintain or train personnel specialized in the use of chemical and bacteriological appliances of all kinds, even as personnel of other undertakings; not to publish even for purely theoretical purposes regulations or instructions dealing with the use of the said aggressive appliances.
>
> (5) Not to import chemical and bacteriological appliances of any kind specifically intended for warlike purposes. [18]

The Danish delegation, referring to its suggestion for the establishment of an international cartel, proposed:

> That the 1925 Convention concerning chemical and bacteriological warfare be supplemented by an undertaking entered into by the states not to engage in any preparatory manufacture or training with a view to the use of such weapons, and to prohibit all private manufacture of means of chemical and bacteriological warfare;
>
> That a technical sub-committee be set up with a view to getting into touch with national cartels for chemical and bacteriological manufacture and studying with them the possibility of organizing an international cartel responsible for

ensuring that such private manufacture shall not be employed for preparation for chemical and bacteriological warfare. [19]

The Turkish delegation supplemented its proposal, which it had presented at the opening stage of the Conference, with the following remarks:

> ... Even after complete prohibition of chemical warfare by international treaties, the chemical industry, restricted to its true purpose, should be internationalized in the same way as the manufacture of war material.
> It would, moreover, be absolutely necessary to take the first step in the direction of prohibition by a complete destruction at the outset of all means of chemical warfare at present in existence. [20]

The Yugoslav representative elaborated the problem of sanctions. He stated that in order to do away with chemical and bacteriological weapons, it had to be decreed that the transgressor country would be excommunicated from the civilized world and all the countries which had signed the convention would be required to proceed against it. It should be known that a country which had used such methods would lose the war and would not attain its object by using them. The proposal called for:

> Prohibition of the employment of and preparation for air bombardment and chemical and bacteriological warfare, even in case of legitimate defence. If, in case of hostilities one of the parties transgressed this obligation, the Council of the League of Nations would have to pronounce its outlawry from the civilized world. In such case all the signatory states would be obliged to render military assistance to the state victim of the transgression. [17]

On 22 April 1932, the General Commission of the Disarmament Conference adopted a resolution by which it declared the "approval of the principle of qualitative disarmament—i.e., the selection of certain classes or descriptions of weapons the possession or use of which should be absolutely prohibited to all states".

In another resolution, it decided that:

> In seeking to apply the principle of qualitative disarmament ... the range of land, sea, and air armaments should be examined by the competent special Commissions with a view to selecting those weapons whose character is the most specifically offensive or those most efficacious against national defence or most threatening to civilians. [17]

On 10 May 1932, upon recommendation of the Bureau—which had been created on 5 February 1932, to assist the President in directing the work of the Conference—the General Commission decided to appoint a special committee to consider chemical and bacteriological weapons from the point of view of qualitative disarmament. It consisted of: Australia, Brazil, Denmark, France, Germany, Italy, Japan, the Netherlands, Poland, Spain, Switzerland, the Union of Soviet Socialist Republics, the United Kingdom and the United States of America. [17]

III. *First report of the Special Committee on CBW*

The Special Committee submitted its report on 31 May 1932.

In Chapter I entitled "Chemical Weapons and Methods of War", the Committee considered the special character of substances that might be used in chemical warfare. These substances only became means of warfare through the use which was made of them in war. This marked a distinction between chemical means of warfare and ordinary weapons.

The same applied to a certain extent to the appliances and devices utilized for the employment of chemical substances in war. There were some that might be used equally well for chemical warfare and for peaceful purposes. Others might be used both for chemical warfare and for other war purposes. A small number were employed specifically for carrying on chemical warfare.

The Committee agreed unanimously that chemical weapons and methods of warfare were the most threatening to civilians. The characteristic of gas employed in chemical warfare was that when once it had been released it was no longer under the control of those employing it. It might reach civilians a considerable distance away from the spot where it was released.

There was some difference of opinion as to whether chemical weapons and methods of warfare were the most specifically offensive in character, and whether they were the most efficacious against national defence.

The Committee specified what should be the definition of chemical weapons and methods of warfare. It included all harmful substances, whether natural or synthetic, whatever their state, whether solid, liquid or gaseous; poisons such as curare or snake-poison were also included.

The Committee was unwilling to undertake an enumeration of the various categories of substances according to their chemical composition. It thought it essential to draw up a definition which should apply to all substances, both those known and those which might be discovered subsequently. It therefore adopted as criterion the physiological effects of the substances on living creatures. All substances having a harmful effect were included in the definition.

No account was taken of the degree of harmfulness. It was thought that if certain gases less pronouncedly harmful in their effects were excluded from qualitative disarmament, the practical value of the system would be considerably weakened. There would be difficulties in ascertaining the nature of the substances employed, and the use of these substances might lead in good faith to an unfounded allegation that a prohibition had not been observed. Furthermore, distinguishing between the various gases on the basis of their relative harmful effects would be a very long and difficult task, and would need to be kept constantly up-to-date.

While admitting the validity of the reasons against permitting the use of certain gases in international warfare, one delegation pointed out that lachrymatory gases, considered separately, did not in fact answer to the criterion of being the most threatening to civilians; that the use of these gases for police purposes could not be open to any objection; and that in some circumstances such a use would even be preferable to other methods which involved bloodshed.[1]

The Committee accepted this view, but was still of the opinion that lachrymatory gases should not be considered separately from the point of view of their use in warfare, since there were serious practical objections to any discrimination between gases.

The Soviet delegation felt that the question of lachrymatory gases for police purposes lay outside the terms of reference of a conference dealing with reduction and limitation of armaments.

The Committee thought that the prohibition should include not merely substances harmful to human beings, but should extend to those harmful to animals. No special reference was made to vegetables, because it was felt that in practice it would not be possible to employ, for the purpose of damaging vegetables, substances which were not also harmful to human beings or animals, or which were not likely to make the vegetables harmful to them.

In the Committee's view, qualitative disarmament applied to chemical substances only if their use was designed to injure an enemy, not necessarily during a military action; the poisoning of wells or springs from which an enemy might possibly draw water would also be prohibited. On the other hand, the use of chemical substances for the maintenance of armies, such as the use of disinfectants and medicaments and the means of destroying harmful animals and parasites, was not in question.

The Special Committee specified what ought to be excluded from the definition:

Explosives. The combustion of explosives may cause a discharge of noxious substances (such as carbon monoxide) which, according to circum-

---

[1] The name of the delegation was not indicated in the report. No detailed minutes of the first session of the Special Committee (18–31 May 1932) were published. In all probability the delegation in question was that of the United States of America.

stances, may have more or less serious destructive effects. That consequence could not be prevented by any means short of prohibiting explosives. At the same time, any practice designed to increase the discharge of noxious substances must be condemned. Such a practice might consist either in introducing certain products into the explosives, or in altering the constitution of the explosives, or in adopting a special method of manufacturing the projectiles.

Smoke and clouds. Smoke could be used as a screen, or for signalling, etc. The smoke and clouds must not be capable, in normal conditions of use, of producing harmful effects upon the organism.

The Committee suggested including among the objects of qualitative disarmament all appliances, devices or projectiles (to which the Soviet and Italian delegations had drawn the attention of the Conference) specially constructed for the utilization of the said noxious bodies with a view to injuring an adversary; it thus condemned material which could only be used for chemical warfare.

The Special Committee ignored the question of the methods whereby qualitative disarmament could be effected.

The Soviet delegation asked that the following observation be inserted in the report:

The Committee on Chemical and Bacteriological Weapons' reply to the General Commission's questions regarding qualitative disarmament is given mainly from the point of view of the prohibition of the use of chemical weapons in wartime. This is tantamount to restating with a few supplementary details the essential ideas contained in the Geneva Protocol of June 17th, 1925, which up to the present is unfortunately still awaiting the ratification of several states.

Such legal prohibitions are, however, inadequate and of merely secondary importance. The Soviet delegation has always attached and continues to attach paramount importance, not to the prohibition of the use of chemical weapons in wartime, but to the prohibition of preparations for chemical warfare in peacetime. Consequently, efforts should be directed not so much to the framing of laws and usages of war as to the prohibition of as many lethal substances and appliances as possible. This is the point of view which the Union of Soviet Socialist Republics will continue to represent in the General Commission.

The Soviet declaration was not discussed.

In Chapter II of the report, entitled "Bacteriological Weapons and Means of Warfare", the Committee stated that bacteriological warfare unquestionably constituted one of the methods of war most threatening to civilians, since epidemics, as they spread, attacked all human beings indiscriminately, and there was no sure method of arresting their effects.

Moreover, bacteriological warfare ought to be included in qualitative disarmament quite irrespective of whether it answered to any of the criteria laid down in the General Commission's resolution. It was so particularly odious that it revolted the conscience of humanity more than any other method of warfare.

The Committee tried to give the fullest possible definition of bacteriological warfare and the methods of such warfare. In the resolution it adopted (see below), reference was made to all methods of disseminating pathogenic microbes, filter-passing viruses or infected substances, wherever such dissemination took place. It was immaterial whether the microbes in question were momentarily innocuous, if they were capable of becoming virulent again.

It was stated that methods of bacteriological warfare may not be employed either against human beings, or against animals or plants.

In Chapter III devoted to "Incendiary Projectiles and Flame-projectors", the Committee found that incendiary weapons possessed a specific means of action which assimilated them to chemical rather than to ordinary weapons. It considered that the use of incendiary projectiles involved a particularly grave menace to civilians, and that the use of flame-throwers was calculated to cause needless suffering. It declared therefore that projectiles specifically intended to cause fires and appliances designed to attack persons by fire should be included in qualitative disarmament.

The Special Committee submitted the following resolutions to the General Commission: [21]

I. Chemical Weapons and Means of Warfare

The Committee considers,
That chemical substances, whether elements or natural or synthetic compounds, as well as appliances or devices for releasing them, can be described as weapons or means of warfare only in virtue of the use that is made of them, for they may be employed or made with a view to entirely different and essentially peaceful uses;
That, when used for the purpose of injuring an adversary, they answer to one or other of the criteria laid down in the General Commission's resolution of April 22, 1932, and, in any case, more particularly to the third of those criteria,

It declares,
That there should be included in qualitative disarmament the use, for the purpose of injuring an adversary, of all natural or synthetic noxious substances, whatever their state, whether solid, liquid or gaseous, whether toxic, asphyxiating, lachrymatory, irritant, vesicant, or capable in any way of producing harmful effects on the human or animal organism, whatever the method of their use.

It also declares,
That appliances, devices or projectiles specially constructed for the utilization

of the said noxious bodies with a view to injuring an adversary should be included in qualitative disarmament.

It observes,
That unless the use of explosives as such is included in qualitative disarmament, the above definition cannot be extended to the noxious substances arising from the combustion or detonation of explosives, provided that the latter have not been designed or used with the object of producing noxious substances.

It further observes that the above definition should not apply to smoke or fog used to screen objectives or for other military purposes, provided that such smoke or fog is not liable to produce harmful effects under normal conditions of use.

II. Bacteriological Weapons and Means of Warfare.

The Committee considers,
That the use of pathogenic microbes for the purpose of injuring an adversary is condemned by the conscience of humanity, quite apart from the fact that it answers to the criteria laid down by the General Commission in its resolution of April 22nd, 1932, and more particularly to the third of those criteria.

It declares,
That all methods for the projection, discharge or dissemination in any manner, in places inhabited or not, of pathogenic microbes in whatever phase they may be (virulent or capable of becoming so), or of filter-passing viruses, or of infected substances, whether for the purpose of bringing them into immediate contact with human beings, animals or plants, or for the purpose of affecting any of the latter in any indirect manner—for example, by polluting the atmosphere, water, foodstuffs, or any other objects—should be included in qualitative disarmament.

III. (a) Incendiary Projectiles.

The Committee considers,
That the use of incendiary projectiles involves a particularly grave menace to civilians.

It declares,
That projectiles specifically intended to cause fires should be included in qualitative disarmament.

It observes,
That this definition does not apply either to projectiles specially constructed to give light or to be luminous and, generally, to pyrotechnics not intended to cause fires, or to projectiles of all kinds capable of producing incendiary effects accidentally.

It considers,
That qualitative disarmament should not extend to projectiles designed specifically for defence against aircraft, provided that they are used exclusively for that purpose.

III. (b) Flame-projectors.

The Committee considers,
That the use of flame-projectors is calculated to cause needless suffering, quite apart from the question whether it answers to any of the criteria laid down by the General Commission's resolution of April 22nd, 1932.

It declares,
That appliances designed to attack persons by fire, such as flame-projectors, should be included in qualitative disarmament.

IV. *Conclusion of the first phase of the Disarmament Conference*

On 22 June 1932, in the General Commission, the US representative read a statement of the instructions issued by President Hoover to the United States delegation to the Disarmament Conference. The programme of disarmament presented to the Conference included a proposal for the abolition of all chemical warfare. [17]

On 20 July 1932, the General Commission considered a draft resolution, submitted by the Rapporteur, on the conclusion of the first phase and preparation of the second phase of the Conference.

Noting that agreement had been reached on a certain number of important points, the draft enumerated measures of disarmament which should form part of the general convention to be concluded. The measures included the prohibition of chemical, bacteriological and incendiary warfare "under the conditions unanimously recommended by the Special Committee".

Among questions to be examined in detail, there were:

Rules of international law shall be formulated in connection with the provisions relating to the prohibition of the use of chemical, bacteriological and incendiary weapons and bombing from the air, and shall be supplemented by special measures dealing with infringement of these provisions.

In presenting the relevant part of the draft resolution, the Rapporteur expressed the view that the Conference had progressed beyond the Geneva Protocol of 1925, and the new provisions framed by the Special Committee had supplemented and improved it; those new provisions constituted a general plan for the total prohibition of chemical and bacteriological warfare. It was necessary—indicated the Rapporteur—to examine the problem of possible violations of the obligations contracted by states.

On 21 July 1932, the representative of the USSR, who took part in the discussion on the draft resolution, pointed out that the Geneva Protocol had been ratified by thirty-four states, including the USSR, the United

Kingdom, France, Italy and Germany, and if the states which had not signed or ratified the Protocol were in agreement with it, they had only to make a formal declaration to that effect, and the instrument would receive universal application. Anything new in the recommendations of the technical bodies with regard to chemical and bacteriological warfare might be made the subject of an additional protocol. But there was no reason for reopening the question of international obligations which were already in existence and encouraging governments to withhold their adherence to or ratification of these obligations.

The Soviet delegate proposed to replace the paragraph on chemical, bacteriological and incendiary warfare in the draft resolution by the following text:

> The States represented at the Conference, in so far as they have not done so hitherto, undertake to sign within three months and to take steps for the speediest possible ratification of the Geneva Protocol of 1925, concerning the prohibition of chemical and bacteriological warfare. . . .

He did not press for a vote on this amendment.

The representative of the Netherlands said that two questions arose in connection with the means of chemical, bacteriological and incendiary warfare: prohibition to use them and prohibition to prepare them. The draft resolution related only to the first of these questions; this did not mean that the second was not covered. As soon as the Disarmament Conference had unanimously prohibited the use of chemical, bacteriological and incendiary warfare, it would be impossible not to conclude that preparation for this form of warfare would also be prohibited. The Special Committee's report dealt with the preparation of chemical warfare only from the point of view of material. It did not deal with personnel. However, if the preparation of material for chemical warfare were banned, training for chemical warfare would also inevitably be prohibited. He was therefore in favour of the draft resolution.

The Norwegian delegate thought that no distinction could be made between barbarous methods of war and less barbarous. If it were desired to make such a distinction, it might be necessary to conclude that chemical warfare was still the least barbarous method and that death by asphyxiation was less painful than that produced by the bursting of shells. However, it would certainly be much easier to abolish the most recent methods of war, as they had become less deeply rooted in military organization, and to proceed by stages and concentrate on abolishing those methods which favoured sudden attacks. It would, nevertheless, be a dangerous illusion to think that the prohibition of air bombardments and chemical warfare would be effective if their preparation were still permitted and states were allowed to remain in possession of those methods of warfare. The technical commissions must be authorized to go to the root of the matter and to study the total abolition of those arms.

On 22 July 1932, the General Commission approved, without amendments, those parts of the draft, submitted by the Rapporteur, which dealt with chemical, bacteriological and incendiary warfare, and on 23 July adopted the resolution as a whole. (The USSR voted against.) [22]

V. *The report of the chairman of the Special Committee*

On 21 September 1932, the Bureau of the Conference considered action to be taken on the decisions of the General Commission regarding the prohibition of chemical warfare. [23] The state of the question was:

The draft convention drawn up by the Preparatory Commission for the Disarmament Conference provided in Article 39 for:

absolute prohibition of bacteriological warfare;
prohibition, subject to reciprocity, of chemical warfare.

The Special Committee had decided in favour of the prohibition of chemical bacteriological and incendiary weapons.

The Bureau had to take a decision on the following questions:

1. Should the requirement of reciprocity be maintained, or should the prohibition of chemical warfare be made absolute?
2. Should this prohibition cover questions of training in time of peace and the manufacture of weapons?

These questions had been raised in a letter from the Norwegian delegation, which was distributed on 1 August to the members of the Bureau and contained amendments to Article 39 of the draft convention.

The proposal read:

Article 39
The High Contracting Parties undertake unreservedly to abstain from the use in war of asphyxiating, poisonous or similar gases, and of all analogous liquids, substances or processes.
They also undertake unreservedly to abstain from the use of all bacteriological methods of warfare.

Article 39 (b)
The High Contracting Parties consequently undertake to destroy within a period of ... months as from the coming into force of the present Convention, all appliances and apparatus for chemical aggression (all asphyxiating gases for use in war, and all appliances for spreading them, such as gas-projectors, pul-

verisers, balloons, flame-projectors and other apparatus) either in use by troops, or in reserve, or in course of manufacture.

They also undertake to destroy any appliances and apparatus for bacteriological warfare at their disposal.

Article 39 (c)
The High Contracting Parties also undertake to abstain from any preparation in time of peace for the use in wartime of the methods of warfare specified in the two preceding articles, and in particular from any training of troops in such methods, and undertake, on the coming into force of the Convention, to take the necessary steps for effectively preventing all preparation on the part of private persons or enterprises within the respective territories for the employment of such methods of warfare. They further undertake to take all necessary steps for effectively preventing the manufacture in, import into, or export from their respective territories of materials or apparatus capable of use for chemical or bacteriological warfare except in so far as such substances or apparatus meet a normal peacetime requirement.

If such substances or apparatus meet a normal peacetime requirement, the High Contracting Parties undertake to keep their import, export and manufacture within the limits of commercial requirements, and to supply the Permanent Disarmament Commission with annual statistics of such import, export or manufacture.

(Suitable provisions for controlling and supervising the observation of the above undertakings shall be inserted in the appropiate chapter of the Convention.

It is understood that defensive apparatus for use against chemical warfare or training in the use of such apparatus is not covered by the above prohibitions.)

To sum up, the Norwegian delegation called for:

Absolute prohibition of the employment of chemical and bacteriological weapons and methods of warfare;

Destruction of all means of CB warfare and of all apparatus used for such warfare;

Prohibition of preparation in time of peace of such methods and of training for their use. [24]

Following a suggestion made by the British representative, a resolution was adopted, reading:

The Secretary-General is requested to report to the Bureau as soon as possible, from the records of enquiries already made:

(1) What is the state of expert opinion as to the practicability of prohibiting the *preparation* of gas as distinguished from prohibiting its use in warfare;

(2) Whether there is any sufficient reason why the preparation and possession of *machines and instruments* for the use of gas should not be forbidden;

(3) The same as to training in use of gas.

On the following day, the Bureau requested the chairman of the Special Committee to submit a report on the same subject and on the problem raised by possible violation of the prohibition of chemical, bacteriological and incendiary warfare. [23]

In compliance with the Bureau's request, the Secretary-General prepared a memorandum reproducing the opinions previously expressed by various League committees of experts (see Chapters 1–3) as to the possibility of preventing preparations and training for the use of chemical and bacteriological weapons in peacetime. [25]

The report of the chairman of the Special Committee, of 25 October 1932, dealt with:
Absolute or relative character of the prohibition.
The prohibition of preparations for chemical, incendiary and bacteriological warfare.
The supervision of the observance of the prohibition of preparations for chemical warfare.
Sanctions in the event of the use of chemical, incendiary and bacteriological weapons. [26]
Here is the summary of the report.

Absolute or relative character of the prohibition

The Geneva Protocol of 17 June 1925, concerning the prohibition of chemical and bacteriological warfare, did not formally exclude the condition of reciprocity. Accordingly, twelve states (Australia, France, India, Iraq, Irish Free State, the Netherlands,[2] New Zealand, Portugal, Romania, Spain, the Union of South Africa and the USSR) had ratified it only subject to that condition.

Article 39 of the draft convention dealing with chemical and bacteriological warfare, prohibited the former on condition of reciprocity, while the prohibition of the latter was absolute. In the course of the Disarmament Conference, however, various delegations declared that any prohibition that was not absolute would lose much of its force. States would contemplate recourse to the chemical weapon as an eventuality if the convention itself admitted it in certain cases. Under such circumstances, any attempt to prohibit preparation for chemical warfare in peacetime would be illusory. Hence the prohibition of chemical warfare must be absolute.

If the prohibition were absolute, the right of resorting to the prohibited means of warfare would be excluded.

---

[2] The condition of reciprocity was attached only to the prohibition of chemical warfare.

Even if an undertaking not to use chemical weapons could not be given by all countries, the undertaking would not be allowed to be given on condition of reciprocity.

It must, nonetheless, be borne in mind that a different position would arise if, in time of war, states which had undertaken not to resort to chemical, bacteriological or incendiary weapons were confronted by other states which had not given such an undertaking.

To ensure that no such situation should arise, the undertaking should be given by all states, or by a very large number of states. If the undertaking could not be made absolutely universal, it might be decided that states accepting the undertaking should be required to take action against states using the chemical weapon, whether or not the latter had given any undertaking on the subject.

Even if a state were a victim of an unlawful war (a war undertaken in violation of the Pact of Paris or the Covenant of the League), it would be bound to refrain from resorting to these prohibited means of warfare.

In the Rapporteur's view, if the absolute character of the prohibition were to be respected, and if states were to regard it as a settled principle governing their conduct, the forms of control provided for must be as satisfactory as possible, and, above all, the penalties must be effective.

The prohibition of preparations for chemical, incendiary and bacteriological warfare

In principle, the prohibition of preparations for chemical, incendiary and bacteriological warfare applied both in peacetime and in wartime. In peacetime, there could be no exception to this prohibition. In wartime, the only exception that might perhaps have to be made would be in the event of an adversary having had recourse to the prohibited weapons. Since use of certain of these weapons by way of reprisal might be allowed under certain fixed conditions, preparations for their use would become lawful in the case contemplated.

The prohibition of preparations for chemical, incendiary and bacteriological warfare had two aspects—the manufacture and possession of material, and the training of troops.

1. Prohibition of the manufacture and possession of appliances and substances.

This material was divided into two classes: defensive material (protective masks) and material for attacking purposes (gases and gas-projecting apparatus).

Defensive material. There seemed to be no reason for prohibiting material solely intended to protect combatants and civilians against chemical and bacteriological weapons should such be employed by a belligerent. The existence of protective material, calculated to lessen the injury caused by the prohibited weapons, would diminish the military advantages that an unscrupulous belligerent might derive from their use, and would therefore make resort to those weapons less tempting to such a belligerent.

Material for attacking purposes. This material would be completely forbidden. The elements covered by the prohibition were:

(*a*) Prohibition to manufacture, import or store appliances exclusively suited to the conduct of chemical warfare (special apparatus for the projection of gases or propulsion of gas-shells).

The prohibition, even if strictly observed, could not form a serious obstacle to the conduct of chemical warfare. In the first place, those were relatively simple appliances which could be fairly quickly manufactured, and, secondly, chemical warfare could be carried on, if necessary, without them.

(*b*) Prohibition to manufacture, import or store chemical substances exclusively suited to the conduct of chemical warfare.

There existed certain gases which could not be used for peaceful purposes. A list of the principal chemical substances falling into this class could be made, but it would not be exhaustive. New substances answering to the above definition might be invented. Moreover, substances which were rightly regarded as exclusively suited to the conduct of chemical warfare might subsequently prove suitable for peaceful uses. In such a case the list would have to be revised. Any stocks of the above-mentioned chemical substances which might be in the possession of states when the convention was put into force should be destroyed.

(*c*) Prohibition to manufacture, import or store appliances and chemical substances capable of being used both for peaceful and for military purposes with the intention of using them in war should the occasion arise.

In this case, it is not the nature of the appliances and substances, but the intention with which they were produced and stored, that would be reprehensible.

This general prohibition might be accompanied by subsidiary prohibitions to issue some listed appliances and substances to the armed forces, to keep them in military establishments (arsenals, fortifications, barracks, etc.) or to store them for army use. If the armed forces need a certain quantity of such appliances or substances for use other than as fighting weapons (e.g., chlorine cylinders, chlorine required for disinfection purposes), the states should be asked to say what quantity they would wish to leave at the disposal of the armed forces. These special prohibitions would, of course, have only a limited value.

Any stocks of chemical substances which might be in the possession of states themselves when the convention was put into force should be disposed of for actual consumption.

It is also possible to fix for each state the maximum quantity of chemical substances capable of use for military purposes which may be stored in industrial establishments of every kind in its territory. That would involve making an estimate of the needs of consumers and the normal reserves necessary to meet those needs. Such an estimate would, of course, present real difficulties.

2. Prohibition of training.

Training might be either for defence or for attack.

Defensive training aimed exclusively at protecting people against the effects of chemical, incendiary or bacteriological warfare must be authorized for the same reasons as were given above in connection with defensive material.

Training for the practice of chemical, incendiary or bacteriological warfare must be absolutely forbidden.

The supervision of the observance of the prohibition of preparations for chemical warfare

The supervision of the observance of the prohibition of preparations for chemical warfare is designed to prevent breaches of the prohibition. The question of supervision arises in peacetime as in wartime, with this difference, that in wartime supervision would be exercised over the belligerents under even more difficult conditions than in peacetime.

Characteristic of the substances used in chemical warfare is that they can easily be prepared in a country having a well-developed chemical industry, and that manufactures of a peaceful character could be used as means of chemical warfare without conversion, or with only very simple and rapid conversion.

In regard to the prohibition of training for chemical warfare, and the prohibition of the manufacture and possession of material exclusively suited to the conduct of chemical warfare, the question of supervision presented difficulties, but they were the same in kind, if not in degree (the manufacture of gas projectors was easier to conceal than the manufacture of heavy guns), as those that were encountered in other fields. This supervision could not therefore be separated from the general system of supervision that would be instituted to enforce other prohibitions laid down by the Convention for the Reduction and Limitation of Armaments.

As regards chemical substances exclusively suited for military use, the attempt to trace their manufacture would entail inspections of factories and a careful study of the processes in progress there. This method of supervision would be of very doubtful efficacy. Concealment would be easy in view of the size and intricacy of the chemical industry, and of the fact that the products in question might not become products exclusively suited for war until they had undergone one slight final transformation at the end of the process of manufacture. Moreover, this form of supervision would have the disadvantage of affording openings for malpractices, or giving rise to suspicion of affording such openings. It would involve a risk of industrial espionage, and might lead to friction, without yielding much in the way of results.

In the case of appliances and chemical substances suited for both peaceful and military use, supervision should establish three facts:

output, and what part of the plant, if any, was not actually used but could be, if desired;

any excess of production over consumption;

that this excess of production over consumption was not a mere accident due, for example, to the invention of new processes of manufacture, or to mistaken economic forecasts, or to a depression, but was intended to allow for the formation of stocks with a view to the conduct of chemical warfare.

In view of the difficulties attendant upon the supervision of the production of chemical substances it would be better to abandon the idea of using such forms of supervision which would entail highly complicated investigations.

A suggestion had been put forward that international agreements between the chemical industries of different countries might permit effective supervision of chemical production with a view to preventing it from being carried out in preparation for chemical warfare. The value of the agreements which would have the effect of rationing production and preventing certain manufactures, would vary according to their nature and the number of parties to them. It did not appear, however, that the Conference for the Reduction and Limitation of Armaments and the governments of which it was composed had any means of deciding upon such agreements. Supervision would gain from their existence, but one could hardly contemplate creating them in order to facilitate the supervision with which the Conference was concerned.

One might consider whether it would not be possible to obtain some results by the following necessarily limited method of supervision:

A specialized section for chemical warfare would be established in the Permanent Disarmament Commission.

It would be the duty of this section to collect all the information that could be found in official trade statistics or private statistics carrying some weight, relating to the manufacture of chemical substances in the territory of each signatory state, and their import into that territory.

Having seen this information and made a preliminary study of it with all the necessary safeguards, the section might ask the governments of the countries concerned to supply it with any further information, if necessary.

Further, any signatory state might, on the strength of any information received by it, apply to the section for explanations regarding chemical substances manufactured in the territory of another state or imported into that territory. The section would make a preliminary examination of such applications, with all due discretion and subject to the necessary safeguards, in order to establish whether it was worthwhile to obtain fuller information. If it were so decided, the state whose position had been impugned would be asked for explanations.

When it was in possession of the information required, the section would decide whether the matter was to be dropped or laid before the Permanent Commission. In the latter case, the Commission would take all proper steps, within the limits of the powers conferred upon it by the disarmament convention.

An idea had been advanced that, to facilitate the provision of information for the specialized section of the Permanent Commission, the signatory states might undertake to send, not only in reply to a request they may receive from the section, but also periodically and regularly, statistics of the manufacture of chemical substances in their territory and the imports of such substances into that territory. It was, however, doubtful whether fuller and more useful information would be obtained in this way than was already available in the trade statistics normally published in most countries.

The limited supervision, as suggested above, would still yield only incomplete results. The work of various technical bodies connected with the League of Nations, which in 1926 studied the question of preparation for chemical warfare, had demonstrated the inevitable imperfection of any form of supervision. It had also been established that for a country with a well-developed chemical industry preparations for chemical warfare would always be easy and could be made very rapidly. One must, therefore, not rely essentially on supervision. A system of penalties to be applied in the event of the employment of chemical weapons must be so severe as to make up for the deficiencies of preventive supervision.

Sanctions in the event of the use of chemical, incendiary and bacteriological weapons

The sanctions meant were those which would follow on a violation, not of the prohibition to prepare for chemical warfare in time of peace or in time of war, but of the prohibition to use chemical weapons in time of war.

The question of sanctions embraced both the attitude to be adopted by the state against which chemical weapons had been employed and the attitude to be adopted by the other parties to the convention.

Establishment of the fact that the prohibition had been infringed

This should be governed by the following rules:

The establishment of the fact of infringement should not be carried out by the state complaining of such infringement, but should be entrusted to a neutral authority.

If the establishment of the infringement were left to the state against which it was alleged to have been committed, it would be attended by no safeguards. Without going so far as to suppose that a state desirous of employing chemical weapons would—in order to provide a semblance of justification for its conduct—accuse its adversary of having had recourse to that prohibited arm, one might simply fear a mistake on the part of the military over-ready to jump to conclusions. They may impute to chemical weapons the asphyxiating effects produced by the normal combustion or detonation of ordinary explosives or by some other cause. But even if the establishment of the infringement were based on fact, it would be lacking in authority if it were effected by the actual victim. It was to the latter's interest that the establishment of the infringement, designed as it was to produce moral, legal and political effects, should be conducted under all desirable conditions of impartiality, so that it would carry conviction with all governments and public opinion.

The establishment of the infringement must be effected within a fairly short time.

Since it must precede any retaliation on the part of the victim and any intervention by third states, it is essential that it should take place rapidly; otherwise the state guilty of resorting to chemical weapons might derive great military advantage from its initiative. If the establishment of the infringement is effected with all requisite speed, it seems unlikely that a few days' delay which might prove unavoidable would really be prejudicial to the victim. As regards the third states, their assistance could not reasonably be given until the facts which called for it have been established with certainty. As to the victim, it was probable that, even if it were to retaliate by itself employing the prohibited weapon, it would require some few days at least to make its preparations.

Practical rules for the establishment of the infringement must be laid down in advance. The injured state must know to whom it should apply in order that the establishment of the infringement may be effected. The person in question, who could be the doyen of the diplomatic corps in the country concerned, would call upon agents such as consuls or military attachés, possibly assisted by doctors or chemists, who would be responsible for actually establishing the infringement. The result of the investigation would be forwarded to the Permanent Commission set up by the disarmament convention and that Commission would draw its conclusions from the investigation and declare that an infringement had or had not taken place. Simultaneously with the person responsible for instituting the investigation, the Permanent Commission would be informed of the occurrence and would have an opportunity of ordering that enquiries be made from authorities of the state accused of the infringement, while it would, in every instance, inform that state of the charge brought against it and request its observations.

It was understood that the states concerned would be obliged to allow measures of supervision to be carried out without restriction, so that the authorities responsible for supervision might move about the country and carry out their observations. A state which placed obstacles in the way of supervision would create an unfavourable presumption against itself.

The actual establishment of the infringement would probably be easy to carry out. There was reason to think that a state which by resorting to prohibited weapons exposed itself to the intervention of the third powers and to retaliation on the part of its adversary would, from the outset, have made use of those arms on a very big scale. It was a fairly simple matter to decide, even without special expert knowledge, in the presence of a certain number of killed and wounded, whether they had been stricken by a chemical weapon or by other arms.

Effects of the establishment of the fact of infringement

The solution found for the specific question of infringements of the prohibition to use chemical weapons must be in harmony with the general system of sanctions laid down in the convention. Nevertheless, the infringement of the prohibition to use chemical weapons would be an offence of a rather special kind. On the one hand, any recourse to chemical weapons would constitute a violation of the convention—a violation of special gravity from the standpoint of its military consequences—and, on the other, since supervision concerning chemical weapons would be of a very limited efficacy in time of peace, as distinguished from the other categories of prohibited arms, the inadequacy of the preventive action should be offset by provision for more drastic penalties.

The attitude of third states in the event of an infringement of the prohibition to have recourse to chemical weapons:

Once the Permanent Commission had established that the prohibition to use chemical weapons had been infringed, other states would be both entitled and bound to take a series of measures against the offending state, which, by gradually increasing the pressure brought to bear upon that state, might induce it to abandon the use of the chemical weapon or make its continued use impossible. (The measures might include diplomatic representations, rupture of diplomatic, economic and financial relations, and blockade.)

After the establishment of the infringement of the prohibition, there would be a consultation between states which would make it possible to determine the methods most appropriate, in the circumstances, to put a stop to the infringement and assist the state which had been a victim of the violation of the convention. That consultation would have to take place as rapidly as possible. To gain time, it might be possible to provide, in the first place, for convening a body smaller in size than the general conference of all states parties to the convention.

Whether the states would be prepared to enter into other obligations in the event of an infringement of the prohibition to use chemical weapons was a political question.

It might be decided that a state guilty of having infringed the prohibition to use chemical weapons would be deemed to have committed an act of war against the other states parties to the convention. Or it might be decided to apply increasingly drastic economic or financial sanctions, or more or less extensive military sanctions.

If states agreed to enter into such undertakings, the prohibition would be accompanied by powerful safeguards and would run little risk of being infringed.

Rights of the state which was victim of the violation of the prohibition:

It is desirable that the state which became a victim of the breach of the prohibition should not retaliate by employing the chemical weapon, and that the assistance given it by third states should compensate it, and more, for the disadvantage resulting from the fact that the chemical weapon had been employed against it. If, however, the state which was the victim of the breach were not assured of receiving such assistance, and in particular, if the assistance were not immediate, but contingent on a consultation between states, which might take some time, the state in question could not be forced to refrain from the use of chemical weapons, when, through being unable to fight its adversary with equal weapons, it might be crushed or suffer great injury. In no case could a state which had honourably fulfilled its obligations under the convention be allowed to be placed in a position of inferiority. It therefore seemed prudent to decide that the establishment by the Permanent Commission of the fact that a belligerent had definitely employed the chemical weapon, should confer on the victimized state the right to retaliate by employing the chemical weapon in its turn, or to use other reprisals, subject to the condition that such reprisals should not occur outside the fighting area.

But the Permanent Commission, after establishing the breach, should have power to recommend the state which had been the victim of that breach to renounce the use, by way of retaliation, of the chemical weapon. In so doing, it would take into account all the circumstances, and more particularly the practical possibility of giving effective aid to the state which had been a victim. The latter may, of course, not follow the recommendation to refrain from using the chemical weapon, but it would do so at its own risk—that is, it might lessen its chances of receiving assistance from third states.

Similarly, and with added authority, the third states consulting together might advise the victim of the breach not to retaliate by using the chemical weapon, or to defer such retaliation.

The possibility of retaliating by the actual use of the prohibited weapon would apply only to chemical and incendiary weapons, and not to the bacteriological weapon. The bacteriological weapon, since it strikes everyone indiscriminately, and cannot be regarded as an appreciable element of military superiority, would still be prohibited, but its use, if duly established, would give the other belligerent the right to employ other prohibited weapons by way of reprisal.

Effects of the application of the Covenant of the League of Nations and of the Pact of Paris:

A complication arose from the fact that the attitude to be adopted towards belligerents by states bound by the Covenant of the League of Nations or the Pact of Paris (the Briand-Kellogg Pact) would be, or should be, determined by the circumstance that one belligerent appeared to them to be supporting a lawful war and another to be supporting an unlawful war. The states members of the League would have to decide whether they should not take, against certain belligerents, the action provided for in Article 16 and lend assistance to states which had suffered aggression in breach of the Covenant.

In such circumstances, if the Covenant-breaking state resorted to chemical warfare, additional action would have to be taken against it. If a state, attacked in breach of the Covenant, resorted to chemical warfare, it would perhaps suffice to refuse or withdraw the assistance to which it would have been entitled.

As the disarmament convention was to be signed also by states which were not members of the League or were not parties to the Pact of Paris, it was not possible for the convention to lay down rules for situations arising out of the existence of other pacts, or to involve in the effects of those pacts states which were not parties to them.

In practice, the application of the disarmament convention, the Pact of Paris and the Covenant of the League might be satisfactorily reconciled. If the states parties to the disarmament convention were called upon to hold a consultation as a result of a breach of the prohibition to use the chemical weapon, they might consider, as one factor in their decision, the lawful or unlawful character of the war from the standpoint of this or that belligerent. The disarmament convention might contain a clause providing that the use of the chemical weapon might deprive states of the assistance laid down in other treaties, to which they would otherwise have been entitled. There would be nothing to prevent states members of the League from deciding among themselves by some appropriate means (special protocol, amendments to the Covenant) that the use of the chemical weapon should produce the same effects as a breach of Article 16 of the Covenant.

Conclusions of the report

The contracting parties should renounce with respect to any state, whether or not a party to the convention, and in any war, however unlawful such war might be on the part of their adversaries, the use of chemical and bacteriological weapons for the purpose of injuring an adversary, the use of projectiles specifically intended to cause fires or the use of appliances designed to attack persons by fire. (Certain substances and weapons used in the normal process of warfare would be explicitly excepted.)

All preparations for chemical, incendiary and bacteriological warfare should be prohibited in time of peace as in time of war, but this prohibition should not apply to material intended exclusively to protect individuals against the effects of such warfare or to the training of individuals in measures of protection.

A special section should be set up in the Permanent Disarmament Commission to deal with questions relating to preparations for chemical, incendiary and bacteriological warfare.

The Permanent Disarmament Commission would establish the fact of the use of chemical, incendiary or bacteriological weapons. It would have the right to carry out for this purpose any preliminary enquiries, both in the territory subject to the authority of the complainant state and in the territory subject to the authority of the state against which a complaint was made.

The declaration of the Commission establishing the fact of the use of chemical, incendiary or bacteriological weapons would entail immediate action on the part of the third states. It would be their right and duty to bring pressure to bear on the offending state. At the earliest possible moment, third states would decide, if necessary, on the punitive or other action to be taken. The state victim of the breach would have the right to retaliate within the fighting area by employing chemical or incendiary weapons.

VI. *Consideration of the chairman of the Special Committee's report*

The report submitted by the chairman of the Special Committee was discussed by the Bureau of the Disarmament Conference at meetings held from 8 to 12 November 1932. [27]

Absolute or relative character of the prohibition

The representative of Switzerland was strongly in favour of absolute and unconditional prohibition of CBW. He thought that it should be universal, and that it should retain this character, even if one or two states did not accept it. What was proposed in regard to chemical and bacteriological warfare simply represented a solemn confirmation of international law which bound even those who did not accede to the convention. In the peace treaties, this form of warfare was considered to be condemned by the conscience of mankind; the prohibition was therefore regarded as existent prior to the peace treaties. Further, the 1925 Geneva Protocol confirmed this virtually pre-existing principle.

The Belgian delegate was also convinced that the existence of the principle of international law prohibiting recourse to chemical weapons was independant of any convention. The purpose of the convention was to organize and add something to this principle, by prohibiting preparations for chemical warfare and by providing a system of sanctions.

The Japanese delegation supported the idea of absolute prohibition without any exception. It proposed that recourse to the use of chemical weapons should be prohibited even by way of reprisals.

The US representative stated the position of his government concerning tear gas. There was no question of its use in time of war, but the United States delegation would have difficulty in undertaking to give up the preparation and employment of this gas for local police purposes.

The French, British and Belgian delegations maintained that the question whether the prohibition of chemical, incendiary and bacteriological methods of warfare should be absolute or relative must necessarily depend on the conclusions to be reached in regard to the prohibition of their preparations, as well as the forms of control or penalties to be provided as guarantees that the prohibitions would be effective.

This was recognized also by other delegations.

Prohibition of preparations for chemical, incendiary and bacteriological warfare

The Rapporteur explained that since defensive measures (manufacture of masks, drill, utilization of masks, etc.) had not been prohibited in the report, the use of defensive appliances presupposed the possibility of manufacturing gases for experimental purposes. However this manufacture would be on such a small scale, being confined to laboratories, that it would not be necessary for the convention to deal with it.

He suggested that it might be expedient to convene the Special Committee to draw up a list of appliances and substances designed solely for the purposes of chemical warfare. This list would not be final but would merely serve as an example. It would allow measures to be taken against certain patented appliances and would make it known that the manufacture of such appliances was illicit.

He was in favour of an explicit prohibition of all appliances and substances connected with chemical warfare, but warned against any illusions in regard to the practical effect of the prohibitions.

The Japanese delegation observed that absolute and universal prohibition of all noxious gases did not mean that populations should be left without any means of defence against possible gas attacks. But a study of the means of defence against gases necessarily involved a study of the methods of attack. If some way could be found of restricting the scope of studies to defence without studying the methods of offence, the Japanese delegation would welcome it. It wondered, however, whether that was feasible.

Referring to the Rapporteur's suggestion that the testing of defensive material might be carried out on a restricted scale, in the laboratory, the representative of Japan pointed out that, in the opinion of experts, laboratory tests were not sufficient to study protection against dangerous gases.

The French representative recalled that, some years before, an accident had occurred which stirred Europe: the whole of one district in a certain city had been poisoned by toxic gases emanating from the stocks of a manufacturer of chemical products, who also sold protective masks. It was possible to obtain at one time, and for a modest sum, a mask and samples of the principal gases. He questioned whether such practices could be tolerated. Without going so far as to prohibit all preparations of defensive material, the French delegate wondered whether it would not be possible to prohibit private manufacture which might lead to such surprising results. The indiscriminate manufacture of defensive appliances and experimental material for those appliances—no matter where, no matter by whom—could hardly be admitted. It would be desirable to provide for strict government supervision and international supervision.

Bacteriological warfare had been totally prohibited, but, in one country represented at the Conference, experiments had been carried out, for purely scientific purposes, which necessitated preparation of tuberculosis bacilli by tens of kilogrammes in order to make possible a chemical analysis of the bacillus. What would happen if the mass production of dangerous bacilli took place in a number of countries?

The appliances peculiar to chemical warfare differed very little from others. Gas shell were very like other shell. Gas reservoirs were quite ordinary receptacles. The form of projector used was very similar to that of appliances for the release of non-poisonous smoke. When it came to the question of prohibiting material which could be used for both peaceful and warlike purposes, the difficulty was even greater. One big firm which had placed on the market a fire-extinguishing apparatus declared that it would serve excellently as a flame-projector.

The French representative further questioned whether there was any difference between instructing a unit in the release of non-toxic smoke for purposes of cover and training the same unit in the release of toxic gases. In both, the same considerations as regards wind direction, humidity of the air, etc. would be taken into account. As a matter of fact, no special training was needed for releasing gases.

He urged that the question of the prohibition of preparations be referred to the Special Committee with a view to formulating more definite rules in the matter.

The United States delegate felt very strongly that before an attempt was made to consider the advisability of accepting the prohibition of preparations, as indicated in the report, the following questions should be discussed in detail:

Whether the means provided for the protection of the individual, i.e., gas masks, were sufficient. Was it not necessary to go further and provide for group protection? If so, not only the question of training individuals and groups for immediate protection would be raised, but also the question of the means to be provided for the purpose. Were the means for the resanitation of gassed areas to be wiped out? If not, by what means might those areas be resanitated? Could groups of men be trained and material maintained for doing away with the effects of gas which had been illegally used? Under what authority should those people be trained? By whom would the material be provided? Where and how would the necessary experimentation be made to provide for such action? Was it proposed to do away with gas hospitals and with the specialized training of doctors to handle what were commonly known as "gas casualties"? If not, where and how should those doctors get their training? What limits should be placed upon laboratory experiments for training purposes?

He then suggested the inclusion in the report of a provision authorizing the training of the police in the use of tear gas for local purposes.

The British delegate said that it was necessary to know exactly what were the appliances and substances suited exclusively for the conduct of chemical, incendiary and bacteriological warfare before any final decision could be taken on the subject of preparations.

The delegate of Switzerland observed that there could be no real distinction between material intended exclusively for war purposes and material which might be used for both peaceful and military purposes and thought that further expert opinion must be obtained. While he had found the idea of a radical prohibition of all offensive or defensive material for chemical warfare very attractive, he wondered whether it would not conflict with the moral sense, with human nature, to claim that an individual or a state should cast prudence to the winds and renounce the idea of self-defence.

The representative of Uruguay spoke about the possible abuses arising out of an authorization to manufacture defensive material. He recalled that during the Arms Trade Conference in 1925, when one delegate had made a survey of the expenditure included in the budgets of the principal states for preparations for chemical warfare, and had stressed the magnitude of the sums devoted to related studies, the representatives of each of the nations in question stated that the studies concerning chemical warfare were for defensive and not for offensive purposes. The difficulty of deciding,

Consideration of the chairman's report

in this sphere, what was offensive and what was defensive was so great that the question should be left open pending an opinion of technical experts.

The Bureau decided to ask the Rapporteur of the Special Committee to draw up a list of questions which would need to be settled before a final decision could be taken on this aspect of the problem. The Bureau, in particular, desired to know whether any technical means existed which could allow defensive preparations without simultaneously allowing preparations for attack, and whether a list could be drawn up of appliances and substances exclusively suited to the conduct of chemical, incendiary and bacteriological warfare.

Sanctions in the event of the use of chemical, incendiary and bacteriological weapons

The representative of France pointed out that since it would not be possible to build up an entirely effective system of prevention—whatever supplementary proposals might be brought forward—the question of sanctions arose in all its gravity. Unless it was made clear to a state wishing to employ the prohibited methods of warfare that the fact of doing so would expose it to very serious consequences, it was to be feared that such a state, which had already accepted the responsibility of breaking the covenants it had signed, would not hesitate to violate one international agreement more and employ so effective a weapon.

As regards the establishment of infringement, the French delegate suggested that the question, whether the measures proposed in the report were adequate for the achievement of the aim in view, should be examined. Would it be easy for the leader of the corps diplomatique to arrange for an enquiry on the spot unless the details were settled in advance? Subject to this reservation, there was no disagreement on the principle.

In the matter of sanctions, which was much more serious, the evil should be balanced against the remedy proposed. It was not to be imagined that, once chemical and bacteriological warfare had been branded as an international crime, any state would lay itself open to the reproach of committing such a crime unless it had decided to bear the consequences of its act. There would be no question of purely localized emission of gas. There would be a large-scale use of gas in order to obtain the advantage of a decisive surprise. Such being the risk, the French delegate doubted whether the scale of sanctions laid down by the Rapporteur would really be very effective. Among the measures of pressure the Rapporteur mentioned the severing of diplomatic relations; but would that be a matter of very great importance to the offending state? It was said, moreover, that there should be an immediate consultation of the other states, which would decide on the measures to be adopted. But these consultations would take time; in the interval, the offending state would pursue its advantage. What would become of a small state which had been attacked by a great power possessing vast industrial resources?

Referring to the paragraph of the report, providing for the right of the state against which chemical weapons had been employed, to use them in retaliation, the French delegate said: if individual retaliation on the part of the attacked state was permitted, it would be impossible to apply the rules for the prevention of the preparation of chemical warfare, and there would be a reversion to the condition of reciprocity. In his opinion, only one solution was possible—recourse to collective retaliation. It would not be the attacked state that would have the right to take justice into its hands, but it would be the community of states which would act on its behalf. The only retaliation which could be admitted must be decided upon by that community under conditions to be laid down. The problem arose not only in chemical warfare. States must realize that as soon as they employed any forbidden weapon they would expose themselves to the use by the community of states of weapons which were not at their own disposal. In other words, it was necessary to make preparations for international punitive action.

The British representative pointed out that no country really anxious to observe the convention would be in a position to undertake immediate reprisals. It was therefore essential that the machinery provided for the establishment of infringement and the denunciation of the aggressor should be able to come into operation even before the attacked country could consider the possibility of reprisals. The matter required careful examination, for a country which was the victim of a breach of the convention must not be allowed to see its cities laid waste and its population decimated and finally find itself in an impossible position owing to the prohibition of reprisals and to a delay in the intervention of other states. Speaking about collective action on behalf of the victim of a breach of the convention, he submitted that measures must include all means of pressure, from moral to active pressure. He was, however, unable to say what sanctions his government might decide upon in certain given circumstances, or what sanctions it would have to reject. It was necessary to avoid any rigid definition.

The Japanese delegate was in favour of the strictest and most severe sanctions, but was not in a position to lay down the extent of the sanctions or the method to be followed for bringing them into play. He was definitely opposed to the employment of gas as a retaliatory measure.

The Greek delegation thought that the crux of the problem lay in framing a system of sanctions sufficiently serious to supplement the inadequacy of preventive measures. If the right of reprisal were rejected, accentuated measures of repression were essential.

The Spanish delegate declared himself opposed to any right of retaliation by the use of chemical weapons and supported the French delegation in demanding effective collective action against states violating the prohibition. He submitted the following suggestion as a guide to the discussion:

If a signatory state has recourse to the use of chemical, incendiary or bacteriological weapons, it shall be, *ipso facto,* considered as having committed an act of war against the other states parties to the Convention; and the said other states shall take repressive action against the state violating the prohibition, which action shall be progressively accentuated with a view to inducing the state in question to forego the use of chemical, incendiary or bacteriological weapons, or preventing it from continuing the use of them, in the last resort employing military sanctions to enforce respect for the obligations under the Convention.

The Permanent Disarmament Commission will organize such action in accordance with the obligation assumed by the signatory states.

The right of retortion against the use of chemical, incendiary or bacteriological weapons is formally forbidden. Any state having recourse to reprisals will thereby place itself outside the Convention.

The representative of Poland said that, while it was important to establish a system of supervision, the main thing was to give a state which had been attacked the certainty that the whole of humanity would back it up by inflicting on the aggressor state effective sanctions without any delay or restriction.

The United States representative agreed that there must be a prompt establishment of the fact of violation. He had some doubts as to the procedure to be followed to establish that fact, and feared that the proposition might prove to be an intolerable burden upon the doyen of the diplomatic corps, particularly if he were representing a small neutral country lying next to one of the great belligerents. That the establishment of fact should be followed up by a consultation of signatory powers seemed to him to be a reasonable and natural corollary to that article of the draft convention which dealt with complaints. But it was not very clear to him as to what further measures should be provided; he was inclined to leave the development of such measures to the Permanent Disarmament Commission.

The Swedish delegate stated that the admission of the right of retaliation would be a retrograde step on the part of the Disarmament Conference and hoped that progress would be made in the direction of collective guarantees.

The Rapporteur then proposed a text whereby the state against which chemical, incendiary or bacteriological weapons had been employed should in no circumstances retaliate by the use of the same weapons, and the third states situated in a given region might decide to undertake jointly, and as rapidly as possible, severe punitive action against the delinquent state, and create for this purpose a joint police force.

A subcommittee was appointed to define the question of sanctions, starting from the principle that the right of retaliation was not admitted.

On 12 November 1932 the Bureau considered the following draft submitted by the subcommittee:

The declaration of the Permanent Disarmament Commission establishing the fact of the use of chemical, incendiary or bacteriological weapons shall have the following effects:

1. Third states shall individually be under an obligation to bring pressure to bear, chosen according to circumstances, and notably according to the special situation in which they are placed in relation to the belligerents, upon the state which has used the chemical, incendiary or bacteriological weapons to induce it to give up the use of the said weapons or to deprive it of the possibility of continuing to use them.

2. A consultation shall be held among third states through the agency of the Permanent Disarmament Commission at the earliest possible moment to determine what joint steps shall be taken and to decide on the joint punitive action of every description to be taken.

These decisions shall be taken by a majority vote (character of the majority to be decided by the General Commission). The minority shall not be bound, but it shall be under an obligation not to hinder the action of the majority.

The Permanent Disarmament Commission shall be entitled to take in advance all preparatory measures with a view to the possible application of the decisions referred to in the foregoing paragraph.

3. Third states situated in a given region may further pledge themselves to undertake jointly and as rapidly as possible severe punitive action against the delinquent state, and, for this purpose, to create beforehand a joint police force.

4. The state against which chemical, incendiary or bacteriological weapons have been employed shall in no circumstances retaliate by the use of the same weapons.

The Polish delegate welcomed the definite prohibition provided for in paragraph 4 of the text. He felt that it would be essential later to define in greater detail the terms of the preceding paragraphs which formed a counterpart to that provision, since the powers signatories to the convention would thereby assume a very great moral responsibility towards the state which was a victim of an aggression of that nature.

The delegate of Italy said that, in a matter which aroused universal reprobation and called for the proclamation of a principle of international law valid alike for the signatory and non-signatory states, it seemed to him questionable to talk of sanctions of a regional character. He doubted whether such a provision would add anything to the general punitive measures and even felt that it might tend to diminish their effect.

He also felt doubtful as regards the efficacy of a regional agreement, as he thought that really universal action, such as a blockade by all the states in the world, would be far more powerful than a demonstration of force on the part of any one group of states. Moreover, a regional agreement backed up by the proposed police force might constitute in reality an alliance against a state which was not a party to that agreement. He, therefore, found himself unable to accept the provision.

The delegate of the United States observed that the discussion on sanctions had assumed proportions not originally anticipated.

The Soviet representative thought that the question of punitive action ought not to form the subject of a partial discussion—that is, it ought not to be settled in connection with the prohibition of the use of chemical weapons. It was a problem which would have to be examined in all its complexity. As this was the first occasion on which the principle of sanctions had been so definitely raised at the Disarmament Conference, the Soviet delegation could not express its final opinion.

The representative of France reiterated his government's opinion that the only means of abolishing chemical warfare was to impress in advance upon possible violators of the prohibition the certainty that they would be made to expiate the breaking of their word. That was why the French delegation had always considered that a system of automatic sanctions was in this case necessary. It had been recognized that absolute prohibition, excluding the right of reprisal by the use of the same weapon, was possible only if states felt sure that collective sanctions would take the place of individual sanctions. The text, however, which was the outcome of the Committee's deliberations, whatever progress it represented, failed to give that certainty. The very principle of consultation contradicted that of automatic sanctions.

The Belgian delegate noted that agreement had not been reached on the question of guaranteeing the victim of an aggression the necessary means of protection.

The Bureau of the Conference decided to forward to the Special Committee on Chemical, Incendiary and Bacterial Weapons (previously called Special Committee on Chemical and Bacterial Weapons) a questionnaire containing a series of points raised in the discussions.

VII. Questionnaire submitted to the Special Committee

A. Prohibition of the preparation of chemical, incendiary and bacterial warfare

I. Defensive material

1. Is it necessary, in order to guard against the effects of chemical arms, to employ devices for individual protection (masks, protective clothing, etc.)?

Is it practicable to entrust the preparation of these devices or some of them to an international body or can it be submitted merely to the technical supervision of an international body? If so, under what conditions?

2. Is the preparation of measures of collective protection (underground shelters, etc.) essential for defence against chemical warfare? Is it practicable to regulate this preparation by means of an international convention? Can it be made subject to the technical supervision of an international body?

3. Does the testing of protective material necessitate the use of poisonous substances? If so, what measures should be taken to prevent this giving rise to abuses?

Should it be made compulsory to declare the quantities of poisonous substances produced with a view to testing protective material? Should these quantities be restricted? Should the results of the tests be made public?

4. Can the preparation of the treatment of victims of chemical warfare give rise to abuse?

II. Offensive material

1. How can the preparation of bacterial warfare be prevented?

2. Is it practicable to prohibit the manufacture, import, export and possession of implements and substances exclusively suitable for use in chemical and incendiary warfare?

   (*a*) Are there such implements and substances?

   (*b*) What are they?

   (*c*) Are they of genuine importance? If the above-mentioned prohibition can be pronounced, would this constitute an effective obstacle to the preparation of chemical warfare?

3. Is it practicable to prohibit the manufacture, importation, exportation or possession of implements and substances capable both of pacific and military utilization?

If not, can the armed forces be forbidden to possess certain stocks of these substances or implements, or can states be obliged to declare those stocks?

4. Can the training of armed forces in the use of chemical weapons be prohibited? What would be the practical effect of this prohibition?

5. Can the Committee suggest other practical forms of prohibiting the preparation of chemical, bacterial and incendiary warfare?

Appendix

Special Case of Lachrymatory Substances

Should lachrymatory substances be included in the category of substances exclusively suitable for use in chemical warfare?

If so, can they be treated separately? Can such treatment give rise to abuse?

Can the limitation of the quantities that may be produced, imported or kept in possession be of practical value?

Is it possible to regulate the treatment of lachrymatory substances otherwise than by limiting the quantities that can be manufactured, imported or exported?

B. Supervision of the prohibition to make preparations for chemical, incendiary and bacterial warfare

1. (*a*) Can the prohibition of such preparations be supervised by consulting commercial statistics of the movements of chemical industries in all countries?

(*b*) Can this supervision be exercised by entrusting to national or international bodies the inspection of chemical factories and by having the following data published:

The nature of the products manufactured therein;
The existing stocks of manufactured products;
The output capacity of the factories?
Is it sufficient to do this for certain factories?

(*c*) Is such supervision of practical value?

2. From what facts will it be possible to deduce that the prohibition to make preparations has been violated?

First system: Supervision based on the existence of regulations concerning production.

(*a*) Limitation of the chemical output capacity of states, or, at any rate, of a certain number of states, so that the chemical warfare potential of certain states should not be too unequal (quotas, industrial agreements, etc.).

(b) Limitation of the quantities of chemical products in stock.

Practical value of this system?

Second system: The freedom of manufactures, imports and stocks is, in principle, complete, but the intention of using these substances for chemical warfare is alone prohibited.

From what facts can this intention be deduced:

(a) From the character of government intervention in the management of production;

(b) From abnormally large outputs;

(c) From abnormal stocks;

(d) From other facts?

Practical value of this system?

3. Can the Committee suggest other practical forms of supervision?

C. Case of a breach of the prohibition to use chemical, incendiary and bacterial weapons against an opponent

Determination of such a breach.

How should the determination of a breach be technically organized?

Who should determine such a breach? Should specialised experts be compulsorily attached to the authority responsible for determining the breach?

Should these experts be designated in advance?

Should two expert investigations be provided for—viz., by the experts of the country attacked and by international experts appointed in advance?

How should the determination of the breach be organized so that it should take place as rapidly as possible?

Penalties

Has the Committee any suggestions to make as regards the penalties to be applied to a state committing a breach of the convention? [28]

VIII. *Summary and comment*

By the time the Disarmament Conference convened in 1932, thirty-three states had ratified the Geneva Protocol which was then generally considered a "no first use" agreement allowing resort to the prohibited weapons in retaliation. But the demand was growing to remove this qualification in the envisaged convention for the reduction and limitation of armaments. There was wide support for the formula suggested by the Preparatory Commission to outlaw unconditionally the use of bacteriological weapons; difficulties arose with regard to chemical weapons. This was the key CBW problem during the first stages of the Conference.

The delegations agreed that the weapons in question belonged to a category of arms most offensive and most threatening to civilians and therefore subject to qualitative disarmament, meaning that their possession or use should be absolutely prohibited. But absolute prohibition was deemed possible only if manufacture and storage of toxic substances and appliances for their employment, as well as training in their use, were also forbidden. Detailed proposals to this effect, including the destruction of stocks, were submitted by Italy and Norway.

All these measures to be valid would have to be supervised so as to provide assurance that they were observed. Suggestions for a centralized international analysis of national statistics of manufacture and trade in chemical substances or for an international cartel to be organized for the purpose of ensuring that private chemical industry was not engaged in the preparation of chemical weapons, were recognized as unsatisfactory. It was accepted, in the light of the technical assessment made for the Disarmament Conference, that no reliable system of verification was possible. The shortcomings of any control methods would consequently have to be offset—argued some delegates—by a system of effective sanctions established in advance. Effective sanctions meant collective reprisals. Reliable machinery for international punitive actions could not be created in the political conditions of the early thirties. Thus, the Conference was faced virtually with the same dilemma as the Preparatory Commission a few years before and turned in a vicious circle.

Much, however, was accomplished in removing uncertainties regarding the scope of the prohibition. The conclusion reached by the Special Committee on CBW was that the prohibition of use of chemical weapons extends to substances capable in any way of producing harmful effects on the human or animal organism (no special reference was made to plants, because it was felt that it would not be possible to employ, for the purpose of damaging plants, substances which were not also harmful to man or animals, or which were not likely to make the plants harmful to them). The use of biological weapons against man, animals and plants was explicitly prohibited.

A resolution adopted by the Special Committee confirmed that tear gas belonged to the category of banned weapons. The USA no longer opposed this interpretation and stated that it was against the use of tear gas in war.

Moreover, it was found that incendiary weapons possessed characteristics assimilating them to chemical rather than to ordinary weapons, and that projectiles specifically intended to cause fires and appliances designed to attack persons by fire should be prohibited.

The legality of defensive measures was not questioned, but there were doubts whether a clear dividing line could be drawn between offensive and defensive preparations.

Much attention was devoted to the elaboration of rules for establishing whether there had been an infringement of the prohibition to use CB and incendiary weapons.

The Conference decided that a more thorough study of the problems involved should be carried out.

Chapter 5. November 1932–1933

I. *The second report of the Special Committee on CBW*

The Special Committee on Chemical, Incendiary and Bacterial Weapons considered the questionnaire, drawn up by the Bureau of the Disarmament Conference, at its meetings held from 17 November to 13 December 1932.

Mr Pilotti of Italy was Chairman, and Professor Rutgers of the Netherlands was Rapporteur. The following ten states were represented on the Committee: Denmark, France, Italy, Japan, the Netherlands, Poland, Spain, Switzerland, the United Kingdom and the United States of America.

The Committee included qualified experts on all subjects submitted to it, and appointed rapporteurs for the different groups of questions: defensive material, Colonel Fierz, Switzerland; medical and bacteriological questions, Professor Di Nola, Italy; chemical questions, Professor André Mayer, France; military questions, General J. G. Benitez, Spain; lachrymatory gases, Mr Davidson Pratt, United Kingdom; establishing the facts and penalties, Mr René Cassin, France.

The rapporteurs were also assisted by other experts in drawing up their special reports.

The replies of the Committee were given under three headings:

A. Prohibition of chemical, incendiary and bacterial warfare.
B. Enforcement of the prohibition against the preparation of chemical warfare.
C. Cases of infringement of the prohibition to employ against an adversary chemical, incendiary and bacterial weapons.

The main arguments advanced in the Special Committee's report, the gist of the conclusions arrived at, and the suggestions put forward, [1] as well as some important statements made during the debate in the Committee, [2] are given here.

A. Prohibition of chemical, incendiary and bacterial warfare

Defensive material

From the technical point of view, it is impossible to guard against the effects of chemical weapons without recourse to devices for individual protection such as respiratory apparatus, masks, protective clothing, gloves, unguents, etc. The possession of such devices by the victims of a chemical attack would be likely to reduce considerably the military advantages obtained by a party violating the prohibition.

On the other hand, it may in certain cases be to the advantage of an armed force to be equipped with protective devices when delivering a chemical attack; masks would be needed whenever the personnel was brought into direct contact with poisonous preparations (e.g., gas clouds, infection of ground). However, protective devices are unnecessary for an attack by means of shells or airbombs. Thus a prohibition to equip armed forces with protective devices would not place any serious barrier in the way of chemical warfare.

Besides, protective devices against poisonous preparations are used in time of peace in a number of industries, and the armed forces would always be in a position to procure them, even if such apparatus did not form part of their normal equipment.

It is important that each country should be able to manufacture such protective devices as it may require. Even if the preparation and manufacture, or the technical testing, of the appliances were to be entrusted to an international body, the countries would still have to conduct experiments on their own account. It would always be necessary to adapt such devices to the special conditions of each country.

Collective protection, such as underground shelters, look-out units, alarms, organization of rescue and disinfection services, stocks of disinfectants, etc., could not be regulated internationally on account of the diversity of circumstances which determine the defensive measures to be taken.

The use of poisonous substances is necessary to test individual devices; it is also required for testing shelters and apparatus for collective protection and for experiments on methods of disinfecting.

In general, the tests cannot be confined to mere laboratory experiments. Tests in the field are indispensable to form an accurate idea of the dangers against which protection is to be provided, and of the efficacy of the means of protection when applied in the conditions likely to arise in practice.

The quantities of poisonous substances employed for the purpose of testing the efficacy of an individual protective device are insignificant. Larger quantities are necessary for tests in the field; but even then, the quantities required are extremely small in comparison with those which would be needed for a chemical attack and those which are in current use in the chemical industry.

Restrictions which might be imposed with regard to the quantities of poisonous substances at the disposal of the armed forces for the purpose of protective experiments would apply only to experiments undertaken by organizations under state control. They would not affect those which might be conducted by private industrial undertakings and which would frequently be the more important. Furthermore, nothing would prevent governments from entrusting their protective experiments to private organizations.

If the governments were requested to publish the results of experiments with protective devices, the information provided would give only an incomplete idea of investigations concerning protection, such research being in large measure carried out by private firms and kept secret.

The treatment to be given to victims of chemical warfare requires instruction and training a staff of doctors, nurses and stretcher-bearers, and the organization of first-aid stations, means of transport and specialized hospitals. Such measures could not reasonably be prohibited.

The experimental study of the treatment of casualties caused by toxic substances resembles laboratory research conducted on poisonings which result from the manipulation of chemicals in industrial undertakings. It is extremely desirable that such research should be continued. Very small quantities of the chemical which could be used in warfare would be necessary. The malpractices to which such research might give rise would be even less serious than those which might result from investigations regarding methods of protection against toxic substances.

The organization in time of peace of a suitable health service represents the most effective means of defence against bacterial infection. It is, however, impossible to guarantee that a health service, however perfectly organized, could unfailingly master all the epidemics which might be disseminated. After causing an epidemic, a country would quickly lose control of it and run serious risks itself.

In the course of the discussion, the United States representative stated that his government would object to any restriction on the preparation of the means necessary to ensure protection. It reserved the right to take suitable measures to train the necessary personnel, either individually or collectively, for this purpose. The US delegation considered that it was the duty of its government to protect US nationals against chemical warfare; in fulfilling this function it should not be subject to international supervision. Nor must there be any interference by anybody other than one responsible to the national government. The US delegation would be unable to agree to an interference in civilian undertakings under the pretext of reduction or limitation of armaments or the prohibition of chemical warfare.

The Committee made these conclusions with regard to defensive material:

Protection against the effects of chemical weapons involves the employment of individual protective devices. In practice, neither the preparation of these devices nor the supervision of such preparation could be entrusted to an international body.

The preparation of collective protection is an essential means of defence against chemical warfare. Such preparatory work could not in practice be governed by an international convention, nor could it be submitted to the technical supervision of an international body.

The testing of protective material involves the employment of poisonous substances, though in quantities insufficient for military action. There is a risk that any obligation to publish the quantities of poisonous substances kept for this purpose or the limitation of such quantities would fail in its object since it would leave out of account the experiments carried out by private industry on its own initiative. For the same reason, only a very partial idea of the work done for protection against poisons would be obtained by imposing an obligation to publish the results of experiments.

The preparation of measures of treatment to be given to the victims of chemical warfare could not give rise to malpractices.

The Committee made the following suggestions regarding the protection of civilians:

To subject certain protective devices, for example, masks, to technical testing by an international body. Such tests might lead to the introduction of standard devices for the protection of civilians.

To set up an international information service for the collection of material regarding protection against chemical weapons. This body would enable all countries and, in particular, those whose technical organizations were inadequate, to keep abreast of methods of preparing the defence of civilians. The service might be attached to the Permanent Disarmament Commission.

Offensive material

Projectiles: The greater part of the projectiles charged with poisonous substances in the First World War were ordinary projectiles. Certain types of projectiles which were invented for the purpose of containing poisonous substances usually differed only very slightly from the ordinary ones. They may be charged either with explosives or with poisonous substances; in particular—and it is this which makes it difficult to characterize them—they are quite suitable for being charged with smoke-producing substances, the use of which is allowed.

Means of projection: To the knowledge of the Committee, there are no means of projection exclusively suitable for chemical warfare. There are ordinary guns firing poisonous shell, and ordinary airplanes capable of transporting receptacles charged with poisonous substances. Mortars intended for this purpose might be used for throwing smoke-bombs. Devices intended for producing clouds of poisonous substances are either ordinary commercial bottles or cylinders or apparatus identical with that generally used for creating artificial smoke-clouds or even therapeutic clouds, for instance to combat diseases of trees. Implements used for spraying the ground with poisonous substances are the same as those used in peacetime for totally different purposes.

Substances: There are substances which, as far as the Committee is aware, have only been used for chemical warfare, such as dichlorethyl sulphide (mustard gas) and certain arsines.

Any list of such substances, however complete it might be at the time it is drawn up, would very soon require amendment, either because new poisonous substances have been discovered by ordinary chemical research not undertaken with a view to chemical warfare, or because the peaceful use of poisonous substances mentioned in the list has been discovered or become practicable.

A prohibition of the manufacture, importation, exportation or possession of substances exclusively suitable for chemical warfare might be proposed. There should, however, be no total prohibition, because a certain quantity of such substances would always have to be prepared with a view to studying the question of protection.

If this prohibition were observed in peacetime, it would give a relative degree of security, inasmuch as the use of toxic substances for aggression might be delayed; only very small quantities would be available at the outbreak of war, while the period necessary for reaching the maximum productive capacity would be prolonged. The prohibition would not, however, constitute a great obstacle to preparations for chemical warfare. The substances in question could be very easily and rapidly produced with the raw materials and intermediary compounds widely disseminated in commerce and easily obtainable.

A prohibition of the possession by the armed forces of certain substances capable both of peaceful and military utilization would be, in practice, ineffective. In countries possessing a chemical industry, there would be nothing to prevent the armed forces from requisitioning the stocks of these substances existing in industrial establishments.

In a country having an important chemical industry, chemical warfare can always be rapidly organized, even though no special preparation has been made in peacetime. Methods for using poisonous substances can be easily adapted. The speed with which all this could be done would depend exclusively on the industrial strength of the country.

The troops can be trained for chemical warfare in the normal course of their training. Poisonous shells are the same as explosive shells; smoke apparatus is the same, whether the fumes are poisonous or not; compressed-gas cylinders are the same whether they are filled with poisonous or inoffensive gases; the throwing of gas bombs from the air does not differ from the throwing of other projectiles by the same means and, in particular, the throwing of illuminating bombs; the spraying of the ground can be carried on by men who are not specialists.

It is not possible to subject bacteriological research to effective supervision. Virulent bacteria are to be found in all bacteriological laboratories—both public and private—and also in hospitals treating contagious diseases. There could be no question of hindering the progress of medical bacteriology (the preparation of sera, vaccines, etc.), the objects of which are humanitarian, by supervising and restricting experiments with virulent cultures. Such supervision, moreover, would never be complete, and therefore always ineffective.

The apparatus and substances used in incendiary warfare are not exclusively suitable for such warfare, except for projectiles and bombs specifically incendiary, which are easy to construct and which can be quickly manufactured. There exists no special training of armed forces in the use of incendiary weapons.

The Committee drew these conclusions with regard to offensive material:

It is not possible in practice to prevent preparation for bacteriological warfare.

It would be possible to prohibit the manufacture, import and possession of apparatus and substances exclusively used for chemical warfare, but such a prohibition would be of only limited value. The substances could be easily obtained by converting ordinary ones; their manufacture could be improvised by any state possessing a chemical industry. There are no projectiles or means of projection exclusively employed for chemical warfare.

It is not possible to prohibit the manufacture, import, export or possession of apparatus and substances capable of employment both for peaceful and for warlike purposes. Any such prohibition imposed upon a state would either be ineffective in practice, in view of the stocks held in industrial establishments, or it would inflict irreparable damage on the chemical industry.

It is possible to prohibit the training of armed forces in the use of chemical weapons, but the practical effect of such a prohibition would be very small as the material used for chemical warfare is not of a specialized character.

It is possible to prohibit the manufacture, import, export or possession of projectiles and bombs specifically incendiary.

It is not possible to prohibit the manufacture, import, export or possession of incendiary substances and apparatus suitable both for peaceful and military uses.

The Committee regretted that it was unable to suggest any other practical measures to enforce the prohibition of the preparation of chemical and incendiary warfare.

Special case of lachrymatory substances

In the Committee's view, the question of lachrymatory substances can not be treated separately as far as the prohibition of the use of poisonous substances in wartime is concerned. But it arises separately in peacetime because some states use lachrymatory implements in police operations.

There are many kinds of lachrymatory substances of varied chemical composition. Some are particularly poisonous—chloropicrin, acrolein, etc. Others have no toxic effects in the conditions under which they have been used for police operations or experiments in protection; such substances are benzyl chloride and bromide, chloracetophenone, etc. It might be possible to draw up a list of "non-poisonous lachrymatory substances", but the Committee did not think it advisable to do so, because it could not state—from the strictly scientific point of view—that compounds designated as non-poisonous may not have dangerous effects under certain conditions of use.

Lachrymatory compounds in general are suitable not exclusively for use in chemical warfare, but, as a rule, are industrial products in common use. There are no special implements for using lachrymatory substances.

The use of lachrymatory substances for police operations may give rise to abuse if, for instance, a state prepared a number of implements charged with lachrymatory substances, out of proportion to the real needs of the police.

In the course of the discussion, the US delegate said that 327 US banks employed lachrymatory substances as a protection against robbery, and convoys of securities and bullion were similarly protected. There was an increasing tendency in the United States to use lachrymatory substances for civilian purposes. Many of the police forces in the USA had been supplied with those substances to disperse gatherings which may disturb peace and to arrest criminals. In some cases members of police forces had been specially trained. The US government, however, was willing to forego the use of lachrymatory substances for military purposes in wartime as this might give rise to abuse. While, therefore, in favour of prohibition in this sense, the US representative strongly urged that the use of lachrymatory gases for police purposes and for protecting private property should be permitted. He felt that special regulations could be introduced which would prevent abuse.

Questioned about possible harmful effects of the lachrymatory substances, the US delegate stated that out of some 30 000 cases investigated in the USA, undesirable effects had been observed only in one case, and even there it was generally believed that the results were due to causes other than contact with lachrymatory substances.

The British delegate feared that, if the use of lachrymatory substances was permitted, those used in war would not be the harmless substances employed in most countries for police purposes, but highly poisonous gases.

The Japanese delegate said that, while not essentially very harmful unless of great density, lachrymatory substances may cause great losses in a general attack.

The French delegate drew attention to the difficulty of regulating the use of lachrymatory substances. For instance, benzyl bromide, a lachrymatory substance used by the French police even before the war in arresting dangerous criminals, was the product most widely employed for charging asphyxiating shells during the war. He stressed that in strong doses, or used under certain conditions, all lachrymatory gases could be poisonous; some were even poisonous in small doses. Finally, he pointed out that the expression "lachrymatory substances" was not a chemical definition, but merely described the physiological effect produced by those substances. The use of lachrymatory substances might lead to the use of sternutatory gas; protection in this latter case was more difficult—masks were required, whereas ordinary goggles were adequate protection against lachrymatory gases.

The representative of the Netherlands suggested that if it were intended to authorize the use of lachrymatory substances by police, private trade in such substances and projection apparatus would have to be forbidden.

The conclusion of the Committee was that lachrymatory substances did not come within the category of substances exclusively employed for purposes of chemical warfare.

The Committee suggested that, in order to avoid abuse, a state wishing to use lachrymatory substances should be obliged to inform the Permanent Disarmament Commission. It should specify the substances, the implements which it proposed to employ and their number. The Commission would examine whether there was any disproportion between the arms notified and police requirements. The information would remain confidential.

The Committee had learned that in some countries industrial firms manufactured or sold implements or devices charged with lachrymatory substances for the protection of private property. It thought that in this case the state should remain responsible for its nationals. If private individuals wished to prepare, sell, purchase or possess implements or devices charged with lachrymatory substances, it would be desirable that they should declare such intention and not be allowed to carry on this industry or trade and possess such implements or devices without being subject to regulations laid down by the state.

B. Enforcement of the prohibition against the preparation of chemical warfare

To supervise the prohibition of preparation for chemical warfare, it would not be sufficient to ascertain the quantity of products manufactured and imported or exported; information would also have to be obtained with regard to their transformation and final use. Existing stocks would have to be known, and the output capacity of factories ascertained. Commercial statistics, however, contain very inadequate information, or no information at all, on these points. In addition, they are often published with delay.

The British delegate noted that in his country there were no statistics containing information on the output of raw materials and manufacture of finished products, as far as chemical products were concerned. The British chemical industry was opposed to publication of output figures for reasons of competition. The French delegate reported that French statistics did not show the output capacity of the chemical industry either.

For fiscal reasons some countries have organized supervision of a certain number of chemical products, such as sodium chloride, alcohol, acetic acid, stearic acid, etc. This fiscal supervision makes it possible to watch not only the manufacture of these products but also their transport to the place of transformation, though it is not always possible to ascertain exactly their final destination. Nevertheless the system works well, but it necessitates the existence of an official body of inspectors permanently present in the factories or authorized to enter them at any moment. Furthermore, the number of products thus supervised, even in countries where the fiscal system is most developed, is comparatively small. Should it be desired to obtain an idea of the nature and quantity of products utilizable for chemical warfare, of the existing stocks, and of the output capacity of factories, the system mentioned above would have to be generalized. Such supervision would completely destroy secrecy in commercial affairs; in many cases it would lead to the divulging of manufacturing secrets, to the detriment of the national industry.

If supervision were entrusted to an international body, chemical factories of the whole world would have to be inspected by foreign inspectors. And even then the measures of inspection might be evaded by the preparation of products similar to those that were really aimed at, or by formation of stocks of semifinished products, or by masking the real capacity of the factories.

The United States representative, taking part in the discussion, expressed the belief that both from the practical and the technical points of view, supervision of the prohibition to make preparations for chemical warfare could be exercised only by governments, and not by an international body. Nor could it be carried out by inspecting output, stocks or output capacity, because this would lead to interference with commercial secrets and would give rise to serious trouble in the chemical industry and perhaps to unfair competition; the composition of products and the nature of manufacturing processes were vital matters to the industry.

The British delegate emphasized the necessity of supervising the employment of products until they had passed through the final processes, if effective supervision were to be exercised. To show the complexity of the problem, he took the example of alcohol. With alcohol it was possible to manufacture acetic acid which was used for many peaceful purposes; chloracetic acid, which could be used both for peaceful purposes and for making a lachrymatory substance, could be manufactured from acetic acid. Alcohol could also be used in manufacturing ethylene, from which chlorhydrin was made; the latter gave ethylene glycol which was used for manufacturing both antifreezing products and explosives. From ethylene glycol it was also possible to prepare thiodiglycol, with which mustard gas was made. But the most important point was that, even if effective supervision could be set up in time of peace, it would cease to be effective as soon as hostilities broke out.

The Committee stated that the inequality of the distribution of mineral deposits among states created between them an inequality of strength in regard to chemical industries. Endeavouring to redress this by limiting the output capacity of countries rich in raw materials or possessing other favourable conditions (power in various forms, such as hydroelectric stations, labour, etc.) was a difficult undertaking, and it was doubtful whether it was economically desirable. Furthermore, the big chemical industries were key industries. The majority of chemical products were used in other industries which could not subsist without them. In order to ensure to some extent their economic independence, certain states, although at a disadvantage from the point of view of raw materials, had developed chemical industries in their own territory. This form of industrialization was necessary to safeguard political independence. Therefore, states would hesitate to give up industries which were of vital importance to them.

Limitation of stocks of products other than those suitable exclusively for chemical warfare would be difficult. Necessity to meet unforeseen demands, accumulation of substances with a view to placing a new product on the market, or obligation to keep substances which have become useless, owing to a technical improvement, but for which it was hoped to find a new application—all these were factors that might justify the existence of stocks.

To illustrate these difficulties, the United States delegate said that in the USA the sulphating of fruit trees, which is carried out in summer, required a large amount of chemical products. During winter the chemical industry concerned accumulated stocks of these products with a view to the increased orders which would be given the following summer. The limitation of stocks would consequently be unfortunate; it would be an interference with an activity which corresponded to perfectly normal industrial needs, and such evil consequences would have no relation to any system intended to prevent the preparation of chemical warfare.

The delegate of France gave still another example: in Argentina ticks attacked cattle, making it impossible to use their hides. To rid the cattle of these insects, they were bathed in ditches filled with a solution of arsenic. The abundance of ticks depended on meteorological conditions. How was it possible in these circumstances to limit and regulate, in advance, the stocks and market of arsenic which was used as a basis for most of the dangerous substances employed as irritants?

The Danish representative, however, thought that an aggression by means of chemical weapons was so serious a matter for the whole of humanity that it was impossible to take private interests alone into account.

The Committee considered that it would always be difficult to discover the intentions of a state wishing to direct its chemical industry towards warlike purposes. The fact that a government intervened in the management of the chemical industry was not sufficient to prove bellicose intentions. The government may encourage the industry to make use of the products of its soil, or induce it to manufacture chemical fertilizers in an agricultural country, or dyestuffs in a country with a textile industry, or arsenic compounds in a country having to fight against tropical diseases, etc. In doing so, the state concerned no doubt increased its war potential—for example, its capacity to produce explosives if it manufactured nitrogenous fertilisers, or to produce poisonous substances if it made arsenical products; but how could it be proved that this was the real purpose? Only when a government had substances which had been prepared for its own account and which were believed to be exclusively suitable for chemical warfare, could it be called upon to prove the legitimacy of its action. It would be more difficult if it confined itself to ordering the preparation, not of the toxic substances themselves, but only of half-finished products in current industrial use, which were of a similar nature.

The Committee reached the following conclusions with regard to the enforcement of the prohibition against the preparation of chemical warfare:

It is not possible to enforce the prohibition against such preparation by examining the commercial statistics of the activities of chemical industries in all countries.

(At the request of the Netherlands a note was inserted in the report expressing the opinion that the fullest possible publicity was essential, not only in regard to apparatus for the protection of the civilian population, but also in regard to military apparatus. It was suggested in the note that the Permanent Disarmament Commission might collect all possible information on chemical industries under government control. As regards private chemical industry, it should see what could be done to supplement that industry's statistics so as to follow its development from the point of view of the prohibition of chemical warfare.)

It is conceivable in theory, but impossible in practice, to exercise control by entrusting national or international bodies with the duty of inspecting chemical factories and of making public the character of the manufactured products, the existing stocks of manufactured products and the production capacity of the factories. If any such control were proposed, it would have to apply to the entire chemical industry. The practical values of such control would be very limited considering the ease with which chemical warfare could be improvised.

It is not possible to base control upon limitation of the chemical production capacity, so that the potential of certain states for chemical warfare should not be excessive compared to that of other states, or upon limitation of the quantity of chemical products in stock. The latter would only be possible in the case of substances used exclusively for chemical warfare.

It is not possible to prohibit the intention to make use of substances for chemical warfare while at the same time leaving the manufacture, import and storage of such material entirely unsupervised. The intention referred to above is not susceptible of proof; it could not be deduced with certainty either from the nature of state intervention in the production, or from the extent of production, or from abnormal stocks, or from other factors.

The Committee made these suggestions regarding supervision:

No manufacture of or trade in poisonous substances suitable exclusively for the conduct of chemical warfare—such as dichlorethyl sulphide (mustard gas), for which no legitimate use was known—should be permissible without government authorization. (Regulations of this kind bearing on various dangerous products already existed in many countries.)

Penal legislation might be introduced in each country providing for punishment of the authors of the preparation of a prohibited form of warfare —for example, chemists or bacteriologists convicted of preparing chemical or bacteriological weapons.

Without going as far as this, several members of the Committee thought that a kind of stigma involving the prohibition to practice their profession might be attached to those engaged in work aimed at the preparation of chemical or bacteriological warfare.

The last two suggestions were not approved by the whole Committee. The French delegate observed that the real difficulty arose not from the fact that scientists themselves could be double-faced, but from the fact that their studies might serve a two-fold purpose. Scientific progress had been used both for good and evil. One of the most terrible features of the Great War was that governments enlisted not only soldiers but also scientists; the latter, when faced with the choice of remaining in a state of indifference in their laboratories when their countries were being attacked by new and terrible weapons, or placing their knowledge at the service of their country, did not hesitate. Besides, research in poisonous substances may be in the interest of mankind. The only means of studying measures for rat-extermination was to make a systematic study of certain poisonous substances. Some of the most dangerous of these substances, hydro-cyanic acid, for example, were required to destroy insects attacking fruit trees, etc.

The Italian delegate added that a chemist's first duty was to his own country; it could not be subordinated to the duty towards the community of nations.

C. Cases of infringement of the prohibition to employ against an adversary chemical, incendiary and bacterial weapons

Establishment of the facts

The Committee felt that the offending state should not derive substantial military advantage, or decisive superiority, from a tardy imposition of penalties due to delay in establishing the facts.

Certain substances, like chloropicrin, brominated ketones, etc., leave traces on the ground for only a relatively short space of time (a few hours at the most). There are volatile substances, like phosgene, the use of which can only be proved by medical examination of the victims. Such examination would have to be made before certain clinical or anatomo-pathological phenomena have disappeared or undergone transformation. (The effects of lachrymatory substances disappear immediately.)

In case of the use of bacteriological weapons, the establishment of the fact of infection is particularly urgent, not only in order to take effective measures against epidemic contamination, but also to discover any evidence of a deliberate character of the contamination and to determine the persons who have taken part in the prohibited acts on behalf of a state at war. The difficulties of this investigation are greatly aggravated by the fact that the effect of a bacteriological contamination does not make itself felt until the end of the period of incubation, while a deliberate attempt at contamination is not necessarily successful.

The facts should be established in such a way as to carry weight with all governments and with public opinion.

In the course of the discussion the Italian delegate remarked that the evidence submitted by the experts of the country attacked, even if not altogether conclusive, must not be rejected. However speedy the investigation, the main element on which the finding would be based was the evidence of the victim state.

The Committee reached the conclusion that the establishment of facts of infringement should fulfill three conditions: it should be extremely rapid, it should afford the greatest possible guarantees of impartiality, and it should be carried out by persons of recognized qualifications and of high moral standing.

*Body by which the facts of infringement are
to be established*

The Permanent Disarmament Commission would not be in a position itself to determine with the necessary speed whether a breach has been committed.

The aggrieved party, however, should be able to have the fact of the use of prohibited weapons established, without delay, by qualified persons already in or near its territory.

The conclusions of the Committee with regard to the organs responsible for establishing the facts of infringement were as follows:

The duty of collecting evidence would in normal circumstances be entrusted to an international "Commission for urgent initial investigation".

Evidence collected by the experts of the complainant state is unilateral in character and serves chiefly to justify the complaint.

The Commission for urgent initial investigation may be constituted in

Second report of the Special Committee on CBW

peacetime or may be composed of representatives of the Permanent Disarmament Commission accredited to the belligerent states.

Failing these arrangements, the complainant state should apply to the doyen of the diplomatic corps, who would appoint as members of the investigation commission: military attachés, members of the diplomatic corps or consuls de carrière, technical experts preferably of foreign nationality, selected from a list drawn up in advance by the Permanent Disarmament Commission. (For the purpose of establishing the fact of the use of bacteriological weapons, technical experts might include clinical doctors, veterinary surgeons, biologists and bacteriologists.) Under certain circumstances, it should also be possible to apply to magistrates.

The Permanent Disarmament Commission should not be required to undertake a supplementary enquiry, but should have the right to do so.

Procedure for establishing the facts of infringement

The complainant state should immediately inform the Permanent Disarmament Commission and at the same time see to it that the initial investigations are carried out.

The Commission for urgent initial investigation should forthwith enquire into the matter and report to the Permanent Disarmament Commission.

The latter should inform the accused state of the complaint and should, if necessary, order an enquiry in its territory.

When the evidence of the states involved had been heard, the Permanent Disarmament Commission would state whether the prohibited weapon had been used.

The states involved and, if necessary, all the other signatory states should take all necessary steps to enable the commissioners to perform their duty. The executory regulations should also deal with the facilities to be given to the commissioners in regard to transport, communications and the technical work.

Penalties

The Special Committee realized that the subject of penalties was primarily political. It confined itself to formulating, among the possible suggestions, those that related to technical measures, and to examining only the technical aspects of other measures.

It ignored the question of breaches of the prohibition to make preparations in time of peace or war, which did not fall within its terms of reference.

The Committee made these suggestions:

States signatory to the Convention would give the attacked state scientific, medical and technical assistance in repairing, attenuating or preventing the effects of the use of the prohibited weapons. Preparations for the granting of assistance may be entrusted to an international information and documentation service for protection against chemical weapons, the establishment of which was contemplated in the report. (See page 150.) These measures, if not organized on a universal basis, may be of regional or continental character, however they would then lose much of their efficacy.

It is possible that a state may not be in a position to furnish to the attacked state its share of the necessary scientific, medical and technical assistance. In such a case, the state concerned should be asked to make a financial contribution to the treatment of the victims of chemical, incendiary or bacteriological warfare, and to the protection against such warfare.

The means of pressure varying from diplomatic representations to military measures are mainly of a political nature. This also applies to economic measures, such as the breaking off of commercial and financial relations between the signatory states, their nationals or residents, and the guilty state, its nationals or residents. However, there is one technical suggestion which deserves to be considered: the stoppage of supplies to the guilty state of raw materials, products and appliances necessary for chemical, incendiary and bacteriological warfare.

Certain materials, such as sulphur, arsenic, bromine and iodine are very unequally distributed throughout the world. In some cases, the prohibition to dispatch the raw materials, as well as products and appliances necessary for chemical or incendiary warfare, would hamper the continuance of that warfare.

No rapid or practical effect could be expected if the guilty state possessed a highly developed chemical industry. In addition to its own supplies, such a state may have considerable stocks of raw materials imported in advance from abroad, or may find substitutes or manufacture other types of poisonous gases or liquids. With regard to bacteriological warfare the suggested measure would not be effective.

As regards the question of reprisals, the following conclusions were reached:

The recognition of the right of individual reprisals would compromise the prohibition to make preparations for chemical or incendiary warfare, unless the preparation of such reprisals was made conditional on the previous establishment of the fact of infringement, and the victim state was assured of concrete assistance from other states in making the said preparations.

There can be no doubt that the transgressor state would necessarily be in a position of inferiority if the countries not directly concerned agreed to exert collective reprisals by means of chemical or incendiary weapons.

If the Convention excluded all reprisals, the transgressor state could derive very important advantages from the use of prohibited weapons. The temptation to have recourse to such weapons would be great in the absence of individual or collective retaliation.

It would be desirable for the Permanent Disarmament Commission and the states signatories to the Convention to avail themselves of the short period between the submission of a complaint by a state and the establishment of the fact of a breach, to make preparations with a view to the possible application of penalties.

II. *Consideration of the second report of the Special Committee by the Bureau of the Disarmament Conference*

The Bureau of the Conference considered the report of the Special Committee at its meetings held on 25 and 30 January 1933. [3]

The Rapporteur, supplementing the report of the Special Committee, submitted a series of draft conclusions to serve as a basis for articles to be embodied in the Disarmament Convention. [4]

The draft conclusions provided:

That the use in warfare of chemical, incendiary and bacterial weapons should be prohibited;

That all preparations for such warfare should be prohibited in time of peace as in time of war, it being understood that this prohibition would not apply to material and installations to ensure individual or collective protection against the effects of such warfare;

That, in order to enforce the prohibition, the manufacture, import, export or possession of appliances and substances exclusively suited to the conduct of such warfare should be forbidden;

That the manufacture, import, export or possession of chemical appliances and substances suitable for both peaceful and military purposes, with intent to use them in war, should be prohibited;

That the instruction and training of armed forces in the use of such weapons should be forbidden;

That the Permanent Disarmament Commission should examine complaints put forward by states alleging a violation of the prohibition to prepare for such warfare;

That a procedure for establishing the fact of the use of such weapons should be organized under the supervision of the Permanent Disarmament Commission and include the constitution and functioning of a commission for urgent initial investigation and the institution of supplementary enquiries in the territory of the state attacked, at the discretion of the Permanent Disarmament Commission;

That, following the establishment of the fact of violation, third states should individually be under an obligation to supply the state attacked with scientific, medical and technical assistance, to bring pressure to bear upon the offending state and, in the first instance, to cut off supplies of raw materials, products and appliances necessary for such warfare, and to consult, through the Permanent Disarmament Commission, as to what joint steps or joint punitive action might be taken;

That an international information and documentation service concerning protection against chemical weapons should be established under the supervision of the Permanent Disarmament Commission.

Prohibition of preparations for chemical and bacterial warfare

The British representative said that the report made somewhat depressing reading. It showed that the countries rich in certain raw materials and with a highly developed industry were very well equipped for carrying on chemical warfare and there were no practical means of preventing such warfare. Moreover, it was clear that, for the most part, appliances and substances suitable for chemical warfare existed in industry for perfectly legitimate purposes. It had, therefore, to be admitted that everything depended on the good will of states in carrying out their undertakings. He warned against inserting in a convention provisions which, under existing circumstances, could not be applied with any certainty and the realization of which was impracticable.

The United States representative thought that the greatest difficulties would inevitably be encountered if it were desired to go beyond an undertaking not to prepare appliances and substances for chemical warfare in time of peace. It should be borne in mind that ratifications of a convention on the subject would be more easily obtained, if it contained only a simple statement of main principles, without entering into so many controversial details. Referring to that part of the report which dealt with lachrymatory substances for police operations, he questioned the requirement to submit a list of such substances and appliances. These elements were constantly changing. Lachrymatory substances and appliances were used in the United States throughout the entire country by the police, banks and various undertakings, and the models frequently varied. To ask the states to carry out a census among a large number of departments and private organizations would be to impose on them an extremely arduous task. The USA was prepared to state, in the name of humanity, that it renounced the use of chemical weapons, but there was no reason to demand in addition that the government should engage in impracticable and extremely tiresome investigations.

The Rapporteur (Netherlands) agreed that it was impossible in time of peace to provide guarantees against the possible use of chemical weapons in case of war. This was because chemical warfare could be improvised and could be prepared without there being any visible sign of such preparation. In reply to the US remarks regarding lachrymatory appliances and substances, he pointed out that police might, in fact, be equipped with arms suitable for use in time of war. The Special Committee considered that this was not a matter to be treated with indifference from the point of view of disarmament and of the prohibition to prepare for chemical warfare.

The US delegate made a reservation with regard to the above-mentioned points of the report.

Penalties for the use of chemical or bacterial weapons

The German delegate considered that the employment of the prohibited weapons should not be countenanced even by way of retaliation. He recalled that the international provisions relating to respect for the Red Cross, the treatment of prisoners, etc., did not allow for reprisals in the case of violation. He wondered, in any case, whether it was indispensable to settle immediately the question of penalties to be applied to the state which had recourse to chemical, incendiary and bacterial weapons. The penalties contemplated were not, in fact, peculiar to this kind of warfare. They were equally applicable to other methods of war. The German delegation was, therefore, of the opinion that a special chapter of the disarmament convention should be devoted to action in the event of violation of its provisions.

The Soviet representative supported the German proposal to postpone the discussion until the Disarmament Conference dealt with the question of penalties in general.

The US representative said that the problem of violations must be studied as a whole. Unless and until it were found that the general clauses of the convention were insufficient, it was premature to examine special measures applicable to any one part of it.

The French delegate pointed out that, whatever steps might be provided for in the general case of a breach of the convention, it would be necessary, in the special case of violation of the prohibition to use chemical, incendiary and bacterial weapons, to make provision also for special measures. The problem, if adjourned, would still call for solution. What was the sense of admitting that, in the case of the use of chemical, incendiary or bacterial weapons, the establishment of the fact of use was of special value, unless it was thereby recognized that immediate repressive action was called for?

The representative of the United Kingdom thought that it would be easier to take a decision with regard to special penalties for the employment of prohibited weapons when the Conference came to discuss action to be taken in the case of recourse to war. He added that it was essential in considering the question to remain within the limits of what was possible and not to ask more of human nature than human nature could bear. There was no country which, when subjected to chemical attack, would agree to wait for authority before exercising its right of reprisals. Public opinion would not accept such a limitation.

The Rapporteur believed that the use of chemical weapons would produce a strong reaction throughout the world, which could be felt even in the state using the prohibited arm, and bring about a change of government there. Other states would probably hasten to ask the guilty state not only to promise not to resort again to that kind of warfare but also to give pledges ensuring that that promise would be kept. Lastly, it might be expected that every effort would be made by third states to bring about the cessation of that kind of warfare, and that measures of conciliation would be taken. In certain cases immediate reprisals would be not only useless but even harmful. Reprisals must be made subject to a preliminary establishment of the facts. Such a solution would be a compromise between absolutely prohibiting reprisals and unreservedly allowing them.

The Italian delegate said that the penalties suggested by the Special Committee were not sufficiently severe. He thought that the question was not yet ripe for discussion by the Bureau in view of the divergencies of opinion which had been revealed.

On 30 January 1933, the Bureau adopted the following resolution:

> The Bureau agrees to the principle of special measures being taken in case of a violation of the prohibition of the use of chemical, incendiary and bacterial weapons.
>
> It decides to elaborate the relevant articles with regard to such special measures after the general penalties for the case of the violation of the Convention have been examined by the Conference.

On the same day, as a result of consideration of the Special Committee's proposals, the Bureau instructed the Drafting Committee to frame articles for embodiment in the draft disarmament convention.

The Drafting Committee prepared the text requested by the Bureau, on 7 March 1933. [5]

III. *The British draft disarmament convention*

On 16 March 1933, the delegation of the United Kingdom submitted to the General Commission a draft disarmament convention. Part IV of the draft contained provisions concerning chemical, incendiary and bacterial warfare, which were based on the text prepared by the Drafting Committee. [6]

Section I. Prohibition of chemical, incendiary or bacterial warfare

Article 47

The following provision is accepted as an established rule of International Law:

The use of chemical, incendiary or bacterial weapons as against any State, whether or not a Party to the present Convention, and in any war, whatever its character, is prohibited.

This provision does not, however, deprive any Party which has been the victim of the illegal use of chemical or incendiary weapons of the right to retaliate, subject to such conditions as may hereafter be agreed.

With a view to the application of this rule to each of these categories of weapons, the High Contracting Parties agree upon the following provisions:

Article 48

The prohibition of the use of chemical weapons shall apply to the use, by any method whatsoever, for the purpose of injuring an adversary, of any natural or synthetic substance harmful to the human or animal organism, whether solid, liquid or gaseous, such as toxic, asphyxiating, lachrymatory, irritant or vesicant substances.

This prohibition shall not apply:

(*a*) To explosives;

(*b*) To the noxious substances arising from the combustion or detonation of explosives provided that such explosives have not been designed or used with the object of producing noxious substances;

(*c*) To smoke or fog used to screen objectives or for other military purposes, provided that such smoke or fog is not liable to produce harmful effects under normal conditions of use.

Article 49

The prohibition of the use of incendiary weapons shall apply to:
(1) The use of projectiles specifically intended to cause fires.

The prohibition shall not apply to:
(*a*) Projectiles specially constructed to give light or to be luminous and generally to pyrotechnics not intended to cause fires, or to projectiles of all kinds capable of producing incendiary effects accidentally;
(*b*) Incendiary projectiles designed specifically for defence against aircraft, provided that they are used exclusively for that purpose.

(2) The use of appliances designed to attack persons by fire, such as flame-projectors.

Article 50

The prohibition of the use of bacterial arms shall apply to the use for the purpose of injuring an adversary of all methods for the dissemination of pathogenic microbes, or of filter-passing viruses, or of infected substances, whether for the purpose of bringing them into immediate contact with human beings, animals or plants, or for the purpose of affecting any of the latter in any manner—for example, by polluting the atmosphere, water, foodstuffs or any other objects.

Section II. Prohibition of preparations for chemical, incendiary and bacterial warfare

Article 51

All preparations for chemical, incendiary or bacterial warfare shall be prohibited in time of peace as in time of war.

Article 52

In order to enforce the aforesaid general prohibition it shall in particular be prohibited:
(1) To manufacture, import, export or be in possession of appliances or substances exclusively suited to chemical or incendiary warfare.

The quantities of chemical substances necessary for protective experiments, therapeutic research and laboratory work shall be excepted. The High Contracting Parties shall inform the Permanent Disarmament Commission of the quantities of the said substances necessary for their protective experiments.

The manufacture of and trade in these substances may not be undertaken without government authorization.

(2) To manufacture, import, export or be in possession of appliances or substances suitable for both peaceful and military purposes with intent to use them in violation of the prohibition contained in Article 48.

(3) To instruct or train armed forces in the use of chemical, incendiary or bacterial weapons and means of warfare, or to permit any instruction or training for such purposes within their jurisdiction.

Article 53

The provisions of Articles 51 and 52 shall not restrict the freedom of the High Contracting Parties in regard to material and installations intended exclusively to ensure individual or collective protection against the effects of chemical, incendiary or bacterial weapons, or to training with a view to individual or collective protection against the effects of the said weapons.

Article 54

The High Contracting Parties shall inform the Permanent Disarmament Commission of the lachrymatory substances intended to be used by their authorities for police operations as well as of the number of the various appliances by means of which they are to be utilized.

Section III. Supervision of the observance of the prohibition of preparations for chemical, incendiary or bacterial warfare

Article 55

The Permanent Disarmament Commission shall examine the complaints put forward by any Party which may allege that the prohibition to prepare for chemical, incendiary or bacterial warfare has been violated.

Section IV. Establishment of the fact of the use of chemical, incendiary or bacterial weapons

Article 56

Any Party claiming that chemical, incendiary or bacterial weapons have been used against it shall notify the Permanent Disarmament Commission.

It shall, at the same time, notify the authority designated for the purpose by the Permanent Disarmament Commission or, failing such authority, the Doyen of the Diplomatic Corps accredited to it, with a view to the immediate constitution of a commission of investigation.

If the above-mentioned authority has received the necessary powers, it shall itself act as a Commission of Investigation.

Article 57

The Commission of Investigation shall proceed with all possible speed to the enquiries necessary to determine whether chemical, incendiary or bacterial weapons have been used.

It shall report to the Permanent Disarmament Commission.

Article 58

The Permanent Disarmament Commission shall invite the Party against which the complaint has been made to furnish explanations.

It may send commissioners to the territory under the control of that Party for the purpose of proceeding to an enquiry, to determine whether chemical, incendiary or bacterial arms have been used.

Article 59

The Permanent Disarmament Commission may also carry out any other enquiry with the same object.

Article 60

The Parties involved in the above-mentioned operations, and, in general, all the Parties to the present Convention, shall take the necessary measures to facilitate these operations, particularly as regards the rapid transport of persons and correspondence.

Article 61

According to the result of the above-mentioned operations, the Permanent Disarmament Commission, acting with all possible speed, shall establish whether chemical, incendiary or bacterial weapons have been used.

Article 62

The details of the application of the provisions of this chapter shall be fixed by regulations to be issued by the Permanent Disarmament Commission.

The draft convention was adopted as a basis of discussion.

IV. *Amendments to the draft convention*

The provisions of the British draft convention relating to chemical, incendiary and bacterial warfare were read in the General Commission on 30 May 1933. [7]

The representative of the Netherlands drew attention to the fact that the right to retaliate was allowed to a victim of the illegal use of chemical or incendiary weapons, subject to conditions to be agreed upon. He recalled that the Special Committee had taken the view that the exercise of the right to retaliate must be contingent on the previous establishment of the use of chemical and similar weapons by the adversary.

The British delegation, however, argued that considerable delay may be involved in establishing the fact of use of prohibited weapons by the adversary.

The United States delegate referred to Article 52 of the draft convention, concerning the prohibition of preparations for chemical, incendiary and bacterial warfare, and particularly to paragraph 2 which dealt with "intent to use them [appliances or substances suitable for both peaceful and military purposes] in violation of the prohibition contained in Article 48". He did not see any way of adopting legislation based upon intent. He asked the Commission to consider the problem and see whether all of Article 52 was really necessary. With regard to Article 54, the US representative repeated that in his country the use of lachrymatory gases for police purposes was very widespread. If the furnishing of information on this matter were made a contractual obligation in accordance with Article 54, the USA would be totally unable to carry it out. He did not believe the federal authorities would be able to obtain the information short of taking a complete census of the United States. If his country signed the obligation, it would run the grave risk of being reproached for bad faith in case it was unable to communicate the information.

The German delegation expressed the opinion that the use of chemical weapons and gas should be completely prohibited, even as a measure of retaliation.

The representative of Yugoslavia, speaking on behalf of the delegations of the Little Entente, stated that the prohibition of chemical bombardment must be absolute, and that only collective sanctions should be applied to enforce the prohibition.

The French representative supported the Yugoslav statement. He added that sanctions were necessary, not only to enforce the prohibition of chemical warfare and of preparations for that warfare, but also in case of any breach of the convention. Collective sanctions were, in his opinion, the only ones which should be contemplated.

Despite a number of reservations, the UK draft was unanimously accepted as the basis of the future convention.

The president of the Conference invited the delegations to submit their amendments with a view to a second reading of the draft at a later stage.

The provisional text of the draft convention, prepared in the light of the modifications adopted in the first reading and of the amendments presented by the delegations, and published on 22 September 1933 [8], contained the following proposals and observations in Part IV dealing with chemical, incendiary and bacterial warfare:

Amendment, proposed by the Little Entente, to delete paragraph three of Article 47. [9]

Amendment by the United States to replace Article 54 by:

The High Contracting Parties undertake to inform the Permanent Disarmament Commission annually of the nature of the lachrymatory substances used by their governmental agencies or instrumentalities for police operations, as well as of the number and character of the various appliances by which the said lachrymatory substances are utilized. [10]

V. *Postponement of the Disarmament Conference*

No further action with regard to CBW was taken by the Conference.

In November 1935, the Secretary-General of the League communicated to the members of the Conference for the Reduction and Limitation of Armaments a preliminary report on the progress of the work of the Conference, which the late Mr Arthur Henderson, president of the Conference, had, some time before his death, requested him to distribute. (It was published in July 1936. [11])

On 22 January 1936, the Council of the League of Nations decided to postpone the further convocation of the Conference.

The Conference never reconvened. Its failure brought to nothing the League's efforts to go beyond the prohibition of use of CB weapons contained in the Geneva Protocol of 1925.

VI. *Summary and comment*

The second phase of the Disarmament Conference yielded two important documents concerning CBW: a comprehensive report of the Special Committee on chemical, incendiary and bacterial weapons, and a United Kingdom draft convention for general disarmament, which included provisions for the prohibition of those weapons.

The main conclusions reached by the Special Committee were:

It was not possible to prohibit the manufacture, import, export or possession of apparatus and chemical substances capable of employment both for peaceful and warlike purposes.

It was possible to prohibit the manufacture, import and possession of apparatus and substances suited exclusively for chemical warfare, as it was possible to prohibit the training of armed forces in the use of chemical weapons, but the means which might be employed to ensure observance of those prohibitions would be of very limited value; no effective control could be devised.

Lachrymatory substances could not be treated separately from poisonous as far as the prohibition of use in war is concerned. The employment of tear gas for police operations in peacetime was not questioned, but some sort of supervision over the stocks of those substances and apparatus for their use, should be exercised.

It was not possible to prevent preparations for bacteriological warfare.

One could prohibit the manufacture, import, export and possession of projectiles and bombs specifically incendiary, but not of incendiary substances and apparatus suitable both for peaceful and military uses.

The establishment of the fact of violation of the prohibition to use CB and incendiary weapons should be very rapid, offer a guarantee of impartiality and be carried out by competent persons of high moral standing.

The attacked state should be given scientific, medical and technical assistance, while pressure should be brought to bear on the offender to make him stop the use of prohibited weapons.

It would be advisable to establish an international information and documentation service for protection against chemical weapons.

Notwithstanding the discouraging conclusions concerning the possibility of enforcing the prohibition of preparations of chemical, incendiary and bacteriological warfare, there were strong demands to include such prohibition in a convention for the reduction and limitation of armaments.

The relevant provisions of the UK draft, accepted as the basis of the future convention, reflected this attitude. The proposals put forward were to a great extent declaratory in character, no effective means having been provided to secure their implementation. In one respect they fell short even of the Committee's requirements, namely, in affording the victim of the illegal use of chemical or incendiary weapons the right to retaliate without prior official establishment of the use of similar weapons by the adversary.

With all its shortcomings, the text of the convention, if adopted, would have constituted a considerable advance, when compared to the Geneva Protocol of 17 June 1925. In particular:

It would have established an absolute prohibition of bacteriological weapons, i.e., prohibition even to retaliate with the use of these weapons;

It would have provided a broad definition of bacteriological and chemical weapons, explicitly including lachrymatory substances—a point discussed and not contested in the debate (the USA was even ready to accept some restrictions regarding tear gas used for internal police operations);

It would have prohibited incendiary in addition to CB weapons;

It would have prohibited preparations for chemical, incendiary and bacteriological warfare in time of peace as in time of war, including the training of armed forces in the use of banned weapons;

It would have instituted some international supervision and government control of substances used for protective experiments;

It would have enabled states to submit complaints with regard to the possible breaches of the prohibition to prepare for chemical, incendiary or bacteriological warfare;

It would have introduced some procedure for establishing the fact of the use of the prohibited weapons.

The withdrawal of Germany and German rearmament brought about the breakdown of the Disarmament Conference and of all attempts to achieve universal reduction and limitation of armaments as well as CB disarmament.

Much, however, can be learned from the record of the debate. It will be noted that many proposals advanced during the League's period have been revived in recent years, and a number of points made then remain topical now.

Chapter 6. 1935–1938

I. *The use of gas in Ethiopia*

The Italo-Ethiopian conflict, which started in December 1934 with a frontier incident between Italian Somaliland and Abyssinia, transformed itself into a full-scale war in 1935–1936. Allegations were made by both parties that the opposing forces were using illegal methods of warfare. The Ethiopians complained that the invading Italian troops were employing poisonous gases. The matter was considered in the League of Nations. Here is an account of the Ethiopian charges and of action taken by the League.[1]

In a telegram of 30 December 1935, the Emperor of Ethiopia informed the Secretary-General of the League of Nations that on 23 December 1935 the Italians had made use of asphyxiating and poison gases against the Ethiopian troops in the Takkaze region; he protested "against such inhuman practices". [1]

Following that protest, on 1 January 1936, the Emperor informed the Secretary-General that, on 30 December 1935, the Italians, after bombing the Ethiopian southern army, made use of poison gas "in defiance and violation of . . . international undertakings". [2]

A letter of 6 January 1936, from the Ethiopian representative to the Secretary-General, stated that the Italian military authorities were continuing their policy of terrorism by employing poison gas against the Ethiopian troops in the northern sector of operations. [3]

In a statement of 20 January 1936 the Ethiopian government asked the League to consider whether collective intervention was not desirable and whether energetic steps should not be taken to prevent such atrocities as the systematic bombing of hospitals by Italian aircraft, the use of poison gases, the destruction of open towns, and the burning of churches. [4]

In a memorandum of the Ethiopian National Red Cross Society of 2 March 1936, addressed to the International Red Cross Committee in Geneva and transmitted by the Ethiopian representative to the Secretary-General, details were given regarding the time and place of employment of asphyxiating, poisonous and other gases.

The Ethiopian Red Cross drew attention to the following consignments

---

[1] For a further description of the use of gas in Ethiopia, see Volume I of this study.

of asphyxiating, poisonous and other gases which had been sent from Italy to Eastern Africa:

Between 25 June and 25 December 1935, there passed through the Suez Canal: 45 tons of mustard gas (Jost-yperite); 265 tons of asphyxiants; and 7 483 gas bombs, all consigned to Massawa. On 4 January 1936, at 6.35 a.m. the vessel *Sicilia* passed through the Suez Canal transporting 4 700 bombs containing asphyxiating and tear gases and 3 227 incendiary bombs in cases marked T.U.7, all consigned to Massawa. On 19 January 1936 at 9.25 p.m., the vessel *Madda* passed through the Suez Canal transporting 185 flame-throwers consigned to Massawa. [5]

On 17 March 1936, the Minister for Foreign Affairs of Ethiopia again formally protested against the continued use by the Italians of asphyxiating and similar gases all along the northern front and during the bombardment of open towns in the interior of the country. He considered it the most flagrant breach of the 1925 Geneva Protocol.

On 21 March 1936, the Ethiopian government complained that Italy "is raining down yperite on inhabited centres". [6]

On 23 March 1936, the Chairman of the Committee of Thirteen—a committee entrusted by the Council of the League to examine the situation in Ethiopia as a whole—sent to the Italian government a letter saying that the Committee "would be happy to receive any observations your Government might wish to make in regard to the Ethiopian Government's allegation of the use of asphyxiating, poisonous or similar gases by the Italian army". In its reply of 3 April 1936, the Italian government did not state whether such gases had or had not been employed, although it referred to "the facts of which a tendentious version is put forward in the Ethiopian statements". [7]

On 10 April 1936, the Ethiopian representative in Geneva informed the Secretary-General that the Italians had begun to use asphyxiating gases on the Ogaden front, and that on 8 April they had discharged gas on Dagabur and Sassabaneh. "Eighty persons were affected; they are suffering horribly in the eyes and lungs and from skin burns." [8]

In a telegram of 11 April 1936 to the Committee of Thirteen, the Italian Minister for Foreign Affairs said that the observance of the laws of war must be bilateral. "The Italian military authorities cannot do otherwise than punish every inhuman atrocity committed by its adversary in contempt of every principle of law and morality." [9]

Attached to a letter dated 13 April 1936, from the Ethiopian representative to the Secretary-General, was the evidence given on 23 March 1936, at Geneva, in the presence of two members of the International Red Cross Committee, by Mr Maksymilian Stanislas Belau, a Polish medical doctor who had served with the Ethiopian Red Cross as chief of Ambulance.

At Ambarada, Dr Belau saw yperite casualties (the first yperite casualty he had seen was at Kworam) and then casualties caused by another gas with which he was not familiar. Mr Medynski, his assistant, who had suffered from the latter gas, thought it was phosgene, but Dr Belau was not of the same opinion. It was a gas smelling of hyacinth. A third gas noticed was lachrymatory and sneeze-producing. When Dr Belau was at Ambarada, a large number of yperite casualties came to the Antalo district for treatment. They were chiefly civilians—women and children. A certain place in this district in which yperite had been used, and which was reported to Dr Belau by the population, was disinfected under his direction with calcium chloride. [10]

The Ethiopian representative's letter provided a list of towns attacked with poison gas. [11]

The allegations were catalogued in the "Analysis by the Committee of Jurists of the Documents concerning the Conduct of the War in Ethiopia" annexed to the report of 18 April 1936 of the Committee of Thirteen. [12]

The "Analysis" also contained statements from non-Ethiopian sources regarding gas casualties. A note circulated by the United Kingdom delegation to the Committee of Thirteen, under the date of 8 April 1936, referred to:

Statement by Dr Melly, of the British Red Cross, that on 1, 2 and 3 March he treated about a hundred serious cases of mustard gas burns, a statement supported by photos showing the burns. The photos, taken on 4 March, were in the possession of the British Minister at Addis Ababa.

Statement by Dr J. W. C. Macfie, of the British Ambulance, that between 1 and 8 March 1936 he personally saw and treated several hundred men, women and children suffering from mustard gas burns.

Report that there existed an authenticated account of the Norwegian Ambulance (southern front) of twenty-one cases of mustard gas burns caused by one bomb on 19 March.

Report that the Swedish Ambulance had treated similar cases in December on the southern front.

Report that Dr Junod, of the International Red Cross, and Count von Rosen, a Swedish Red Cross pilot, had seen gas used at Kworam on 17 March and had suffered from it.

Report that Dr Brophil, an Irishman attached to the Ethiopian army sanitary service command and serving with the Ethiopian Red Cross Ambulance No. 3, had stated in London that, in the last week of December and the first week of January, about thirty cases of mustard gas burns were treated by that Ambulance.

In a letter of 18 April 1936, the chairman of the Committee of Thirteen called the Italian government's attention to the desirability of furnishing information with regard to the allegations of the use of asphyxiating, poisonous or similar gases, and referred to the Geneva Protocol of 17 June 1925. The opinion was also expressed that the observation made in the Italian telegram of 11 April (see page 176), could not justify the use of asphyxiating, poisonous or similar gases. [13]

The Italian representative in the League took exception to the Committee's opinion that the use of the chemical weapon could not be justified even for the punishment of inhuman acts of atrocity committed by an adversary in contempt of law and morality. By that statement, said the delegate of Italy, the Committee set itself up as a judge, giving an interpretation of perhaps the most delicate and complex point covered by the Geneva Protocol of 17 June 1925, which contained no provision prohibiting, in derogation of the general principles, the exercise of the right of reprisal for atrocities such as those of which Italian soldiers had been victims, and evidence of which had been brought to the notice of all the members of the League of Nations. [9]

The British delegate pointed out that it was impossible to disregard the evidence which existed and which went to show that poison gas had been used by the Italian armies in their campaign against Ethiopians utterly unprovided with any means of defence against this method of warfare. The Geneva Protocol of 1925, said the British representative, concerned the inhabitants of the whole world. For them it was a charter against extermination. If a convention such as this could be torn up, would not the people, whether living in the crowded cities of Western Europe or in less densely populated areas elsewhere, ask, and ask with reason, what was the value of any international instrument to which the representatives put their name? [14]

The Portuguese delegate formally condemned the use of gases in war, whatever the reasons alleged for their employment. [14]

The representative of Australia recalled that there had been allegations by both sides with regard to the infringement of conventions for the conduct of war. All the charges put forward, if established as being true, were hideous; but there was one that stood out in its menace to humanity and civilization far above everything else, and that was the charge that poisonous and asphyxiating gases were being employed in the war. If that charge proved true, it was impossible fully to realize what menace it constituted to every nation on earth. [14]

The Council adopted a resolution in which it:

Recalls that Italy and Ethiopia are bound by the Protocol of June 17th, 1925, on the use of asphyxiating, poisonous or other gases, and by the Conventions regarding the conduct of war to which these two States are parties, and emphasises the importance which has been attached to these instruments by all the contracting States. [15]

In Circular No. 325 of 27 April 1936, to the central committees of the national Red Cross societies, the President of the International Red Cross Committee stated that the alleged use of asphyxiating, poisonous and similar gases in the Italo-Ethiopian conflict had been given the closest attention by the Committee. On 23 March 1936, the Committee had received a request from the Ethiopian Red Cross asking that national Red Cross societies be requested to send large quantities of gas masks and manuals dealing with technical protection against asphyxiating, poisonous and other gases.

The International Committee did not feel justified in acceding to this request in the form presented. A general appeal for gas masks on behalf of one of the parties, without specifying for what purpose these masks were to be used, would cause the International Red Cross Committee to go outside its proper role. The only masks which the International Committee would be entitled to ask national societies to supply to a sister society were masks intended exclusively for the use by medical personnel or for patients under the care of such personnel. The International Committee therefore confined itself to informing the national Red Cross societies about this request (these societies in varying degrees had already responded to the appeals of the Ethiopian Red Cross). It had also requested its delegation in Addis Ababa to ascertain how many masks the Ethiopian Red Cross required for the exclusive use by persons of the above category. The president of the Committee explained that the reserve which the Committee had felt bound to show in this circumstance must not be interpreted as tacit acceptance of a method of warfare which it had never ceased to condemn. [16]

On 30 April 1936, the Italian government replied officially to the letter of 18 April 1936, from the chairman of the Committee of Thirteen (see page 178). It stated that the 1925 Geneva Protocol contained no provision excluding the exercise of the right of reprisal, by way of exception to the general principles which admitted that right. By the clause in the Protocol referring to the use of chemical weapons, the signatory powers merely declared that they recognized the prohibition of the use of gases therein mentioned as embodied in international law, but they added no clause modifying the existing legal situation in regard to the right of reprisal. In any case, continued the letter, neither the Committee of Thirteen nor any other organ of the League of Nations would be competent to give an interpretation of the 1925 Geneva Protocol. The Protocol, as it appeared from its formal clauses, had no connection with the League and had also been signed by powers that were not members of the League. While repeating the assurance that it intended to comply with the provisions of the 1925 Protocol, the Italian government could not agree that this Protocol precluded the exercise of the right of reprisal "in punishment of such abominable atrocities as those committed by the Ethiopian forces (torture and decapitation of prisoners; emasculation of the wounded and killed; savagery towards, and the killing of non-combatants; systematic use of dum-dum bullets, etc.)". It concluded by saying that Ethiopia, having been the first to violate the conventions and rules of war, and having done so time and again, had no right to refer to them. [17]

By letter of 2 May 1936 to the Secretary-General, the Ethiopian representative forwarded a series of documents and photographs concerning violations of the laws of war by the Italians. [18] They included:

(*a*) Statement of 9 April 1936, by Dr Schuppler, head of Ambulance No. 3, on the use of poison gases.

Dr Schuppler informed the Foreign Office in Addis Ababa that on 14 January 1936 battle gas bombs had been used by Italian flyers. Twenty country people were killed; he treated about fifteen cases of gas-bombing, two children among them. The burning was caused by mustard gas used south of the Pass Alaghi.

Five miles westward of Amba Aradam, Dr Schuppler himself had been lightly gassed; it was also mustard gas. He found a gas bomb (mustard gas) 16 kilometres west of the plain Kworam. In this district there were only civilians. The bomb measured 1.30 m by 10 cm [19].

(*b*) Report of 11 April 1936, on the use of poison gases, signed by Mr Walter M. Holmes, correspondent of the Nordisk Telegraph Agency of Copenhagen and the *Times* of London.

Mr Holmes, in a letter to the Minister for Foreign Affairs in Addis Ababa, described aerial bombardment in an area of the northern front, where the use of mustard gas was a frequent occurrence. His first personal experience of gas bombardment was on Sunday, 1 March 1936, in the bush between Alamata and Kobbo, about ten miles south of Kworam. The Italian bombardment was carried out indiscriminately. The dropping of several large containers of corrosive liquid was noted; the presence of gas was quickly felt through the impregnation of the atmosphere over considerable zones with vapour, whose pungent biting character left no doubt that it was a substance known in the Great War as "mustard".

Later in the same day, he visited a base which the British Ambulance had then established at Alamata and saw there persons suffering from burns which were undoubtedly inflicted by a liquid of the "mustard" type. Some of these cases, he was informed, had been brought in during that day; others had arrived on the two or three preceding days. While a number showed burns on the head and shoulders obviously caused by falling liquid, a much larger number were severely injured in the legs and lower parts of the body. In several cases, large areas of skin had been removed from the legs and thighs; some of these men had also suffered extremely severe and painful burning of the genital organs. The cause of this particular form of injury was as follows: the gas was being dropped in large containers, one of which had actually been brought into the ambulance compound, inspected and photographed. It was a torpedo-shaped object of a total length of about four feet. On striking the ground, the nose of the container became detached and from a bottle within a quantity of liquid amounting to about forty lbs was released. Falling in dense bush, this liquid was scattered over ground and foliage and remained there for a considerable period. Its corrosive quality persisted for some two or three days. Not only troops, but also peasants passing through the bush on their usual occupations and coming into contact unawares with contaminated foliage, sustained the injuries described above.

Subsequently, at Kworam and in the neighbourhood of Lake Ashangi, Mr Holmes witnessed bombardments by Italian airplanes; gas bombs were frequently used. In addition to dropping the containers already described, the Italians, flying over Kworam Plain at relatively low altitudes, used the method of spraying. This method caused more widespread injury and was certainly more difficult to escape from than the dropping of gas shell. There seemed little protection from the fine rain of corrosive liquid which descended from the plane, unless possibly something in the nature of a diving suit were devised; but Ethiopian soldiers and peasants were not provided with even the most elementary forms of mask or protective clothing. Consequently, large numbers of them, subjected to this form of attack, received ghastly injuries to the head, face and upper parts of the body. One evening when Mr Holmes was riding across Kworam Plain, shortly after such a gas attack, he came upon the British Ambulance warrant officer, Mr Atkinson, administering first aid to victims. It had been necessary to send him down from the cave in which the ambulance was then located because many of the victims had been blinded by the gas-spraying and could not go up into the hills for treatment. After this it became a daily occurrence for the Ambulance to send officers down to the plain to treat victims thus incapacitated. On the evening in question, Atkinson treated fully 100 cases of burning by corrosive liquid.

Another case of injury by gas occurred when the British Ambulance officers, Captain Townshend Stephens and Dr Empey, went to the assistance of the crew of an Ethiopian Red Cross plane which had been bombarded by the Italians while lying on the open ground at Kworam. The officers found themselves passing through a zone of mustard gas and both shortly afterwards showed marked indications of inhalation of the vapour, while Captain Townshend Stephens suffered slight but distinct burns on the throat. Among the wounded, who during Mr Holmes' stay in the region of Kworam were streaming back from the battles south of Makale and in the Tembien, there were a great proportion of gas victims. Many were suffering from gangrened wounds due to the lack of facilities and materials for treating the effects of gas at the front. [20]

(c) Report of 10 April 1936, on the use of poison gases, signed by Dr John M. Melly, Head of the British Red Cross Ambulance in Ethiopia.

Dr Melly reported that in the latter half of February, while the British Ambulance Service was situated at Waldia, several cases of severe burns from mustard gas had been treated. These persons had made their own way down from the front. On 28 and 29 February and 1 March, about 150 cases of severe burns from mustard gas were treated by the advance unit of the British Ambulance Service in Ethiopia near Alamata. Many of the patients were women, children and infants. The burns of the large majority had been contracted locally. During the three days that the unit was situated near Alamata, several mustard gas bombs were dropped in the region.

Between 7 and 22 March, while the unit was situated in the region of Lake Ashangi, between 200 and 300 cases of burning by mustard gas were treated. Many had been temporarily blinded and a special clinic had to he held about a mile away from the unit, where the gassing had been most severe, as the victims being blind were unable to find their way to the Ambulance. When the unit was in this location, mustard gas was frequently used in the vicinity.

A large number of the burns treated were of a terrible nature. [21]

(d) Report of 19 March 1936, on the use of poison gases, sent to the International Red Cross in Addis Ababa by Messrs Gunnar Ulland and Vale, doctors attached to the Norwegian Red Cross at Inga Alem. The report was accompanied by an extract from a letter, dated 20 March 1936, from Mr Smith, a missionary with the Sudan Interior Mission.

Mr Ulland made the following statement regarding the use of gas bombs by the Italians on 19 March 1936:

At 8 a.m. two trimotor Italian bombing planes went over Yerga Alem.

At 5 p.m. two patients were brought along to the camp, both suffering from severe irritation of the eyes with epiphora and blepharospasmus and strong irritation of the mucous membranes of the nose and throat. One of them also had a bulbous burn of the skin on both feet. Four more persons were found suffering from exactly the same symptoms. All six were put in the hospital.

On the spot where the bomb exploded, there was a hole—three metres in diameter and one and a half metre deep. Judging by the three pieces found in the hole, the bomb must have been 75 cm long and about 30 cm in diameter, and made up of 2 mm thick sheet-steel. The grass around the hole for about five metres was faded. There was a distinct smell of mustard up to 80 metres from the spot. The injured persons were located ten to twenty metres from the exploding bomb.

In a postscript to the above statement, Mr Vale said that on 20 March 1936 they had treated fifteen more patients injured by a gas bomb on the previous day. All these patients had bulbous burns on the feet and legs, and a few in the face. The injuries had all the characteristics of burns from mustard gas.

Mr Smith, a missionary, wrote that on 19 March 1936 gas had been dropped in two containers; one exploded, the other did not. Over thirty people were affected. It seemed to be mustard gas mixed with another gas. [22]

(e) List of places bombed with poison gases during the period 22 December 1935–7 April 1936. The document also reproduced figures showing the consignments of poison gases to East Africa, which passed through the Suez Canal. (See page 176.)

The list of gas bombardments was as follows.[2]

Takkaze	22 December 1935
Amba Alaji	26 December 1935
Borana	30 December 1935
Makale	31 December 1935
Sokata	10 January 1936
Makale	21 January 1936
Megalo	16 February 1936
Waldia Road	27 February 1936
Kworam	16 March 1936
Ylan Serer	17 March 1936
Kworam	17 March 1936
Kworam	18 March 1936

[2] Different spelling of localitites appeared in different documents. It is presumably attributable to the fact that at that time a uniform official transcription of geographical names in Amharic did not exist.

Irga Alem	19 March 1936
Irga Alem	21 March 1936
Inda Mehoni	29 March 1936
Inda Mehoni	30 March 1936
Kworam	4 April 1936
Kworam	5 April 1936
Kworam	6 April 1936
Kworam	7 April 1936

In the last four bombardments the gas was sprayed on the town.

It was noted that the list was far from complete, because since the beginning of March 1936 the Italians had been systematically bombing with gases every day on the fronts and on centres of civilian population. [23]

(f) Five photographs taken on 19 March 1936, showing the effects of the poison gases on the victims at Irga Alem. [24]

On 4 May 1936, the Swedish Minister for Foreign Affairs informed the Secretary-General of the League that the Swedish government had received certain information pointing to the use of gas in Ethiopia, but had not felt called upon to open an enquiry into the matter. [25]

The Italian forces continued their advance and on 5 May 1936, the Emperor of Ethiopia having fled from Addis Ababa, the capital was occupied by the Italians.

In a letter of 9 May 1936, to the Secretary-General, the Italian government reiterated that it regarded the Geneva Protocol of 1925 as not precluding the exercise of the right of reprisal in punishment of "such abominable atrocities as those committed by the Ethiopian forces, which would be inconceivable in civilized countries". [26]

On 12 May 1936, the Italian delegation left Geneva refusing to discuss the Italo-Ethiopian dispute on the ground that the only sovereignty in Ethiopia was Italian sovereignty. (By that time, the King of Italy had signed a legislative decree in which he assumed, for himself and his successors, the title of Emperor of Ethiopia.)

In a statement of 10 June 1936, a Belgian lieutenant, Armand Frère, formerly military adviser to the army of Ras Desta (the son-in-law of the Emperor), said that during the whole of his stay on the Somali front and during his return by road to Addis Ababa he had never seen any victims of gas attacks. He had been present at at least seventy or eighty bombardments. On many occasions—continued the statement—the Ethiopians tried to make him believe that the bombs were gas bombs. He could definitely say that they were not, for the simple reason that there were no gas victims and because, as he himself had no mask, he could not have escaped the poisonous or other effects of the gases. On many occasions he spoke to the doctors of the Swedish Ambulance; none of them had ever had to attend to a gas victim. What the Ethiopians took for gas bombs were merely incendiary bombs which, after exploding, left greenish yellow traces giving off a smell of slow combustion powder; this had nothing to do with gas and was quite harmless to the touch. [27]

In a memorandum of 15 June 1936, addressed to the International Red Cross Committee, the President-General of the Italian Red Cross commented on the Ethiopian allegations concerning the use of gas:

Complaint was made of the bombing of Makale with asphyxiating gas on 31 December 1935 and 21 January 1936, at which time the town was in the possession of the Italian troops; it had been occupied since 8 November 1935;

Complaint was made of the transport of gas by the S.S. *Sicilia*, a ship which, according to information received, was used exclusively for the transport of troops and therefore could not carry explosives for which special equipment is necessary;

Complaint was also made of the transport of gas[3] by the S.S. *Madda*, which was alleged to have passed through the Suez Canal on 19 January 1936, whereas it had been ascertained that the vessel in question sailed on her first voyage to East Africa from Naples on 27 February 1936, carrying nothing but motor vehicles and foodstuffs.

The examples were meant to provide illustration of the unreliability of Ethiopian statements. [28]

On 30 June 1936, the Emperor of Ethiopia, in his address to the League's Assembly stated:

At the outset, towards the end of 1935, Italian aircraft hurled tear gas bombs upon my armies. They had but slight effect. The soldiers learned to scatter, waiting until the wind had rapidly dispersed the poisonous gases.

The Italian aircraft then resorted to mustard gas. Barrels of liquid were hurled upon armed groups. But this means too was ineffective; the liquid affected only a few soldiers, and the barrels upon the ground themselves gave warning of the danger to the troops and to the population.

It was at that time when the operations for the encirclement of Makale were taking place that the Italian command, fearing a rout, applied the procedure which it is now my duty to denounce to the world.

Sprayers were installed on board aircraft so that they could vaporise, over vast areas of territory, a fine, death-dealing rain. Groups of nine, fifteen, eighteen aircraft followed one another so that the fog issuing from them formed a continuous sheet. It was thus that, from the end of January 1936, soldiers, women, children, cattle, rivers, lakes and fields were constantly drenched with this deadly rain. In order to kill off systematically all living creatures, in order the more

---

[3] Actually, the Ethiopians alleged that the vessel *Madda* had transported flame-throwers consigned to Massawa (see page 176).

surely to poison waters and pastures, the Italian command made its aircraft pass over and over again. That was its chief method of warfare.

The very refinement of barbarism consisted in carrying devastation and terror into the most densely populated parts of the territory, the points farthest removed from the scene of hostilities. The object was to scatter horror and death over a great part of the Ethiopian territory.

These fearful tactics succeeded, men and animals succumbed. The deadly rain that fell from the aircraft made all those whom it touched fly shrieking with pain. All who drank the poisoned water or ate the infected food succumbed too, in dreadful suffering. In tens of thousands the victims of the Italian mustard gas fell. It was to denounce to the civilised world the tortures inflicted upon the Ethiopian people that I resolved to come to Geneva. None other than myself and my gallant companions in arms could bring the League of Nations undeniable proof. The appeals of my delegates to the League of Nations had remained unanswered; my delegates had not been eyewitnesses. That is why I decided to come myself to testify against the crime perpetrated against my people and to give Europe warning of the doom that awaits it if it bows before the accomplished fact. [29]

II. *Sanctions against Italy*

On 11 October 1935, the Co-ordination Committee, set up by the Assembly of the League of Nations to coordinate the action of the governments of member states in applying economic and financial sanctions against Italy under Article 16 of the Covenant,[4] adopted the following proposal for an arms embargo:

[4] Article 16 of the Covenant reads as follows:
"1. Should any Member of the League resort to war in disregard of its covenants under Articles 12, 13 or 15, it shall *ipso facto* be deemed to have committed an act of war against all other Members of the League, which hereby undertake immediately to subject it to the severance of all trade or financial relations, the prohibition of all intercourse between their nationals and the nationals of the covenant-breaking State, and the prevention of all financial, commercial or personal intercourse between the nationals of the covenant-breaking State and the nationals of any other State, whether a Member of the League or not.
"2. It shall be the duty of the Council in such case to recommend to the several Governments concerned what effective military, naval or air force the Members of the League shall severally contribute to the armed forces to be used to protect the covenants of the League.
"3. The Members of the League agree, further, that they will mutually support one another in the financial and economic measures which are taken under this Article, in order to minimize the loss and inconvenience resulting from the above measures, and that they will mutually support one another in resisting any special measures aimed at one of their number by the covenant-breaking State, and that they will take the necessary steps to afford passage through their territory to the forces of any of the Members of the League which are co-operating to protect the covenants of the League.
"4. Any Member of the League which has violated any covenant of the League may be declared to be no longer a Member of the League by a vote of the Council concurred in by the Representatives of all the other Members of the League represented thereon."

Proposal I

Export of Arms, Ammunition and Implements of War.

With a view to facilitating for the Governments of the Members of the League of Nations the execution of their obligations under Article 16 of the Covenant, the following measures should be taken forthwith:

(1) The Governments of the Members of the League of Nations which are enforcing at the moment measures to prohibit or restrict the exportation, re-exportation or transit of arms, munitions and implements of war to Ethiopia will annul these measures immediately;

(2) The Governments of the Members of the League of Nations will prohibit immediately the exportation, re-exportation or transit to Italy or Italian possessions of arms, munitions and implements of war enumerated in the attached list;

(3) The Governments of the Members of the League of Nations will take such steps as may be necessary to secure that arms, munitions and implements of war, enumerated in the attached list, exported to countries other than Italy will not be re-exported directly or indirectly to Italy or to Italian possessions;

(4) The measures provided for in paragraphs (2) and (3) are to apply to contracts in process of execution.

Each Government is requested to inform the Committee, through the Secretary-General of the League, within the shortest possible time, of the measures which it has taken in conformity with the above provisions.

The Annex, attached to the Proposal, listed among the categories of arms subject to embargo:

Category VI
1. Livens projectors and flame-throwers.
2. Mustard gas, Lewisite, ethyldichlorarsine and methyldichlorarsine.
3. Powder and explosives. [30]

On 16 October 1935, a revised list of articles considered as arms, ammunition and implements of war was substituted for the list attached to Proposal I (now named Proposal IA). It included:

Category V
1. Flame-throwers and all other projectors used for chemical or incendiary warfare.*
2. Mustard gas, Lewisite, ethyldichlorarsine, methyldichlorarsine and all other products destined for chemical or incendiary warfare.*
3. Powder for war purposes and explosives.

The following remark was made:

*It should be observed that the utilization of these articles has been, and still is, prohibited under the Convention of June 17th, 1925. These articles are only mentioned above because their manufacture being free (the more so, as in many instances they serve various purposes), the Committee desires to emphasize that the export of such products could in no circumstances be tolerated. [31]

By 13 December 1935 the League had received information from the following fifty members that measures prohibiting export of arms, ammunition and implements of war to Italy were in force: Afghanistan, Argentina, Australia, Belgium, Bolivia, Bulgaria, Canada, Chile, China, Colombia, Cuba, Czechoslovakia, Denmark, Dominican Republic, Ecuador, Estonia, Finland, France, Greece, Haiti, Honduras, India, Iran, Iraq, Irish Free State, Latvia, Liberia, Lithuania, Luxemburg, Mexico, Netherlands, New Zealand, Nicaragua, Norway, Panama, Peru, Poland, Portugal, Romania, Siam, Spain, Sweden, Switzerland, Turkey, United Kingdom, Union of South Africa, Union of Soviet Socialist Republics, Uruguay, Venezuela, Yugoslavia. In Austria and Hungary the export of arms was illegal; it had been prohibited under the Treaties of Peace. Luxemburg and Switzerland prohibited the export of arms, ammunition and implements of war both to Ethiopia and Italy. Thus they had not accepted that part of Proposal I which related to Ethiopia (abolition of measures prohibiting or restricting the exportation, re-exportation or transit of arms, munitions and implements of war to Ethiopia). Luxemburg based its attitude upon its policy of neutrality; Switzerland referred to the Hague Convention of 1907, concerning the rights and duties of neutral powers and persons in war on land, and to its neutral status. [32]

The French representative thought that the attitude of Switzerland may lead to very serious consequences in view of the important part Switzerland played in transit matters in Europe. He stated that his government could not admit the validity of the Swiss government's arguments and considered that they ran counter to Article 16 of the Covenant and the London Agreement between the League Council and Switzerland regarding the latter's entry into the League. The Polish, Greek, Romanian, Soviet and British representatives endorsed the French position. [33]

A few countries had somewhat modified and extended the embargo list as compared to that annexed to Proposal IA.

For instance, the United Kingdom included in the schedule of goods prohibited to be exported to Italian territory:

Flame-throwers and all other projectors and machines (including smoke-producing apparatus) used for chemical or incendiary warfare;

Mustard gas, Lewisite, ethyldichlorarsine, methyldichlorarsine, ethyliodoacetate, chloro-acetophenone, chlorosulphonic acid, diphenylaminechloroarsine, bromobenzylcyanide, diphenylchlorarsine, diphenylcyanoarsine, phosgene, chlorpicrin and all other noxious substances whatsoever, intended for offensive or defensive purposes in warfare. [34]

The Swedish list of goods prohibited for exportation comprised:

Phosgene (oxychloride of carbon);

Diphenylaminochlorarsine (Adamsite), diphenylcyanarsine, diphenylchlorarsine, ethyldichlorarsine, phenyldichlorarsine, chlorovinyldichlorarsine (Lewisite) and methyldichlorarsine;

Benzyl bromide, benzyl iodide, benzyl chloride, bromacetone, benzyl cyanobromide, cyanogene bromide, methyloethylbromide, ketone, ethylbromacetate, methylbromacetate, methylic ether of cyanocarbonic acid, dibromated dimethylidether, methyldibromothylketone, dichlorethylsulphide (mustard gas, yperite), dichlorated dimethylic ether, dichlorated phenylcarbylamine, iodacetone, ethyliodacetate, chloracetophenone, chloracetone, chlorocyaniccyanogen chloride, formate of dichloromethylochlorine, ethylchloroformate, monochlorous methylchloroformate, trichloromethyl chloroformate (disphosgene, chloroformate of methylperchlorate), chloropicrin (chloroform nitrate), ethylchlorosulphonate, methylchlorosulphonate, methylbromoethylketone, nitrobenzyl chloride (ortho-), palite (chloroformate of chloromethyl), carbon tetrachlorosulphide, thiophosgene and bromides of xylyl.[5] [35]

The League sanctions were applied half-heartedly and never extended to include military measures. They failed to stop aggression and the use of chemical weapons.

Following a recommendation adopted by the Assembly which had been convened on the initiative of the Argentine government on 4 July 1936, [37] the Co-ordination Committee proposed on 6 July 1936 that the members of the League should abrogate on 15 July 1936 the restrictive measures taken by them. [36] The League thus recognized its inability to assure the application of the provisions of the Covenant and respect for the 1925 Geneva Protocol.

III. *The use of gas in China*

In the late thirties, allegations of the use of poison gas were also made during the hostilities between Japan and China. In this case, the League was even slower to react than in the case of Ethiopia. Listed below are a few instances of Chinese official complaints on which some action was taken.[6]

On 9 May 1938, the Chinese government, in an appeal addressed to the League of Nations Secretary-General, stated that the Japanese forces on the Shantung front had employed poison gas on a number of occasions and, according to confirmed reports, were intensifying their preparations for the

---

[5] The chemical terminology used here is reproduced from the original League of Nations documents.

[6] The use of gas in China in the thirties is described more fully in Volume I.

use of such gases on a large scale "in a desperate attempt to turn the tide of the war". Two battalions of chemical warfare troops, which constituted part of the newly-formed independent mechanized unit under the command of Lieutenant-General Motoma, had left Kobe for the Shantung front via Tsingtao on 19 April. Another mechanized unit, under the command of Major-General Kikuchi, which had just left Japan for South Shantung, also included chemical warfare troops. More forces for chemical warfare were being despatched from Japan.

The appeal stressed that recourse to such a method of warfare was contrary not only to existing international conventions but also to all considerations of humanity. [38]

On 10 May 1938, the representative of China asked the Council to take steps immediately "with a view to forestalling the perpetration of a heinous crime". [39]

On 14 May 1938, the Council adopted a resolution in which it:

Recalls that the use of toxic gases is a method of war condemned by international law, which cannot fail, should resort be had to it, to meet with the reprobation of the civilized world; and requests the Governments of States who may be in a position to do so to communicate to the League any information that they may obtain on the subject. [40]

On 19 September 1938, the Chinese representative again asked the League to take effective measures to deter Japan from continuing to pursue "such illegal and inhuman methods of warfare as the use of poison gas and indiscriminate air-bombing of undefended towns and civilian populations". [41]

On 30 September 1938, the Council adopted yet another resolution [42] by which:

Taking note of the information contained in the various communications of the Chinese representative on the subject; and of his statement on the urgent need for the constitution of a Commission of Neutral Observers in China to investigate the cases of the use of poisonous gas in China, to watch the situation in respect thereto, and to make reports for examination;

Invites the Governments of the States represented on the Council and on the Far-Eastern Advisory Committee having official representatives in China to investigate through the diplomatic channel, so far as this may prove practicable and by the most appropriate method, such cases as may be brought to their notice and to submit all relevant reports for examination and consideration.

The British representative, while accepting the resolution, noted that the evidence received on the subject had been very diverse and conflicting. He therefore maintained an attitude of some reserve in regard to certain of the Chinese representative's observations. [42]

During subsequent months, the deterioration of the international situation rendered it impossible for the League to give even moral assistance to China. The centre of international tension shifted back to Europe.

IV. *Summary and comment*

The first major breach of the Geneva Protocol occurred in 1935–36, during the Italo-Ethiopian War. Although no official fact-finding machinery was set in motion, conclusive evidence was submitted to the League of Nations (detailed accounts by eye-witnesses, including medical doctors, photographs of victims, etc.) to the effect that Italy, a party to the Protocol, had used chemical weapons against Ethiopia, another party to the Protocol.

The gas employed was first tear gas and then mostly mustard gas, dropped in bombs or other containers and sprayed from aircraft over vast areas. According to testimony by the Emperor of Ethiopia, chemical warfare, waged indiscriminately against troops and civilians not provided with any means of defence, produced tens of thousands of casualties and determined the issue of war in favour of Italy.

Italy did not explicitly deny that it had had recourse to the prohibited weapons. It tried to explain its action by alleging that the Ethiopians had been the first to violate the established rules of conduct in war, and claimed that the Geneva Protocol had not modified the "existing legal situation in regard to the right of reprisal".

In the opinion of a League committee dealing with the situation in Ethiopia, the Italian allegations did not justify the use of gases. It may be noted that in adhering to the Geneva Protocol Italy had made no reservation regarding its applicability, and that no one in the course of the CBW debate in the League had referred to a possibility of using chemical weapons in retaliation against a state which had infringed the rules of warfare other than those prescribed by the Protocol.

Some economic and financial sanctions were applied by the League against Italy for committing aggression. The list of arms subject to embargo included products and appliances used for chemical or incendiary warfare. But the steps taken were obviously inadequate. The League was unable to stop aggression and ensure respect for the Geneva Protocol.

In the 1930s gas was also alleged to have been used by Japan in China. In this case there was less documentary evidence available to the League of Nations. Nevertheless, apart from the British delegation which had some reservation, there were few doubts that gas had actually been employed.

Although Japan was not party to the Geneva Protocol, the use of gas was regarded as a violation of the accepted customs of war. The League's response to China's appeals was even more sluggish than in the case of Ethiopia. In 1938, the international situation rapidly deteriorated and the European nations were too preoccupied with events near home to take action in defence of China.

Chapter 7. 1946–1953

The United Nations Charter, which came into force on 24 October 1945, accorded a fairly low priority to the establishment of a system for regulating armaments. In actual practice, however, the United Nations has made disarmament one of its primary objectives. The first task it set itself in this field was to deal with weapons of mass destruction, mainly atomic weapons which had been used in the final stage of World War II.

I. *Chemical and bacteriological weapons defined as weapons of mass destruction*

The United Nations General Assembly resolution of 24 January 1946, establishing a Commission to deal with the problems raised by the discovery of atomic energy, charged that Commission to make specific proposals "for the elimination from national armaments of atomic weapons and of all other major weapons adaptable to mass destruction". [1]

The UN General Assembly resolution of 14 December 1946, which set up principles governing the general regulation and reduction of armaments, asked for the elimination of "... all other major weapons adaptable now or in the future to mass destruction...." [2]

In 1947, when the United Nations Commission for Conventional Armaments was discussing the programme of its work, the need arose to define the categories of armaments falling within its terms of reference.

The majority of the Commission was of the opinion that the best way to arrive at a definition of conventional armaments would be to start by defining weapons of mass destruction.

On 5 September 1947, the United States submitted a draft resolution by which weapons of mass destruction would be defined as including "atomic explosives, radioactive material, lethal chemical and biological weapons, and any weapons developed in the future which have characteristics comparable in destructive effect to those of the atomic bomb or other weapons mentioned above". [3]

The matter was considered at the third meeting of the working committee of the Commission for Conventional Armaments.

The UK representative suggested that weapons of mass destruction include atomic, chemical and biological weapons, while all other weapons should be considered as conventional armaments.

The Australian delegate, however, thought that radioactive material should also be regarded as a weapon of mass destruction.

The representative of the United Kingdom then asked the US representative why in his draft resolution he had added the word "lethal" to "chemical and biological weapons".

The reply was that this had been done in order to distinguish between deadly chemical and biological weapons and those which were not deadly such as tear gas and smoke screens, etc. The use of the word "lethal", continued the US delegate, might be subject to discussion, but if it were not adopted it would be necessary to replace it by another word expressing the same idea.

Australia pointed out that it would be desirable "in order to avoid confusion" to suppress the word "lethal" in the US draft.

The Soviet delegate said that the US proposal failed to define atomic weapons. [4]

On 8 September 1947, the United States tabled a revised draft resolution according to which weapons of mass destruction included "atomic explosive weapons, radioactive material weapons, lethal chemical and biological weapons, and any weapons developed in the future which have characteristics comparable in destructive effect to those of the atomic bomb or other weapons mentioned above". [5]

In the course of the discussion which took place on 9 September 1947, at the fourth meeting of the working committee of the Commission for Conventional Armaments, the Soviet representative criticized the US definition as too restrictive. During World War II, he said, the Germans had used weapons other than atomic, with mass destructive effect on population and cities far from the field of battle. He instanced the flying bomb and rocket as weapons of mass destruction, though they contained neither radioactive nor lethal chemical or biological components. It was his view that an attempt to establish two lists, one consisting of atomic weapons and other weapons of mass destruction and another consisting of conventional weapons, would be a wrong method.

The statement reflected the Soviet position of that time that general regulation and reduction of armaments should cover all kinds of armaments.

The British delegate stated that "V" weapons would definitely fall within the category of conventional armaments since their destructive effect, statistically considered, had not been large. The number of persons killed by the

14 000 "V" weapons dropped on England was 56 000, or four persons per weapon.[1]

The US revised draft resolution was adopted by seven votes to two with two abstentions. The delegations which abstained (China and Brazil) voiced no objection with regard to the substance of the resolution, but had some reservations as to the procedure. Those who voted against were probably the USSR and Poland. (There was no roll-call.[2]) [6]

The matter was raised again in 1948 during the consideration of the report of the working committee by the Commission for Conventional Armaments.

On 11 August 1948, the Ukrainian representative defined weapons of mass destruction as weapons directed primarily against peaceful populations and weapons not of defence but of aggression. [7]

On 12 August 1948, however, the Commission approved the definition as given by the working committee, with the USSR and the Ukrainian SSR opposing. [8]

In his Introduction to the Annual Report on the Work of the Organization, 1 July 1947 – 30 June 1948, the UN Secretary-General said that the prolonged debate on the control of atomic energy and the demonstrations of the tremendously destructive power of atomic weapons that the United States had given to the world had distracted attention from developments in the field of bacteriological and lethal chemical weapons. Whatever the situation regarding atomic weapons—continued the Secretary-General—there had never been any effective monopoly of bacteriological and chemical weapons; some of these weapons were probably potentially as destructive of human life as atomic weapons but not a single proposal had been made by the member nations for a system of preventing or controlling their manufacture, nor had there been any discussion or study of the problem in the United Nations. Meanwhile, it could be assumed that, as in the case of atomic bombs, stocks of these weapons were piling up and new discoveries were constantly being made that rendered them more deadly. He then expressed the belief that the Assembly should give special attention to the problem and suggested that "... it might be fruitful to begin a study of some of the problems involved in the control of bacteriological and lethal chemical weapons". [9] The suggestion was not followed up.

The UN General Assembly resolution of 11 January 1952 directed the

---

[1] On 21 October 1969, the British delegate at the Conference of the Committee on Disarmament said: "Nobody who was in southeast England during the V-bomb attacks in the last war would, at that time at any rate, have had any hesitation, I think, in describing them as weapons of mass destruction."

[2] The committee consisted of states members of the Security Council: France, Australia, Belgium, Brazil, China, Colombia, Poland, Syria, the USSR, UK and USA.

Disarmament Commission to prepare proposals to be embodied in a draft treaty (or treaties) for the regulation, limitation and balanced reduction of all armed forces and all armaments, "for the elimination of all major weapons adaptable to mass destruction, and for effective international control of atomic energy ...". [10]

II. Allegations of the use of bacteriological and chemical weapons in Korea and China

On 25 June 1950, hostilities broke out between North and South Korea, each side claiming to have been attacked by the other. The United Nations Security Council resolved that a breach of the peace had occurred (at the time of the attack the Council was being boycotted by the USSR because of the presence of a representative of Taiwan) and the UN members were called upon to render assistance to South Korea. Sixteen states contributed armed forces; some others provided supplies and services. A unified UN command was formed with a US commander, the US troops constituting more than half of the UN forces present in Korea. In October 1950, China intervened by sending volunteers to fight alongside the North Koreans.

The UN forces, and more particularly the US forces, were accused of waging bacteriological and chemical warfare.[3] In a cablegram of 8 May 1951, addressed to the President of the UN Security Council, the Minister of Foreign Affairs of the Korean People's Democratic Republic protested against the use of bacteriological weapons by the US forces in Korea. The following facts were given, relating to the period from December 1950 to January 1951.

Several areas were simultaneously infected with small-pox, seven to eight days after their liberation. Sicknesses broke out in the town of Phenian, in the provinces of northern and southern Phenian, Kanvon, southern Hamhenz, and Hvanhe.[4] The number of persons suffering from small-pox mounted rapidly and by April there were more than 3 500 cases, 10 per cent of which were fatal. The disease was particularly widespread in the provinces of Konvon, 1 126 cases; southern Hangen, 817 cases; Hvanhe, 192 cases. No case of small-pox was discovered among the combatants of the People's Army and the Chinese Volunteer Units, who thanks to timely measures were protected against the small-pox epidemic. According to information received from Japan a number of small-pox cases were also noted

---

[3] The allegations are also dealt with in Volume I.
[4] The transcription of Korean and Chinese geographical names is given as in the United Nations documents.

there in the month of January, the infection having been brought in by US Army men who had taken part in the battles in North Korea and had been infected with small-pox as a result of the use of bacteriological weapons by their units. [11]

On 22 February 1952, the Minister issued an official statement, subsequently transmitted to the United Nations Secretariat, on the continued use of bacteriological weapons in Korea. He alleged that US troops had dropped large quantities of infected insects which spread bacteria of infectious diseases. On 28 January 1952, the statement continued, in Nonsodon and Ensudon districts southeast of Ichkhon, enemy military aircraft dropped three types of insects never seen in Korea before the war, similar to black flies, fleas and bugs. On 29 January military aircraft scattered flies and bugs over Ichkon district. On 11 February large numbers of paper packets containing fleas, spiders, mosquitoes, ants, flies and other types of small insects were dropped on military positions in the Chorwon district. A large quantity of flies was dropped in the Sanoyan district, and fleas, flies, mosquitoes and other insects were dropped in the Pyongyang district. On 13 February, an aircraft dropped flies, mosquitoes, spiders, fleas and other small insects in the Kymkhu district. On 15 February various insects were dropped in the Pyongyang district. On 16 February similar insects were dropped in the vicinity of two villages, Khansu and Okenri, on the banks of the Pukkhanjgan River. On 17 February flies and fleas were dropped in the vicinity of Sansinri and Khasinri, north of Pyongyang. Bacteriological research established that the insects dropped on the positions of troops and behind the lines carried plague, cholera and other infectious diseases.

The US military were also accused of having tried out bacteriological weapons on the island of Kochzhedo on prisoners of war, members of the Korean People's Army and the Chinese People's Volunteers, in collaboration with the Japanese "bacteriological warfare criminals." [12]

On 24 February 1952, the Foreign Minister of the People's Republic of China supported the North Korean protest and called for action "to stop these crimes against humanity". On 8 March 1952 he stated that between 29 February and 5 March the US forces had sent sixty-eight formations of military aircraft, in a total of 448 sorties, to invade China's territorial air in the Northeast, to spread large quantities of germ-carrying insects at Fushun, Sinmin, Antung, Kwatien, Linkiang and other areas, and to bomb and strafe the Linkiang and Changtienhokow area. The details of these incidents, included in an official statement, were as follows:

1. On 29 February, US aircraft, in a total of 148 sorties in fourteen formations, flew over Antung, Fushun and Fengcheng and spread insects over Fushun. Upon investigation on the spot insects of a black colour were

found within an area of fifteen to twenty kilometres in Fushun county covering Takow, Lijen and Fangsiao villages and Lientaowan.

2. On 1 March, US aircraft, in a total of eighty-six sorties in fourteen formations, flew over Fushun, Tatungkow, Changtienhokow, Kwantien and Chi An and spread insects of a black colour resembling fleas over Makinchwang and other places in Fushun county. Of these planes, eight in one group strafed a point five kilometres northeast of Changtienhokow.

3. In March,[5] US planes, in a total of seventy-two sorties in twelve formations, flew over Fushun, Antung, Tatungkow, Changtienhokow, Kiuliencheng, Chi An, Kwantien and Changpai. They dropped large quantities of flies, mosquitoes, fleas and other types of insects over Takow and other parts of Fushun county and areas between Fushun and Mukden.

4. On 3 March, five formations of US aircraft, in a total of thirty-two sorties, intruded into and spread insects over Antung, Langtow and Chi An.

5. On 4 March, thirteen formations of US aircraft in a total of seventy-two sorties, intruded and spread insects over Antung, Langtow, Tatungkow, Kiuliencheng, Changtienhokow, Sinmin, Chi An, Hungkiangkow and Kwantien. At 11 a.m. of the same morning, six US aircraft were observed over Langtow. They dropped from a height of 5 000 metres two cloth receptacles which burst open some 2 000 metres from the ground and then a swarm of flies was found near the highway. At two o'clock in the afternoon, a US aircraft was observed over Paikipao and Yaoyangho in Sinmin county. It dropped a load of flies. On the same day, US aircraft were active over Dwatien, and afterwards flies, mosquitoes, crickets and fleas dropped by US aircraft were immediately found east of Kwantien city and at Hungsheklatze.

6. On 5 March, ten formations of US aircraft, in a total of thirty-eight sorties flew over Antung, Anpingho, Changtienhokow, Hunkiangkow, Tunghua and Linkiang. Of these, one group of eight planes at about 8 a.m. indiscriminately bombed and strafed Lindiang, wounding two inhabitants and destroying five houses.

The Chinese Foreign Minister warned that members of the US Air Force who invaded China's territorial air space and used bacteriological weapons would be dealt with as war criminals on capture. [12]

Two outside groups produced reports on the allegations. These were circulated to the Security Council. The first group was a commission of the International Association of Democratic Lawyers.[6] On 31 March 1952, it

---

[5] The exact date was not indicated.

[6] The composition of this commission was as follows: Heinrich Brandweiner, Professor of International Law, University of Graz, Austria, President; Luigi Cavalieri, Advocate at the Supreme Court of Rome, Italy, Vice-President; Jack Gaster, solicitor, London,

issued a *Report on U.S. Crimes in Korea.* The report stated that the United States had used chemical warfare as well as biological warfare. On 2 April 1952 the same commission published a *Report on the Use of Bacterial Weapons in Chinese Territory by the Armed Forces of the United States.* The second group was an "International Scientific Commission for the Investigation of the Facts Concerning Bacterial Warfare in Korea and China".[7] It was created under the aegis of the World Peace Council. It produced its report in October 1952. [14]

What was said about chemical warfare by the commission of lawyers in their first report was as follows:

On 6 May 1951, three B-29s attacked the city of Nampo with gas bombs. Four hundred and eighty persons died of suffocation and 647 others were affected. According to eye-witnesses' reports, the gas spread immediately after the explosion of the bombs which produced a smoke, first black, then green-yellow, and yellow, and then colourless; the smoke had a disagreeable smell, resembling the smell of chlorine; its poisonous effect lasted until it evaporated within about two hours; people in the shelters (including many children) were particularly affected. The symptoms of the victims: difficulty in breathing, hoarse voice, fainting, coughing, watering of the eyes, running nose, headache, feeling of exhaustion, the skin feeling hot, vomiting, spitting foam and blood, fever, cyanosis, feeble pulse, symptoms of acute bronchitis and photophobia. The blood showed considerable augmentation of white and red corpuscles and of haemoglobin. Post-mortem examination showed an augmentation of volume and weight of the lungs which carried on the periphery marks of the ribs; a red-black liquid mixed with exudates flowed out of the parenchyma. The surface of the bronchial tubes became clear grey in colour and came away easily. The kidneys and the heart showed

---

United Kingdom; Marc Jacquier, advocate at the Court of Appeal, Paris, France; Ko Po-nien, Director of the Research Department of People's Institute of Foreign Affairs, Peking, China; Marie-Louise Moerens, advocate, Brussels, Belgium; Letelba Rodrigues de Britto, advocate, Rio de Janeiro, Brazil; Zofja Wasilkowska, judge of the Supreme Court, Warsaw, Poland.

[7] The composition of this commission was as follows: Dr Andrea Andreen, Director of the Central Clinical Laboratory of the hospitals board of the city of Stockholm, Sweden; Jean Malteree, *Ingénieur-Agricole,* Director of the Laboratory of Aninal Physiology, National College of Agriculture, Grignon, France; Dr Joseph Needham, F.R.S., Sir William Dunn Reader in Biochemistry, University of Cambridge, United Kingdom; Dr. Oliviero Olivo, Professor of Human Anatomy in the Faculty of Medicine of the University of Bologna, Italy; Dr Samuel B. Pessoa, Professor of Parasitology at the University of São Paulo, Brazil; Dr N. N. Zhukov-Verezhnikov, Professor of Bacteriology at, and Vice-President of, the Soviet Academy of Medicine, USSR; Dr. Franco Graziosi, Assistant in the Institute of Microbiology, University of Rome, Italy; Dr. Tsien San-Tsiang, Director of the Institute of Modern Physics, Academia Sinica (Chinese National Academy), China.

signs of loss of blood and points caused by dilation of the capillary structure; the meninges were smoothened and their structures dilated; the spinal marrow which was dissected showed ecchymoses in the white matter.

In the affected area of the city grass became yellow-brown, objects containing an alloy of copper became blue green and rings of silver became black.

On 6 July 1951, two US jets spread gas or some other unidentified chemical product over an area of about 100 to 200 metres east of the village of Poong-Po Ri (south of Wong San). Shortly after the planes had passed, two farmers felt an itching on the exposed parts of the body (face, hands and feet); they observed red spots growing to a size of haricot beans, which then swelled and filled with pus. The injuries resembled second degree burns but with much more serious erosive action and took a longer time to recover. While the victims were in hospital, the blisters containing the pus or liquid broke and the dead skin fell off after the application of Pomatum, leaving light scars.

Where the gas had spread, 10 per cent of agricultural plantations and more, especially the leaves of beans, showed round white spots which were about 1 cm apart, similar to the red spots found on the exposed parts of the victims' bodies. The witnesses testified that there had been no sound of explosion of bombs or machine-gunning.

On 1 August 1951, two bombs were dropped on the villages of Yen Seug Ri and Won Chol Ri (Province of Hwanghai). They broke in the air with an unusually feeble sound and produced a black smoke; a yellow-green cloud spread on the ground. Four civilians were killed and forty poisoned with symptoms and results similar to those found during the bombardment of Nampo. The leaves of trees fell, cereals on the ground were damaged; brass objects became black.

On 9 January 1952, the village of Hak Seng north of Won San was bombarded by two planes. After bombardment, eighty-three persons were poisoned and showed symptoms similar to those found after the Nampo bombardment. They had, in addition, itching in the throat and felt a sweet taste in the mouth. [13]

What both groups had to say about biological warfare is described in Volume V in connection with the problems of verifying allegations of use of BW.

On 5 March 1953, the head of the Soviet delegation to the United Nations forwarded to the President of the UN General Assembly statements by two captured United States Air Force officers who described the bacterial warfare in Korea. [15]

By a cablegram of 30 March 1953, the North Korean acting Minister of Foreign Affairs transmitted to the UN Security Council an investigation report of a commission of the Central Committee of the United Democratic Fatherland Front, which in the chapter "Use of Weapons of Mass Destruction" listed some new cases of bacteriological warfare, relating to the last months of 1952 and early 1953. [16]

III. Discussion in the UN Disarmament Commission of bacteriological and chemical warfare in the light of Korean and Chinese allegations

On 14 March 1952, the representative of the USSR in the Disarmament Commission suggested that, in view of "such latest facts as the use by the American forces in Korea and China of bacterial weapons", the Commission should consider without delay the question of the violation of the prohibition of bacterial warfare. The Soviet plan of work for the Disarmament Commission, submitted on 19 March 1952, included: "Consideration of the question of violations of the ban on bacterial warfare, the question of the impermissibility of the use of bacterial weapons and the question of calling to account those who violate the ban on bacterial warfare". The Soviet delegate recalled that in July 1949, at the Geneva Conference on the protection of war victims, the US government had rejected a proposal for a declaration that the use of bacterial and chemical substances, atomic or any other weapons of mass destruction was incompatible with the elementary principles of international law. He referred to reports that the United States was preparing for bacterial warfare, and that Japanese war criminals who had used bacterial weapons during the Second World War, were working in the US Army as consultants and experts. [17]

The US delegate stated that no bacterial warfare was being waged by the United Nations in Korea, and asked that the Soviet government exercise its good offices to prevail on the Chinese and North Korean authorities to accept the proposal of the International Committee of the Red Cross for an impartial investigation of the charges. [18]

The Chinese (Taiwan) delegate pointed out that China had suffered from epidemics from time to time throughout centuries. Under the present conditions, in particular, it was natural that epidemics should break out and the United Nations Command could not be blamed for that. [19]

The representative of the Netherlands said that the charge raised by the USSR did not belong in the Disarmament Commission which was a technical body instructed to prepare draft treaties. [19]

The British delegate also deplored that the charge should have been made in the Disarmament Commission which was not competent to discuss and even less to investigate any such alleged incidents. He reminded the Commission that the failure of the potato crop in Czechoslovakia had also been ascribed to the Americans who, it was asserted, had dropped potato bugs from aeroplanes on the fields of that country. [20]

The Brazilian representative remarked that if the USSR really had a case, it should avail itself of normal channels and take the matter to the competent organs of the United Nations. [20]

The delegate of Canada considered willingness to accept impartial investigation as a crucial test of sincerity in this matter. [20]

With regard to the proposal that investigation into the charges of bacterial warfare in Korea be conducted by the International Committee of the Red Cross, the USSR took the following position: the International Red Cross Committee is not competent to settle such problems; it is not an international organization but a Swiss national organization and as such cannot act objectively and without bias; it compromised itself in the past by protecting fascist war criminals. The investigation which had already been carried out by an authoritative commission confirmed the facts of use of bacterial weapons by US troops. [21]

The Soviet draft plan of work for the Disarmament Commission calling, among others, for consideration of the question of violations of the ban on bacterial warfare, was rejected on 28 March 1952. [22]

The Commission adopted a programme of work which included: "Elimination of weapons of mass destruction and control with a view to ensuring their elimination." [23]

On 28 May 1952, the US representative on the Disarmament Commission set forth the elements of his government's position on the general subject of bacterial warfare. He said that for various reasons, some of which were not at all connected with the substance of the 1925 Geneva Protocol, the United States Senate had not considered the document to be the best way of meeting the problem of gas warfare. As to the USSR, continued the US delegate, it had not agreed to stop manufacturing weapons for gas warfare or bacterial warfare. It had not even promised not to use gas and bacterial warfare; it merely promised not to use such weapons first. Bacterial weapons could be eliminated only if certain states were willing to establish an effective system of safeguards. The safeguards connected with bacterial weapons would differ somewhat from those connected with atomic energy, and also from those connected with other types of non-atomic weapons, in that different materials and techniques would be involved. The first and all-important safeguard against bacterial warfare was an open world, where no state could develop the military strength necessary for aggression without other states having ample warning and ample opportunity to protect themselves. Other feasible safeguards, such as industrial safeguards, would be desirable and could be developed. As long as the present situation prevailed "we must—as all other nations must—continue to be prepared to defend ourselves". He added that it was certainly true that at least all great powers were capable of launching bacterial warfare attacks. [24]

The Soviet delegate noted that the USA had not officially condemned the use of bacterial weapons and had not declared that it would not be the first to use these weapons. [24]

The French representative said that although France had ratified the Geneva Protocol of 1925, subject to reservation that any future enemies should also have renounced the employment of such weapons, and should not violate their undertakings and treacherously resort to them in the course of hostilities, it recognized that the Protocol had weaknesses, since it instituted no form of control. France believed that the prohibition of weapons of this type should be universal, and that it should at the earliest possible date be brought under a system of international control as strict as can be devised in matters of this kind. The measures required would include publication and verification of laboratory research, control of media of bacterial dissemination and emission, and also, as far as possible, control of the most dangerous special chemical products capable of being used to produce the most destructive type of emission. [25]

On 15 August 1952, the US representative reverted to the question of the elimination of germ warfare. He emphasized that the matter must be included as an essential part of a comprehensive and balanced disarmament programme and could not be satisfactorily dealt with as a separate or isolated problem. It would be a mistake to assume that, because the United States had not ratified the Geneva Protocol, it was opposed to the effective outlawing of poisonous gas and biological weapons directed against human beings. The USA was not unmindful of the fact that the ratification of the Geneva Protocol by forty-two states was a significant manifestation of mankind's desire, which the USA shared, to see these weapons, along with all other weapons adaptable to mass destruction, banned from national armaments. When the Geneva Protocol was submitted to the Senate for ratification in the 1920s, the United States was retreating rapidly into isolationism and neutralism and feared any involvement with the League and any treaties originating from Geneva. The intervening events had demonstrated the Protocol to be inadequate and ineffective to achieve its objective. The hope entertained at the end of the First World War that states could rely upon treaty promises and treaty declarations without safeguards to ensure their observance had turned out to be illusory. Mussolini was no more deterred from using poisonous gas in Ethiopia in the 1930s by the Geneva Protocol which Italy had accepted, than was Germany deterred from using poisonous gas in the First World War by the Hague Conventions which Germany had accepted. It was shocking to hear the Soviet representative in the Security Council (see below) suggest that Hitler was deterred from using poisonous gas and bacterial warfare in the Second World War by the Geneva Protocol. Would the men who consigned defenceless women, old men and children to gas chambers have respected the Geneva Protocol, save for their fears of reprisals, asked the US delegate.[8] He then pointed out that the USSR and other states which ratified the Geneva Protocol reserved the right to employ poisonous gas and germ warfare in reprisal. In view of the proved inadequacies of the Protocol, the USA did not criticize the USSR or other states parties to the Protocol for conducting research and making other preparations for the use of poisonous gas and bacterial warfare as precautionary measures. But it did criticize the USSR for attacking the USA for taking the same precautionary measures and for making false charges that the USA was using bacterial warfare in Korea. The USA believed, as the Soviet delegation had maintained in 1932 at the Conference for the Reduction and Limitation of Armaments, that paramount importance should be attached not to the prohibition of chemical weapons in wartime, but to the prohibition of preparation for chemical warfare in peacetime, and that efforts should be directed not so much to the framing of laws and usages of

---

[8] On 8 June, 1943, the President of the USA made the following statement:

"From time to time since the present war began there have been reports that one or more of the Axis powers were seriously contemplating use of poisonous or noxious gases or other inhumane devices of warfare.

"I have been loath to believe that any nation, even our present enemies, could or would be willing to loose upon mankind such terrible and inhumane weapons. However, evidence that the Axis powers are making significant preparations indicative of such an intention is being reported with increasing frequency from a variety of sources.

"Use of such weapons has been outlawed by the general opinion of civilized mankind. This country has not used them, and I hope that we never will be compelled to use them. I state categorically that we shall under no circumstances resort to the use of such weapons unless they are first used by our enemies.

"As President of the United States and as Commander-in-Chief of the American armed forces, I want to make clear beyond all doubt to any of our enemies contemplating a resort to such desperate and barbarous methods that acts of this nature committed against any one of the United Nations will be regarded as having been committed against the United States itself and will be treated accordingly. We promise to any perpetrators of such crimes full and swift retaliation in kind, and I feel obliged now to warn the Axis armies and the Axis peoples, in Europe and in Asia, that the terrible consequences of any use of these inhumane methods on their part will be brought down swiftly and surely upon their own heads. Any use of gas by any Axis power, therefore, will immediately be followed by the fullest possible retaliation upon munition centers, seaports, and other military objectives throughout the whole extent of the territory of such Axis country." (*Documents on American Foreign Relations*, volume V, Boston: World Peace Foundation, 1944.)

war as to the prohibition of as many lethal substances and appliances as possible.

But the USA did not intend, before such measures and safeguards had been agreed upon, to invite aggression by committing itself to would-be aggressors and Charter-breakers that it would not use certain weapons to suppress aggression. To do so in exchange for mere paper promises would be to give would-be aggressors their own choice of weapons. It was not the function of the Disarmament Commission to codify the laws of war, but if it attempted to do so, it would have to deal with the whole range of weapons and methods of warfare to be proscribed, the machinery necessary to secure the observance of the rules and the matter of sanctions, reprisals and retaliation in case of violation.

As to safeguards to ensure the elimination of bacterial warfare, the US delegate remarked that it might be true that there were no theoretically fool-proof safeguards which would prevent the concoction of some deadly germs in an apothecary's shop. But bacterial weapons, to be effective in warfare, required more than the dropping at random of a few infected spiders, flies or fleas. They required industrial establishments, facilities for maintaining the agents, transport containers and disseminating appliances. Such arrangements and facilities would not readily escape detection under an effective and continuous system of disclosure and verification of all armed forces and armaments which was a necessary prerequisite of any comprehensive disarmament programme. The US proposal was that, at appropriate stages in such a system, agreed measures should become effective providing for progressive curtailment of production, progressive dismantling of plants, and progressive destruction of stockpiles of bacterial weapons and related appliances. Under this proposal, with cooperation in good faith of the principal states concerned, all bacterial weapons, and all facilities and arrangements connected therewith, could by completely eliminated from national armaments and their use prohibited. [28]

On 18 August 1952, the representative of Greece quoted a resolution adopted on 7 August 1952 by the Eighteenth International Red Cross Conference at Toronto which, considering that bacterial weapons constituted the greatest danger to humanity, urged that all governments which had not as yet acceded to the Geneva Protocol of 17 June 1925, or ratified it, should do so in the shortest possible time, and without reservation. He interpreted the resolution to mean that the Geneva Protocol must be ratified and applied equally by all without reservation, for it would be absurd to ask a great power to ratify the Protocol now, without any reservation, when in the judgment of the most qualified experts it had become obsolete or, at any rate, had been bypassed by events, and at the same time to accept the legality of reservations made years ago by another great power. [27] (Greece had ratified the Protocol without reservations.)

In response to the US statement, the Soviet delegate re-submitted the proposal for the "consideration of the question of violations of the ban on bacterial warfare, the question of the impermissibility of the use of bacterial weapons and the question of calling to account those who violate the ban on bacterial warfare", and asked that it should be taken up by the Commission immediately. [28]

He considered the US representative's statement of 15 August 1952 as a belated attempt on the part of the US government to justify its refusal to ratify the 1925 Geneva Protocol. He compared the line taken by the USA with that of Japan which had abstained from ratifying the Protocol in order to be free to use the bacterial weapon. At the trial of the Japanese war criminals in Khabarovsk in December 1949, continued the Soviet delegate, it was established on the basis of documentary evidence that the Japanese had for many years been preparing for bacterial warfare against the USSR, China, the USA and the UK. It was also established that the Japanese had actually used bacterial weapons against the Mongolian People's Republic, the USSR and China. In the areas where bacterial weapons were used, epidemics of plague and typhus broke out.[9]

Further, the representative of the USSR questioned the reasons given by the USA for not having ratified the Protocol. The reference to the alleged "retreat into isolationism" was, in his view, out of place. The US owners of the large chemical firms and monopolies producing poison gases and lethal substances were the principal opponents of ratification. It was opposed by the American Legion, the leaders of which were reported to have been in the pay of the owners of the chemical firms of that time. It was also opposed by the US militarists on the ground that "no other state or people could be trusted" and that with the help of the weapons of mass destruction prohibited by the Geneva Protocol more people could be killed at less cost, while preserving the material property belonging to those people.

The Soviet delegate also challenged the assertion that the absence of control prevented the United States from ratifying the Geneva Protocol. He quoted the League of Nations Special Committee's opinion that the super-

---

[9] In December 1949, Japanese military prisoners, including the former commander-in-chief of the Kwantung Army, appeared before a military tribunal at Khabarovsk, USSR, on charges of having prepared and employed bacteriological weapons. The proceedings of the trial were published in 1950. The trial record contains charges concerning the Japanese production and test facilities at Harbin, as well as the use of Chinese and Soviet prisoners as subjects in BW experiments, resulting in thousands of fatalities. Some sections deal with the use of bacteriological weapons in China, USSR and Mongolia. (*Materials on the Trial of Former Servicemen of the Japanese Army Charged with Manufacturing and Employing Bacteriological Weapons.* Moscow, 1950.)

vision of preparation for bacterial warfare would never be complete, and therefore always ineffectual.

As to the International Red Cross Conference, the resolution of which had been read out and interpreted by the Greek representative, the delegate of the USSR pointed out that the essence of that resolution was to call upon the governments of all states which had not yet ratified the Geneva Protocol to do so, and it was self-evident that the appeal was directed at the US government. [29]

On 27 August 1952, the Soviet proposal was rejected. The delegations of Chile, France and Turkey then proposed to amend the programme of work, which had been adopted previously, by the addition of the words "including bacterial weapons" after the words "weapons of mass destruction", so as to read: "Elimination of weapons of mass destruction including bacterial weapons, and control with a view to ensuring their elimination." [30]

This proposal was adopted with the USSR abstaining.

IV. Discussion in the UN Security Council of bacteriological and chemical warfare in the light of Korean and Chinese allegations

On 18 June 1952, following a request by the USSR, the UN Security Council included in its agenda the question of an appeal to all states which had not yet done so, to accede to and ratify the Geneva Protocol of 1925. [31]

The Soviet representative said that the 1925 Geneva Protocol had been of outstanding importance in the history of international relations and that the political, legal and moral obligations assumed by states under this international agreement had proved an effective restraining influence on the aggressive states. Not a single one of those aggressive states, continued the Soviet delegate, dared to ignore the importance of the Protocol prohibiting the use of chemical and bacterial weapons in war. He observed that there was some difference of opinion among statesmen and public figures in various countries as to the admissibility of using bacterial weapons. This circumstance, and the fact that in a number of countries preparations were made for bacterial warfare creating a threat to international peace and security, made it imperative that the United Nations should adopt appropriate measures to prevent the use of such weapons.

The representative of the USA referred to the campaign being waged by the USSR concerning the use of bacterial warfare in Korea, which he qualified as false and malicious. In appraising the merits of the proposal regarding the Geneva Protocol, it was absolutely essential, continued the US delegate, to keep in mind the motive of those who made that proposal. He recalled that the USA had signed but had not ratified the 1925 Protocol for reasons which might be of interest to a historian of US attitudes of that period, but which were not relevant to a consideration of the problem today. In 1947 the President of the United States had withdrawn the Geneva Protocol from the Senate calendar, along with eighteen other treaties which had become just as obsolete as the Geneva Protocol. The world had moved since 1925, and the question of ratification had to be viewed in the light of new facts. One of those facts was that the USSR, in acceding to the Protocol, had made reservations according to which it felt free to use poison gases or germ weapons against any state which had not ratified the Protocol, and against any state which it decided to label an enemy, and which it declared had used these weapons.[10] By charging the United Nations Command with the use of bacterial weapons, the USSR had set the stage for using the weapons itself, if it should decide to declare that the states resisting aggression in Korea were its enemies. The US delegate characterized the Soviet draft resolution as a fraud, and said that the world was concerned not about the announced intentions of states, but about the known abilities of states.

The USA had never used germ warfare in the Second World War or at any other time. It had not used and was not using germ warfare of any kind in Korea. The USA stood ready to eliminate weapons of mass destruction through the establishment of an effective system based upon effective safeguards so that their use would indeed be impossible. But it was unwilling to place reliance solely upon paper promises which permitted the stockpiling of unlimited quantities of germ warfare or other weapons.

In the course of the discussion, the Greek representative said that the propaganda which had preceded the formulation of the appeal and the all-too-apparent designs of its sponsors could not but confirm the opinion that the Geneva Protocol was obsolete and outstripped by subsequent events.

---

[10] The Soviet reservations to the Geneva Protocol are as follows:
"(1) The said Protocol only binds the Government of the Union of Soviet Socialist Republics in relation to the States which have signed and ratified or which have definitely acceded to the Protocol. (2) The said Protocol shall cease to be binding on the Government of the Union of Soviet Socialist Republics in regard to any enemy State whose armed forces or whose allies *de jure* or in fact do not respect the prohibitions which are the object of this Protocol."
Many other parties to the Geneva Protocol made similar reservations.
On 22 February, 1938, the Soviet War Commissar stated:
"Ten years ago or more, the Soviet Union signed a convention abolishing the use of poison gas and bacteriological warfare. To that we still adhere, but if our enemies use such methods against us, I tell you what, we are prepared—and fully prepared—to use them also and to use them against aggressors on their own soil." (*New York Times*, 23 February 1938).

He disagreed with the Soviet thesis that the Geneva Protocol had served during the Second World War as a deterrent against the use of bacterial warfare by the aggressors. If Germany and Japan did not resort to bacterial and chemical weapons, it was because they were fully aware of the preparations of the democratic powers in that field. [32]

The representative of the Netherlands was willing to see what could be done to improve on the 1925 Geneva Protocol, and was prepared to cooperate in the Disarmament Commission in designing effective methods to eliminate bacterial warfare. [32]

The Brazilian representative said that efforts should be directed towards finding a more adequate instrument than the Geneva Protocol for doing away with those means of mass destruction, and not towards trying to utilize again an international instrument that had lost its usefulness. [32]

The British representative, without suggesting that the 1925 Geneva Protocol had been of no value, expressed the view that the USSR greatly exaggerated its actual influence on events during the preceding twenty-five years and particularly during the Second World War. The value of the Protocol and indeed of any declaration banning or limiting the use of any particular weapon or method of warfare, must rest entirely on the good faith of the governments which were parties to it, so long as the declaration was not supported by any system of control which would ensure that its provisions were duly carried out. In a case of aggression—continued the British delegate—the aggressor was not likely to be over-scrupulous and it would be very foolish to suppose that he would observe any pledges which he might have undertaken if he thought he could profitably violate them. Having violated the principal obligation under the Charter not to aggress, why should he not violate any other obligation? This was surely the case with the Nazi government during the last war and Premier Stalin himself obviously did not suppose that Hitler would be restrained by the mere fact that Germany was a party to the Geneva Protocol.[11]

[11] In a letter to Premier Stalin, on 20 March 1942, Prime Minister Churchill wrote:
"Ambassador Maisky lunched with me last week, and mentioned some evidences that the Germans may use gas upon you in their attempted spring offensive. After consulting my colleagues and the Chiefs of Staff, I wish to assure you that His Majesty's Government will treat any use of this weapon of poison gas against Russia exactly as if it was directed against ourselves. I have been building up an immense store of gas bombs for discharge from aircraft, and we shall not hesitate to use these over all suitable objectives in Western Germany from the moment that your armies and people are assaulted in this way.

"It is a question to be considered whether at the right time we should not give a public warning that such is our resolve, as the warning might deter the Germans from adding this new horror to the many they have loosed upon the world. Please let me know what you think about this, and whether the evidence of German preparations warrants the warning.

"There is no immediate hurry, and before I take a step which may draw upon our

The United Kingdom intended to observe the Geneva Protocol most scrupulously, though it would do its best to improve upon it in so far as it could, for example, by working out plans for the control of bacterial warfare in the Disarmament Commission. [32]

The representative of China (Taiwan) said that the difficulty with the Geneva Protocol was that it provided for prohibition without guarantees, safeguards or controls. It was a defective instrument. Subscribing states tried to protect themselves with reservations; the reservations in turn weakened the instrument. Thus, for example, according to one type of reservation, the signatory state was free to use germ warfare if its enemy used germ warfare; a unilateral charge or accusation would automatically absolve a signatory state from the obligations of the Protocol. [33]

The USSR again stressed the force of the obligations which the Protocol placed on States and said that "... The existence of the Protocol and the fact that the United Kingdom government had ratified it made it impossible for Mr Churchill to use any chemical weapons at all, inasmuch as he and the United Kingdom government were bound by its provisions. Had it not been for the Protocol, it is not certain that Mr Churchill would not have decided to drop chemical bombs on Germany. But the obligations arising out of the Geneva Protocol compelled the United Kingdom government to ob-

---

cities this new form of attack I must of course have ample time to bring all our anti-gas precautions to extreme readiness."

In his reply of 30 March 1942, Premier Stalin said:

"I wish to express to you the Soviet Government's gratitude for the assurance that the British Government will look upon any use by the Germans of poison gas against the USSR in the same light as if this weapon had been used against Great Britain, (and) that the British Air Force will immediately use against suitable objectives in Germany the large stocks of gas bombs held in England."

In a letter of 1 April 1942, Prime Minister Churchill wrote to President Roosevelt:

"Having heard from Stalin that he was expecting the Germans would use gas on him, I have assured him that we shall treat any such outrage as if directed upon us, and will retaliate without limit. This we are in a good position to do. I propose, at his desire, to announce this toward the end of the present month, and we are using the interval to work up our own precautions. Please let all the above be absolutely between ourselves." (Winston S. Churchill, *The Second World War*, volume 4).

On May 11, 1942, Prime Minister Churchill stated:

"The Soviet Government have expressed to us the view that the Germans, in the desperation of their assault, may make use of poison gas against the armies and peoples of Russia. We are, ourselves, firmly resolved not to use this odious weapon unless it is used first by the Germans. Knowing our Hun, however, we have not neglected to make preparations on a formidable scale. I wish now to make it plain that we shall treat the unprovoked use of poison gas against our Russian ally exactly as if it were used against ourselves and if we are satisfied that this new outrage has been committed by Hitler, we will use our great and growing air superiority in the West to carry gas warfare on the largest possible scale far and wide against military objectives in Germany.

"It is thus for Hitler to choose whether he wishes to add this additional horror to aerial warfare." (*New York Times*, 11 May 1942.)

serve it and merely to resort to the legitimate reservation it had made on signing the Protocol to warn Hitler not to use chemical weapons. Such was the binding power of the Geneva Protocol. ... The fact that Mr Churchill, with, in his own words, an enormous stock of chemical bombs at his disposal, did not make use of them at a time of bitter conflict, shows that the Geneva Protocol acted as a deterrent, prevented the launching of bacterial and chemical warfare at that time, and saved millions upon millions of lives." [35]

The French representative recalled that France had made its ratification of the 1925 Geneva Protocol subject to reservations which in no way weakened the effectiveness of the engagements undertaken. The reservations merely constituted an explicit formulation of the implicit conditions for carrying out any bilateral or multilateral engagements entered into in good faith. Such engagements could have validity only in relation to contracting parties which themselves respected the engagements. A violation by one of the signatories set free the other signatories from the obligations which they may have contracted in relation to that signatory. That was an elementary principle of private as well as of international law. In taking cognizance of this principle in the form of reservations, the French government, like the other states which had acted in the same way, was merely doing what regard for clarity and honesty suggested. The reasons which had led France to sign and ratify the Geneva Protocol remained valid. The Protocol was the culmination of a long period of work in which France played a leading role, and at the same time a confirmation of the principles and ideas to which the French people remained deeply attached. France did not regard it as out-of-date or old-fashioned; it had retained all its legal value and moral authority. Although perhaps an effort should be made to improve certain points and to strengthen its practical effectiveness by merging it in a wider system for the control and abolition of weapons of mass destruction, pending the achievement of this desirable result, the Geneva Protocol remained the chief international instrument which could, if respected, strip war of some of its more barbarous aspects. Its provisions were as binding in 1952 as they had been in 1925 on those states which had signed and ratified it, and the states which abstained from doing so had never challenged its principles or disputed its moral value.

However, the French delegation did not feel able to call upon states which had not yet done so to ratify the Protocol or accede to it, unless they were at the same time offered certain guarantees that they would not be immediately and insultingly accused of violating it, without any means of recourse whereby they could defend themselves against such accusations. [34]

The representative of Pakistan thought that the 1925 Protocol, at the time when it was signed, had been probably the best thing that could be contrived in order to keep certain awful aspects of war in abeyance. But it was not a protocol to end bacterial warfare or to end poison gas warfare. It was a protocol in which retaliation and reprisals were regulated. In this context the Pakistani delegate recalled that the Protocol had not stopped Italy from visiting horrors upon the Ethiopian people, while both states were parties to it without any reservations. Therefore, although the USA— and certain other states—may not have ratified the Protocol at some time for reasons best known to themselves, today it was useless to ask them to ratify it. For hundreds of millions of people this would not be enough. Those nations which for a long time to come could not physically be in a position to use bacteria or poison gases on other nations, could only be the victims of such things, and it would not satisfy them if the Protocol "were signed ten times over" because it had been seen to be broken; poison gas had been used and there was a possibility of bacteria being used in any major world conflict in the future, in spite of the Protocol. A settlement must be reached whereby the arms race would stop, the major powers would actually begin to share with the rest of the world the abhorrence of using arms of mass destruction, and the agreements arrived at would not be on paper only but would take a shape and form which would assure that there were practical checks on all the major powers, checks not only for their own safety but for the peace and safety of the rest of the world. [35]

On 26 June 1952 a vote was taken on the Soviet draft resolution appealing for the ratification of or accession to the 1925 Geneva Protocol by those states which had not yet done so. There was one vote in favour (USSR) with ten abstentions (Brazil, Chile, China [Taiwan], France, Greece, Netherlands, Pakistan, Turkey, UK, USA). The draft was thus not adopted, having failed to obtain the affirmative vote of seven members. [36]

In explaining his vote, the US representative pointed out that the USA had not ratified the Geneva Protocol because there was another effort in which it was engaged—to achieve genuine disarmament and genuine control of weapons of mass destruction, which would make it possible to eliminate these weapons. [36]

On 1 July 1952 the Security Council included in its agenda a US proposed item: "Question of a request for investigation of alleged bacterial warfare." [37] The Soviet proposal [38] to invite representatives of the People's Republic of China and the People's Democratic Republic of Korea to participate in the consideration of the item, having been rejected, the Soviet delegation decided not to take part in the discussion on the subject. [39]

In presenting his proposal, the United States representative recalled the history of the germ warfare charges, pointing out that the way they had been spread testified to their purely propagandistic nature.

He also recalled that on 4 March 1952, the US Secretary of State had stated that the charges were false, that the United Nations had not used and was not using any sort of bacterial warfare. Similar denials had been made by the Secretary-General of the United Nations, by the United Nations Commander-in-Chief, by the US Secretary of Defense, and by other officials of UN members, including those contributing forces in Korea. On 11 March the US Secretary of State had requested the International Committee of the Red Cross to determine the facts, emphasizing the need for an investigation on both sides of the battle lines in Korea, and issuing specific invitation to the Red Cross investigators to cover the areas behind the UN lines. The International Committee of the Red Cross agreed to set up a committee to make such an investigation, provided both parties agreed to it and offered their cooperation; the committee would consist of persons of moral and scientific independence, and would also include scientific experts proposed by Far Eastern countries not taking part in the conflict. The US Secretary of State accepted the offer of the International Red Cross.

Independent scientists, stressed the US delegate, including at least ten Nobel prize winners, had publicly expressed complete scepticism of the charges, such as the spreading of typhus and plague through the medium of infected fleas and lice in the freezing winter temperatures of Korea. They had pointed out that in that part of the world diseases such as typhus and plague may be expected to assume epidemic proportions unless the authorities were controlling their natural carriers. But the World Health Organization's offer to provide technical assistance in controlling the reported epidemics in North Korea, transmitted by the UN Secretary-General, was rejected.

In proposing an impartial investigation, the US representative expressed confidence that any such investigation would wreck the germ warfare campaign. But if the investigation was rejected, he added, the campaign would just as surely be wrecked, for it would be a confession that the charges would not bear the light of day. [39]

The delegate of the Republic of China (Taiwan) said that during the Japanese occupation there had been large Japanese laboratories in Manchuria devoted to experimentation in connection with bacterial warfare. Japan had assembled many of its scientists there for that purpose. After the war, the USSR brought to trial a number of person as war criminals charged with having worked on germ warfare, most of whom were military people, not scientists or doctors of medicine. According to a private communication received by the Chinese delegate, these Japanese scientists had been working in cooperation with Soviet and Chinese scientists to conduct further experiments and certain counties in the northeast provinces of China had been used as areas for controlled experimentation; some of that control had failed, and, as a result, man-made epidemics spread. [40]

On 3 July 1952, a vote was taken on a draft resolution, submitted by the USA, which requested the International Committee of the Red Cross, with the aid of such scientists of international reputation and such other experts as it may select, to investigate the charges and to report the results to the Security Council as soon as possible; called upon all governments and authorities concerned to accord to the International Committee of the Red Cross full cooperation, including the right of entry to, and free movement in, such areas as the Committee may deem necessary in the performance of its task; requested the Secretary-General to furnish the Committee with such assistance and facilities as it might require.

There were ten votes in favour and one against. The draft was not adopted, the opposing vote being that of a permanent member of the Security Council (USSR). The US representative then submitted another draft resolution, by which the Security Council would conclude, from the refusal of the governments and authorities making the charges to permit impartial investigation, that these charges must be presumed to be without substance and false; and would condemn the practice of fabricating and disseminating such false charges. There were nine votes in favour of the draft, one abstention (Pakistan) and one against. Since the vote against was that of the USSR, this draft resolution was not adopted either. [42]

On 13 July 1952, the Minister of Foreign Affairs of the People's Republic of China notified the members of the United Nations and the Security Council that his government had decided to recognize the Geneva Protocol of 17 June 1925, acceded to in the name of China on 7 August 1929. [43]

V. Discussion in the UN General Assembly of bacteriological and chemical warfare in the light of Korean and Chinese allegations

On 8 April 1953, the UN General Assembly adopted a resolution by which it requested the Disarmament Commission to continue its work for the development by the United Nations of comprehensive and coordinated plans providing, among other things, for "the elimination and prohibition of all major weapons, including bacteriological, adaptable to mass destruction". The whole programme was to be carried out "under effective international control in such a way that no State would have cause to fear that its security was endangered".

The seventh session of the UN General Assembly also discussed, at the initiative of the USA, the "question of impartial investigation of charges of use by United Nations forces of bacteriological warfare". The Soviet proposal to invite the representatives of the People's Republic of China and the People's Democratic Republic of Korea to participate in the discussion was again rejected. [45]

The US delegate called the charge of use of bacteriological warfare by the United Nations forces a "big lie". He said that it was Stalin who had first invented the lie of bacteriological warfare in a speech on 7 January 1933, when he was attacking certain resistance elements of the Soviet population. Again, in the purge trials of 1937 and 1938, the charge of bacteriological warfare had been raised. One defendent was reported to have confessed to manufacturing virulent bacteria in order to destroy herds of swine, and another to having connived with Japanese intelligence services to infect the Red Army with highly virulent bacilli in the event of war.

In the campaign against the United Nations in Korea, continued the US representative, use was being made of the so-called investigations, carried out by different commissions. Thus, for example, the International Scientific Commission for the Investigation of the Facts Concerning Bacterial Warfare in Korea and China, composed of members who had formed their views before going to China, never cared to examine the evidence on the spot. One of its members, Dr Andrea Andreen, had stated upon her return to Sweden in 1952 that the Commission felt so sure of the integrity of their Chinese hosts that they entirely trusted statements which the Chinese made regarding US use of germ warfare. Another device was to extort confessions from prisoners by various techniques. Dealing with the case of alleged confessions by US military personnel, which the USSR delegation had circulated to the members of the General Assembly's First Committee, the US representative stated that they were totally false in their general assertions and in their specific allegations. He read out a statement of 25 March 1953 by General Omar N. Bradley, Chairman of the US Joint Chiefs of Staff, denying that the Joint Chiefs had ever made a plan for bacteriological warfare in Korea or had ever sent a directive to that effect to the Commanding General, Far East Command, and that any US military forces had engaged in bacteriological warfare. He also quoted statements by Major General C. F. Schilt—who had been Commanding General of the First Marine Air Wing in Korea from July 1951 to April 1952— and Major General Clayton C. Jerome—who had replaced Major General Schilt in that capacity and served up to January 1953—denying categorically all charges of bacteriological warfare in Korea and declaring that no orders concerning such warfare had ever been received by the First Marine Air Wing, which had been referred to in the alleged confessions as having dropped germ bombs in Korea, or issued to subordinate units of the Wing, and that no plans had ever been prepared for bacteriological warfare by that Wing.

In the draft resolution submitted by the USA and other nations whose forces were engaged in Korea, it was proposed that the General Assembly should call upon the governments and authorities concerned to cooperate with a commission to be established by the Assembly to conduct an impartial investigation of the charges that had been made. The proposed commission should be allowed to travel freely throughout such areas of North and South Korea, the Chinese mainland and Japan as it might think necessary in the performance of its task. It should also examine all prisoners of war who were alleged to have made confessions regarding the use of bacteriological warfare but, prior to their examination by the commission, those prisoners of war should be taken to a neutral area. The prisoners would remain under the responsibility and custody of the commission until the conclusion of hostilities, in order to preclude fear of reprisals. [46]

The Australian delegate quoted from a report furnished to him by Sir MacFarlane Burnet, F.R.S., an Australian scientist and microbiologist, in which the latter stated that, after examination of the evidence, he was quite certain that bacteriological warfare methods had never been used by the United Nations forces in Korea.

The representative of the Netherlands remarked that the governments of China and North Korea could easily hamper any impartial investigation by refusing access to their territory. [47]

The New Zealand delegate referred to an opinion expressed in August 1952 by the President of the New Zealand Association of Scientific Workers. That body had studied the reports of scientists and of the committees of investigation, all of which were in sympathy with the Chinese-Korean cause, and had noted a striking contrast between those documents and the statements of politicians and newspaper correspondents; while the latter referred to hundreds of incidents involving sixty different objects, the scientists mentioned only about a dozen cases involving about seven kinds of insects. Moreover, the evidence of the presence of strange insects was based entirely upon the statements of people of no scientific training. All the insects described belonged to species that were commonly found in that area. The allegations regarding the carriage of disease germs by those insects had been examined by various experts who had shown that the reports conflicted in several respects with all that was known of the transmission of disease by insects. The investigation reports stated that Western scientists had made great technical advances and had discovered means of infecting insects with disease germs; at the same time, however, the same persons were described as using methods of distribution which were not only inefficient to the point of absurdity, but had been an almost complete failure.

With regard to the report of a group of scientists selected by the World Peace Council, the New Zealand representative said that there was no indication that the group in question had ever found any tangible evidence in the places where bacterial weapons were alleged to have been used. The scientists claimed to have interrogated four prisoners of war who allegedly confessed to having participated in germ warfare, but there was no indication that they had been allowed to visit the camps and talk to any prisoner they chose. [47]

The delegate of the Republic of China (Taiwan) said that two scientists, Mr C. H. Curran of the American Museum of Natural History, and Mr René Dubos of the Rockefeller Institute, had studied photographs of bombs, insects and germs published on 15 March 1952 by the *Peking People's Daily,* and their conclusions appeared on 3 April 1952 in the *New York Times*: the photographs were a childish forgery; the bombs illustrated were leaflet bombs, and the germs would not be able to resist the air pressure when the bombs were dropped; the insects photographed were not capable of carrying germs; the germs shown did not correspond to the names given to them, and none of them could be transmitted by insects.

Three Canadian scientists, W. H. Brittain, A. W. Baker and C. E. Atwood, after having studied the documentation on bacteriological warfare, concluded that none of the evidence furnished could be regarded as scientific proof. As to the allegation that the insects discovered were out of season, the Canadian scientists believed that the existence of such insects and spiders at the beginning of spring in a temperate climate was quite normal. If the searchers had thought that the species in question were new, the reason was that whole communities had, for the first time, turned out to search for insects. As to conditioning insects to abnormally low temperatures, nothing of the kind had ever been accomplished. [47]

The representative of Belgium thought that the government in Peking had invented the charges to excuse the inadequacy of its medical services. [47]

The British delegate referred to the alleged confessions of two US prisoners of war. He said that they were far too circumstantial to carry conviction. According to them, hundreds, even thousands, of members of the US armed forces would have known if germ warfare was being used. It was quite inconceivable, regardless of security precautions, that no whisper of the use of bacterial weapons, if such weapons were being employed, should have come from any of those men, except of course those who were alleged to have made confessions after being taken prisoner. The confessions spoke of the horror and disgust shared by those who had been told about the use of germ weapons. It was quite incredible that all those men should have made no attempt either to protest or to make the facts known. [48]

The French delegate was of the opinion that nothing would contribute more to the strengthening of a climate of cooperation than if the accusers accepted verification of their charges. [48]

The representative of Indonesia felt that the only acceptable solution would be to set up a commission which had the endorsement of the accuser and the accused. The establishment of a commission by a majority of votes could not lead to a satisfactory outcome. [49]

The Czechoslovak representative expressed the view that the establishment of an organ to investigate the matter without the consent of the governments concerned would constitute a serious infringement of the sovereignty of the People's Republic of China and the People's Democratic Republic of Korea, as well as a violation of the United Nations Charter. [47]

The attitude of the Soviet Union was considerably softer than in the previous years. It was the time of intensive negotiations in Panmunjom, Korea, with a view to ending the hostilities. The representative of the USSR stated that the important issue now was the United States ratification of the Geneva Protocol of 1925. Such ratification would constitute the best proof that the USA did not intend to resort to the use of bacteriological weapons in the future. [49]

On 23 April 1953, the draft resolution sponsored by the USA and its allies was adopted by the General Assembly. According to the resolution, after the President of the General Assembly had received an indication from all the governments and authorities concerned of their acceptance of the proposed investigation, a commission composed of Brazil, Egypt, Pakistan, Sweden and Uruguay, would be set up and would carry out immediately an investigation of the charges that had been made. [50]

In opposing the resolution, the USSR said that an investigation had already been carried out by authoritative organizations and commissions; besides, the Korean-Chinese side should have participated in the discussion of whether any new investigation was desirable and necessary. [51]

In a draft resolution submitted on 9 April 1953, Poland called for the adoption of a decision on the unconditional prohibition of atomic weapons and other weapons of mass destruction and the establishment of strict international control over the observance of that decision by all states; it also called on all states, which had not acceded to or ratified the Geneva Protocol of 17 June 1925, to accede to or ratify that instrument. [52]

The USA responded by stating that, while the objectives of the Geneva Protocol were certainly praiseworthy, they were nevertheless inadequate, for not only did the Protocol fail to prohibit the manufacture and stockpiling of chemical and bacteriological weapons, but it made no provision for a system of controls. [53]

The representative of Ecuador said that there were two possible ways of eliminating the use of bacteriological weapons: to ratify the existing instrument (the Geneva Protocol) which, as had already been pointed out, was inadequate under present conditions, or to draft a new instrument for the same purpose. In adopting the resolution on disarmament, the Assembly would appear to have decided in favour of establishing a more up-to-date instrument. [53]

The delegate of Venezuela considered an appeal to states which had not yet ratified the Geneva Protocol of 1925 not to be a matter of extreme urgency. States were entitled to decide whether or not to assume certain obligations, and the refusal to ratify an international instrument did not imply a fixed intention to act in a manner contrary to the provisions of that instrument. [54]

The Korean War ended on 27 July 1953, by the signing of an Armistice Agreement. Nevertheless, at the eighth session of the UN General Assembly the US representative brought up again the question of impartial investigation of charges of use of bacterial warfare by United Nations forces. By then, the President of the General Assembly had reported that the USA, Japan and South Korea had accepted the terms of the UN resolution of 23 April 1953, and that there had been no reply from the Chinese People's Republic and North Korea who had made the charges. [55]

The United States also submitted sworn statements made by US airmen upon their repatriation denying that they had ever waged bacterial warfare, and declaring that their confessions had been false and had been extracted by coercive methods. [56]

The Soviet representative claimed that the attempts to prove that the prisoners' confessions had been secured under duress were without foundation. It was the USA that had extorted repudiations. He submitted a draft resolution inviting those states which had not yet ratified the Geneva Protocol of 1925 to do so, [57] and stressed that it was the duty of the United Nations to devise measures which would ensure that this major norm of international law should be safeguarded and affirmed. [56]

The delegate of the United Kingdom felt that the question of the elimination and prohibition of the use of bacterial weapons was a matter which the Disarmament Commission was competent to examine. [58]

The Australian delegate said that the USSR wished to use its draft resolution as an indirect implication of guilt on the part of those states which had not ratified the Geneva Protocol prohibiting the use of bacterial weapons. He considered it advisable that the First Committee of the General Assembly declare that the charges brought by the USSR were without foundation. [58]

The representative of Greece said that only if the USSR recognized the fallacy of its charges could its appeal for the ratification of the Protocol be received without suspicion. [59]

The French delegate pointed out that the documents of the commissions of inquiry selected by the accusers had failed to show any sign of critical spirit, scientific exactitude or logic, or even intellectual honesty. It had never been possible to establish a link of cause and effect between the appearance of certain aircraft and the presence in China of insects and rodents. The evidence was often second- or third-hand and nothing had been done to examine it seriously and verify the theories propounded. [60]

The Soviet draft resolution calling for the ratification of the Geneva Protocol was not put to the vote. The General Assembly decided to refer the draft to the Disarmament Commission for such consideration as was deemed appropriate under the plan of work and pursuant to the terms of reference of the Commission. [61]

On 28 November 1953, the General Assembly adopted a resolution in which it recognized the general wish and affirmed its earnest desire to reach agreement as early as possible on a comprehensive and coordinated plan, under international control, for the regulation, limitation and reduction of all armed forces and all armaments, for the elimination and prohibition of atomic, hydrogen, bacterial, chemical and all such other weapons of war and mass destruction, and for the attainment of these ends through effective measures. [62]

On 21 December 1953, the Soviet government issued a statement (subsequently included in the Disarmament Commission Official Records), [63] in connection with President Eisenhower's speech on atomic armaments to the United Nations General Assembly, of 8 December 1953. [64]

The statement contained the following passages:

Nearly thirty years ago the governments of 49 countries reached agreement and signed[12] the 1925 Geneva Protocol forbidding the use of chemical and bacteriological weapons and recognizing as a crime the use of such weapons of mass extermination of peoples. This agreement between those Governments, which was also signed at the time by the Soviet Union, has produced positive results.

Everyone knows that during the First World War wide use was made of such

---

[12] The number of signatories to the Geneva Protocol was actually thirty-eight. See page 342 for the list of signatories.

weapons for the mass extermination of peoples as asphyxiating and poisonous gases, and other forms of chemical weapons, the use of which met with the determined condemnation of the peoples. Even at that time there also existed the danger of the use of harmful bacteriological weapons designed to infect the civilian population of towns with very serious diseases; and the conscience of the overwhelming majority of people could not tolerate this. This was the factor which produced the need for an international agreement in the form of the above-mentioned Geneva Protocol, which condemned and prohibited the use of chemical and bacteriological weapons in warfare.

Had it not been for this Protocol, which was signed by 49 States though it has not yet been ratified by all these States, it is quite obvious that there would have been no restriction on the use of chemical and bacteriological weapons in the Second World War also. The fact that not a single Government engaged in the Second World War dared to use chemical and bacteriological weapons proves that the aforesaid agreement of the States against chemical and bacteriological weapons had positive significance. Nor, naturally, should one underestimate the circumstance that on the basis of this international agreement the countries of the anti-Hitlerite coalition declared with determination that attempts by the enemy to use chemical weapons in the war would meet with crushing reprisals.

The considerations which have been adduced also apply fully to atomic and hydrogen weapons. The United Nations, of course, does not include these weapons amongst conventional armaments but regards them as a special type of weapons—weapons of mass destruction.

VI. *Summary and comment*

Chemical and biological weapons were not employed on the battlefields of World War II. The use of atomic bombs in the final stage of the war overshadowed the importance of other weapons, and no debate in depth on CB disarmament was held during the first years after the war.

In 1947–48 the question of CBW was brought up in the United Nations in connection with discussions on the definition of weapons of mass destruction.

Different characteristics of weapons of mass destruction were given at different times: great destructive and annihilating power; most specifically offensive; favouring sudden attacks; affecting large areas; producing a surprise effect; causing unnecessary suffering of the combatants; directed specifically against civilians or most threatening to civilians; involving indiscriminate use; producing unpredictable results; uncontrollable as to consequences. The United Nations did not succeed in elaborating a comprehensive formula acceptable to all. Chemical and biological weapons were generally considered as belonging to the mass destruction category (along with atomic bombs), but the definition provided by the United States was restricted to the lethal kinds of those weapons. There was also some uncertainty as to whether other weapons, having a great destructive effect on civilian population and usually referred to as conventional, should not fall under the same category.

A suggestion made by the UN Secretary-General in 1948, to study the problems involved in the control of CBW, was not followed up.

The discussion of CBW became animated during the Korean War (1950–53), when the United States was charged with using bacteriological and chemical weapons in Korea and China. The accusations specified the names of the localities affected, the dates of attacks, the number of casualties and the damage caused; testimonies as well as findings of special enquiries were presented and widely circulated. The evidence produced was considered unconvincing by the majority of UN members. They argued that the investigating commissions were appointed by organizations which in advance sided with the accusers, and that the reports were based on second-hand information. The USA hotly refuted the charges as slanders; its proposals for impartial investigation were rejected by the accusing side.

The consideration of the allegations of use of chemical and bacteriological weapons, though inconclusive, produced some indirect results. It emphasized the force of the international customary law prohibiting the use of CB weapons; everyone, including the United States which was formally not bound by the 1925 Geneva Protocol, expressed repulsion at the idea that any country should use the banned weapons. It also highlighted the problems involved in the verification of allegations and demonstrated the need for an agreed machinery to conduct such verification.

The discussion shed some light on the attitude of various states towards the Geneva Protocol. In the heat of cold war polemics the great powers took rather extreme positions, rating the value of that instrument very high or very low.

As previously in the League of Nations, the USSR repeatedly called for universal adherence to the Geneva Protocol. However, draft resolutions to this effect were not accepted by the United Nations; the context in which they were presented may have implied the guilt of those states which had not ratified the Protocol, i.e., chiefly the United States. The USSR contended that the Protocol had fully proved its binding force by stopping the Germans from using poisonous gases during World War II.

The USA asserted that the Protocol was obsolete and ineffective and that the reservations attached to it by a number of states added to its weaknesses. The Protocol had not prevented the Italians from using gas in the 1930s, and if Germany had not resorted to chemical warfare it was only because of the Allies' warnings that immediate reprisals would be taken against it. (Reference was made, in this connection, to statements by Churchill and Roosevelt in 1942 and 1943.)

Even those who minimized the importance of the Geneva Protocol did not contest the principles contained therein.

It should be noted that assertions made in the UN that poisonous gas had not been used at all during the last war, were strictly speaking inaccurate. Gas was not used in combat, but it was employed by the Germans against the civilians of enemy countries in occupied territories and on a mass scale.

In the course of the debate the USSR often referred to the trial of the former servicemen of the Japanese army, who had been sentenced by a Soviet tribunal for manufacturing and employing biological weapons during World War II. Later, however, in 1953, the Soviet Union officially declared that not a single government engaged in the last war had dared to use CB weapons. The UN Secretary-General's report on CBW, issued in 1969, stated that there was no evidence that biological agents had ever been used as modern military weapons. (See chapter 9.)

In the 1950s, the US opposition to committing itself to a prohibition of use of chemical and bacteriological weapons by ratifying the Geneva Protocol was probably due, not so much to the intrinsic combat value of the weapons in question, as to the reluctance to create a precedent for a ban on the use of nuclear weapons, weapons which it regarded as an essential element in its military posture vis-à-vis the USSR. The USA insisted on solving the problem in conjunction with other disarmament measures, through disclosure and verification of all armed forces and armaments, followed by progressive reduction and elimination of prohibited weapons, under appropriate control and safeguards—a proposition hardly realistic at that time.

Chapter 8. 1954–1967

I. *CBW prohibition in Germany and Austria*

A number of restrictions in the field of armaments, which had been imposed upon Germany as a result of its defeat in World War II,[1] were gradually relieved. However, in the mid-fifties, when the Federal Republic of Germany was about to regain sovereignty and join Western military and political organizations, its future partners were anxious to receive guarantees that weapons of mass destruction and some other major arms would not be produced in West Germany. Renunciation of those weapons was considered an important condition for Germany's membership in the Western European Union and the North Atlantic Treaty Organization. It was carried out through a unilateral German declaration.

On 3 October 1954, the Chancellor of the Federal Republic of Germany announced: [64]

The Federal Republic undertakes not to manufacture in its territory any atomic weapons, chemical weapons or biological weapons;

The Federal Republic agrees to supervision by the competent authority of the Brussels Treaty Organization to ensure that these undertakings are observed.

Chemical and biological weapons were defined as follows:

Chemical weapons

(*a*) A chemical weapon is defined as any equipment or apparatus expressly designed to use, for military purposes, the asphyxiating, toxic, irritant, paralysant, growth-regulating, anti-lubricating or catalysing properties of any chemical substance.

(*b*) Subject to the provisions of paragraph (*c*), chemical substances, having such properties and capable of being used in the equipment or apparatus referred to in paragraph (*a*), shall be deemed to be included in this definition.

(*c*) Such apparatus and such quantities of the chemical substances as are referred to in paragraphs (*a*) and (*b*) which do not exceed peaceful civilian requirements shall be deemed to be excluded from this definition.

---

[1] At the Tripartite Conference of 1945, the Heads of Government of the UK, USSR and the USA decided that the production of metals, chemicals, machinery and other items that are directly necessary to a war economy should be "rigidly controlled and restricted to Germany's approved post-war peacetime needs". [65]

An agreement, signed in 1949 between France, the UK and the USA, established a number of prohibitions designed to prevent the development of a new war potential by West Germany, and to remain in effect until a peace settlement was concluded. Among the agents proscribed, poison war gases and toxic products from bacteriological and plant sources were listed. [66]

Biological weapons

(*a*) A biological weapon is defined as any equipment or apparatus expressly designed to use, for military purposes, harmful insects or other living or dead organisms, or their toxic products.

(*b*) Subject to the provisions of paragraph (*c*), insects, organisms and their toxic products of such nature and in such amounts as to make them capable of being used in the equipment or apparatus referred to in (*a*) shall be deemed to be included in this definition.

(*c*) Such equipment or apparatus and such quantities of the insects, organisms and their toxic products as are referred to in paragraphs (*a*) and (*b*) which do not exceed peaceful civilian requirements shall be deemed to be excluded from the definition of biological weapons.

A precise list of items banned was established a few years later.[2]

The unilateral West German renunciation was subsequently transformed into an international commitment. The Chancellor's declaration was annexed as Annex I and the definitions of chemical and biological weapons as Annex II to Protocol No. III modifying and completing the Brussels Treaty of 1948, in view of the Federal Republic of Germany's accession to the Treaty. The contracting parties, members of the Western European Union (Belgium, France, the Federal Republic of Germany, Italy, Luxembourg, Netherlands and the UK) took note of and recorded their agreement with the above declaration by signing Protocol No. III on 23 October 1954. The Agency of the Western European Union for the control of armaments was to satisfy itself that the undertakings not to manufacture certain types of armaments were being observed.

In practice, the control is carried out through inspections, test-checks and information visits to industrial plants and laboratories.[3] (The functioning of the WEU verification machinery is described in Volume V.)

---

[2] On 3 September 1968, at the Conference of Non-Nuclear-Weapon States, the Minister for Foreign Affairs of the Federal Republic of Germany said:

"The Geneva Protocol of 1925 does not define chemical and bacteriological weapons. Should the problem of B and C weapons be discussed, they should be specifically determined.

"In this respect the definitions laid down when Germany renounced production in 1954 could be of value. We offer our assistance and support for all efforts aiming without discrimination at effectively remodelling the prohibition of B and C weapons with the object of banishing man's fear of them."

[3] In 1969, the USSR transmitted a letter from the Minister for Foreign Affairs of the German Democratic Republic to the UN Secretary-General with a memorandum accusing the Federal Republic of Germany of secretly preparing for chemical and bacteriological warfare. The East German Minister demanded the withdrawal of US bacteriological and chemical weapons stockpiled on West German territory, and an end to the development and production of these weapons in firms in the Federal Republic of Germany and in their subsidiaries abroad. [67]

The same subject was brought up in two other documents—a message from the President of the People's Chamber of the German Democratic Republic, transmitted to the United Nations by the representative of Poland, [68] and a statement of

The Austrian Treaty of 15 May 1955 also contained clauses prohibiting biological and chemical weapons: [70]

(*i*) Austria shall not possess, construct or experiment with (*a*) any atomic weapon, (*b*) any other major weapon adaptable now or in the future to mass destruction and defined as such by the appropriate organ of the United Nations, ... (*j*) asphyxiating, vesicant or poisonous materials or biological substances in quantities greater than, or of types other than, are required for legitimate civil purposes, or any apparatus designed to produce, project or spread such materials or substances for war purposes.

(*ii*) The Allied and Associated Powers reserve the right to add to this Article prohibitions of any weapons which may be evolved as a result of scientific development.

II. CBW prohibition in the proposals for the reduction and limitation of armaments

The cessation of hostilities in Korea created a more favourable climate for disarmament talks between the great powers.

On 2 April 1954, the Prime Minister of India said:

We have maintained that nuclear (including thermonuclear), chemical and biological (bacterial) knowledge and power should not be used to forge these weapons of mass destruction. We have advocated the prohibition of such weapons by common consent and immediately by agreement amongst those concerned, which latter is at present the only effective way to bring about their abandonment.

The statement was included in the letter of 8 April 1954, from the representative of India to the UN Secretary-General. [1]

On 19 April 1954, the UN Disarmament Commission decided [2] to establish a subcommittee, with a mandate to discuss disarmament proposals in detail, pursuant to General Assembly resolution of 28 November 1953.

---

the government of the GDR, transmitted by the representative of Hungary. [69] The latter statement drew attention to the fact that the renunciation of atomic, biological and chemical weapons by the government of the Federal Republic of Germany in 1954 applied only to the production of these means of warfare in West Germany, and left open the possibilities of production abroad; production of parts or substances in West Germany and their final assembling or mounting abroad; purchase of these weapons abroad; possession or joint possession of these weapons; control or joint control over these weapons; and scientific and technological preparation of their production.

The government of the German Democratic Republic asserted that the Federal Republic of Germany had not even observed the renunciation qualified this way, and that the observance of the ban on producing atomic, biological and chemical weapons on West German territory had never been checked up.

The allegations were not discussed in the United Nations.

(See page 220.) The subcommittee consisted of Canada, France, UK, the USA and the USSR.

During the meetings of the subcommittee, various proposals, draft resolutions, memoranda and working papers were submitted, referring to the prohibition of weapons of mass destruction. Some of these documents specifically mentioned chemical and bacteriological weapons.

The UK memorandum of 21 May 1954, dealing with the categories of weapons to be covered by a disarmament convention, suggested that the weapons to be prohibited should include chemical and biological weapons. [3]

Another UK memorandum, of 13 September 1955, dealing with methods, objects and rights of inspection and supervision, stipulated that all states should supply the control organ, set up under a disarmament convention, with all the information required on "plants making chemical and biological weapons". The control organ would have the right to analyse and check that information, and, at a later stage of the disarmament programme, to establish resident inspection posts at those plants or inspect them through routine visits. [4]

The French working paper of 6 October 1955 listed among objects which should be subject to control, "chemical industries producing or capable of producing gunpowder, explosives, poison gases, etc". [5]

The working paper submitted by France and the United Kingdom on 3 May 1956 included in the list of objects to be controlled in stage one of disarmament: chemical and bacteriological armaments; all military establishments and installations which use or store chemical or bacteriological armaments; and all documents, such as estimates and accounts, necessary to verify expenditure on chemical and bacteriological armaments. In stage two, factories and other installations in which armaments are being manufactured or assembled or in which essential components are being manufactured, or which can be readily adapted to the manufacture of such armaments or components, would be subject to control. A note attached to the paper explained that a definition of armaments, to be agreed, may be expected to include bombs and equipment for discharging or disseminating chemical or bacteriological substances. The definition of essential components would include major components of armaments which are manufactured elsewhere than at the place of assembly of the whole armament, such as the filling of chemical or bacteriological weapons. Any alleged breach would be considered by the executive committee, consisting of China, France, the USSR, the UK and the USA, which would be permament members of the committee, and of ten other states which would be elected for two-year terms. If the executive committee considered that an irregularity covered by Chapter VII of the UN

Charter had occurred, it should immediately inform the UN Security Council or General Assembly and suspend such of the prohibitions, limitations and reductions provided for in the disarmament treaty as it may consider necessary, pending a decision under the Charter by the Security Council or General Assembly. The executive committee would equally have the right to suspend such of the prohibitions, limitations and reductions as it might consider necessary if it decided on the basis of a report from the director-general of the international disarmament organization that a participating state was markedly and consistently late in fulfilling its obligations. [6]

III. *CBW prohibition in the programmes for general and complete disarmament*

In an address to the UN General Assembly, on 17 September 1959, the British Foreign Secretary presented a programme of comprehensive disarmament under effective international control. Among the objectives to be reached at the final (third) stage of the programme were: a ban on the manufacture of nuclear, chemical, biological and other weapons of mass destruction; and a ban on the use of such weapons. The United Kingdom also suggested that, "There should be a re-examination of the possibility of controlling and then eliminating the remaining stocks of nuclear and other weapons of mass destruction." [7]

The USSR, in its "Declaration on General and Complete Disarmament" of 19 September 1959, presented by the Soviet Prime Minister to the General Assembly, stated:

> Stockpiles of chemical and bacteriological weapons accumulated by some States, asphyxiating and poisonous substances and cultures of lethal bacteria which are potential sources of severe epidemic disease will all be finally and irretrievably destroyed.

The Soviet proposed programme included the following measures concerning chemical and bacteriological weapons (to be carried out in the third and last stage of disarmament):

> The entry into force of the prohibition on the production, possession and storage of means of chemical and bacteriological warfare.
>
> All stockpiles of chemical and bacteriological weapons in the possession of States shall be removed and destroyed under international supervision.
>
> Scientific research for military purposes and the development of weapons and military equipment shall be prohibited.

For the purpose of supervising the implementation of the measures proposed, the programme provided for the establishment of an international control organ. Upon the completion of general and complete disarmament, which should include the disbandment of all services of the armed forces and the destruction of all types of weapons, including weapons of mass destruction (nuclear, rocket, chemical, bacteriological), the international control organ would have free access to all objects of control. [8]

Under the UN General Assembly resolution of 20 November 1959, the UK programme of comprehensive disarmament and the Soviet programme for general and complete disarmament were transmitted to the Ten-Nation Disarmament Committee[4] for thorough consideration. [9]

On 16 March 1960, Canada, France, Italy, the United Kingdom and the United States submitted to the Ten-Nation Committee a plan for "General and Comprehensive Disarmament in a Free and Peaceful World". It provided, among other things, for the prohibition of production of nuclear, chemical, biological and other weapons of mass destruction, as well as the reduction of existing stocks of those weapons. [10]

Bulgaria, Czechoslovakia, Poland, Romania and the USSR submitted a document of 8 April 1960, entitled "Basic Principles of General and Complete Disarmament", in which they proposed the prohibition of nuclear, chemical, bacteriological and missile weapons, cessation of their production and destruction of their stockpiles. [11]

In June 1960, the Soviet government put forward proposals for "Basic Provisions of a Treaty on General and Complete Disarmament".[5] According to these proposals, joint studies would be undertaken, in the first stage of the disarmament process, of the measures to be implemented in the second stage relating to the discontinuance of the manufacture of nuclear, chemical and biological weapons and to the destruction of stockpiles of such weapons. In the second stage, there would be a complete prohibition of nuclear, chemical, biological and other weapons of mass destruction, with the cessation of manufacture and the destruction of all stockpiles of such weapons. Representatives of the control organization would conduct on-site inspection of the destruction of all existing stockpiles of nuclear, chemical and biological weapons. [12]

The US delegation on the Ten-Nation Committee submitted, on 27 June

---

[4] On 7 September 1959, the Foreign Ministers of France, the United Kingdom, the United States and the Soviet Union reached agreement on the setting up of a disarmament committee, composed of the four big powers and Canada, Italy, Bulgaria, Czechoslovakia, Poland and Romania. On 10 September, the UN Disarmament Commission welcomed the setting up of the committee.

[5] The proposals were transmitted by the Soviet Premier to other heads of government on 2 June 1960, then submitted to the Ten-Nation Committee on 7 June 1960, and to the UN General Assembly on 23 September 1960.

1960,[6] a "Programme for General and Complete Disarmament Under Effective International Control", which in stage two included the following measure:

> Quantities of all kinds of armaments of each State, including nuclear, chemical, biological and other weapons of destruction in existence and all means for their delivery, shall be reduced to agreed levels and the resulting excesses shall be destroyed or converted to peaceful uses. Agreed categories of missiles, aircraft, surface ships, submarines and artillery designed to deliver nuclear and other weapons of mass destruction shall be included in this measure.

In the third stage of the US proposed programme, an international peace force and agreed contingents of national armed forces would be armed only with agreed types and quantities of armaments. All other remaining armaments, including weapons of mass destruction and vehicles for their delivery and conventional armaments, would be destroyed or converted to peaceful uses. There would be no manufacture of any armaments except for agreed types and quantities for use by the international peace force and agreed remaining national contingents. Verification of each agreed disarmament measure would have to be accomplished in such a manner as to be capable of disclosing, to the satisfaction of all participating states, any evasion of the agreement. [13]

In a declaration of 23 September 1960, describing the situation with regard to the implementation of the 1959 General Assembly resolution on general and complete disarmament, the USSR stated that while the Ten-Nation Committee had been holding talks on disarmament, steps were taken to expand the production of chemical and bacteriological weapons of mass destruction. [14]

The United States, the United Kingdom and Italy, in a draft resolution of 14 October 1960 submitted to the fifteenth UN General Assembly, reiterated that the final goal of a programme of general and complete disarmament must be to achieve, among other things, elimination of delivery systems of all weapons of mass destruction, as well as elimination of all weapons of mass destruction, nuclear, chemical and bacteriological. [15] In addition, Burma, Cambodia, Ceylon, Ghana, India, Indonesia, Iraq, Morocco, Nepal, the UAR, Venezuela, and Yugoslavia specified, on 15 November 1960, that general and complete disarmament should include total prohibition of the manufacture, maintenance and use of bacteriological and chemical weapons, as well as elimination of all equipment and facilities for the delivery, placement and operation of all mass-destruction weapons. [16] The British delegation suggested that scientific, military and administrative

---

[6] The paper was submitted after the delegations of the USSR, Bulgaria, Czechoslovakia, Poland and Romania had left the conference.

experts submit a report on the systems of inspection and control—their capabilities and limitations—which would be effective and fair to all concerned in relation to measures of disarmament, including the prevention of the manufacture of chemical and biological weapons. [17] None of these proposals was put to a vote.

In a joint statement of agreed principles for disarmament negotiations,[7] of 20 September 1961, the USSR and the USA recommended that the programme for general and complete disarmament should contain provisions for the elimination of all stockpiles of nuclear, chemical, bacteriological, and other weapons of mass destruction, the cessation of the production of such weapons and the elimination of all means of delivery of such weapons. Agreements on partial measures of disarmament were also recommended. [18]

On 25 September 1961, the United States submitted to the UN General Assembly a "Programme for General and Complete Disarmament in a Peaceful World" which provided for:

In stage I, the establishment within an International Disarmament Organization (to be set up within the framework of the United Nations) of a Chemical, Biological, Radiological (CBR) Experts Commission for the purpose of examining and reporting on the feasibility and means for accomplishing the verifiable reduction and eventual elimination of CBR weapons stockpiles and the halting of their production;

In stage II, depending upon the findings of the Experts Commission on CBR weapons, cessation of the production of CBR weapons, progressive reduction of existing stocks, and the destruction of the resulting excess quantities or their conversion to peaceful uses;

In stage III, prohibition of the manufacture of armaments except for those of agreed types and quantities to be used by the United Nations Peace Force and those required to maintain internal order. Destruction or conversion to peaceful purposes of all other armaments. [19]

The General Assembly resolution (declaration) of 24 November 1961, concerning the prohibition of the use of nuclear and thermonuclear weapons, recalled that: ". . . The use of weapons of mass destruction, causing unnecessary human suffering, was in the past prohibited, as being contrary to the laws of humanity and to the principles of international law, by international declarations and binding agreements, such as the Declaration of St. Petersburg of 1868, the Declaration of the Brussels Conference of 1874, the Conventions of the Hague Peace Conferences of 1899 and 1907, and the Geneva Protocol of 1925. . . ." [20]

On 15 March 1962, the USSR submitted to the Eighteen-Nation Disarma-

---

[7] The so-called "McCloy-Zorin Statement".

ment Committee[8] a "Draft Treaty on General and Complete Disarmament Under Strict International Control".

The following measures, concerning chemical, biological and radiological weapons, were provided for at the second stage of general and complete disarmament (Article 23):

1. All kinds of chemical, biological and radiological weapons, whether directly attached to the troops or stored in various depots and storage places shall be eliminated from the arsenals of States and destroyed (neutralized). Simultaneously all instruments and facilities for the combat use of such weapons as well as all special devices and facilities for their storage and conservation shall be destroyed.

2. The production of all kinds of chemical, biological and radiological weapons and of all means and devices for their combat use, transportation and storage shall be completely discontinued. All plants, installations, and laboratories that are wholly or in part engaged in the production of such weapons, shall be destroyed or converted to production for peaceful purposes.

3. The measures referred to above in Paragraphs 1 and 2 shall be implemented under the control of inspectors of the International Disarmament Organization. [21]

On 18 April 1962, the United States submitted, also to the Eighteen-Nation Disarmament Committee, an "Outline of Basic Provisions of a Treaty on General and Complete Disarmament in a Peaceful World", which with regard to chemical and biological weapons provided the following:

Objectives ... (b) Elimination of all stockpiles of nuclear, chemical, biological and other weapons of mass destruction and cessation of the production of such weapons; (c) Elimination of all means of delivery of weapons of mass destruction.

In Stage I:

The Parties to the Treaty would agree to examine unresolved questions relating to means of accomplishing in Stage II and III the reduction and eventual elimination of production and stockpiles of chemical and biological weapons of mass destruction. In light of this examination, the Parties to the Treaty would agree to arrangements concerning chemical and biological weapons of mass destruction.

In Stage II:

a. In the light of their examination during Stage I of the means of accomplishing the reduction and eventual elimination of production and stockpiles of chemical and biological weapons of mass destruction, the Parties to the Treaty would undertake the following measures respecting such weapons:

---

[8] At the sixteenth United Nations General Assembly, the USA and the USSR reached agreement on the composition of a new disarmament negotiating body. The membership of the body, called the Eighteen-Nation Disarmament Committee (ENDC), was: Brazil, Bulgaria, Burma, Canada, Czechoslovakia, Ethiopia, France, India, Italy, Mexico, Nigeria, Poland, Romania, Sweden, UAR, UK, USA and USSR. The agreement was endorsed by a General Assembly resolution of 20 December 1961.

(1) The cessation of all production and field testing of chemical and biological weapons of mass destruction.
(2) The reduction, by agreed categories, of stockpiles of chemical and biological weapons of mass destruction to levels fifty per cent below those existing at the beginning of Stage II.
(3) The dismantling or conversion to peaceful uses of all facilities engaged in the production or field testing of chemical and biological weapons of mass destruction.

b. The foregoing measures would be carried out in an agreed sequence and through arrangements which would be set forth in an annex to the Treaty.

c. In accordance with arrangements which would be set forth in the annex on verification the International Disarmament Organization would verify the foregoing measures and would provide assurance that retained levels of chemical and biological weapons did not exceed agreed levels and that activities subject to the foregoing limitations were not conducted at undeclared locations.

In Stage III:

1. Reduction of Armaments

Subject to agreed requirements for non-nuclear armaments of agreed types for national forces required to maintain internal order and protect the personal security of citizens, the Parties to the Treaty would eliminate all armaments remaining at their disposal at the end of Stage II.

2. Method of Reduction

a. The foregoing measure would be carried out in an agreed sequence and through arrangements that would be set forth in an annex to the Treaty.

b. In accordance with arrangements that would be set forth in the annex on verification, the International Disarmament Organization would verify the foregoing measures and would provide assurance that retained armaments were of the agreed types and did not exceed agreed levels.

3. Limitations on Production of Armaments and on Related Activities

a. Subject to agreed arrangements in support of national forces required to maintain internal order and protect the personal security of citizens and subject to agreed arrangements in support of the United Nations Peace Force, the Parties to the Treaty would halt all applied research, development, production, and testing of armaments and would cause to be dismantled or converted to peaceful uses all other facilities for such purposes.

b. The foregoing measures would be carried out in an agreed sequence and through arrangements which would be set forth in an annex to the Treaty.

c. In accordance with arrangements which would be set forth in the annex on verification, the International Disarmament Organization would verify the foregoing measures at declared locations and would provide assurance that activities subject to the foregoing measures were not conducted at undeclared locations. [22]

The Swedish representative to the ENDC, in his statement of 11 May 1962, remarked that in neither of the two documents quoted above was the elimination of chemical and bacteriological weapons envisaged until Stage II. He submitted that there could be good reasons for reduction or elimination of these weapons already in stage I, although he realized that control of such a measure could be difficult in practice. He then asked the two big powers for comments as to the timing of their disarmament measures with regard to the chemical and biological arsenal. [23]

The Soviet representative replied that weapons of mass destruction should be eliminated during the same stage. The USSR had been at one time prepared to agree to the destruction of these weapons at an earlier stage, but the key to the matter was the position of the Western powers. Until nuclear weapons had been destroyed, it was extremely difficult to solve the problem of control over the liquidation of chemical and biological weapons. [24]

The representative of the USA agreed with the Swedish delegation that difficult problems were connected with verifying the elimination of stockpiles and the cessation of production of chemical, and especially biological weapons. The United States had therefore called for an examination of this problem in the first stage. It was, however, ready to participate in an expert study group even prior to Stage I, in order to determine whether measures sufficiently effective and workable could be devised in time to be implemented in the first stage. [25]

The US-Soviet working draft, of 31 May 1962, of part I of a treaty on general and complete disarmament, setting forth the general objectives of the treaty, provided for the elimination of all stockpiles of nuclear, chemical, biological and other weapons of mass destruction, as well as the cessation of the production of such weapons; the elimination of all means of delivery of weapons of mass destruction and the cessation of the production of such means of delivery. [26]

The revised versions of the Soviet "Draft Treaty on General and Complete Disarmament Under Strict International Control", of 22 September 1962 and 28 April 1965, introduced only minor changes to the original article dealing with elimination of chemical, biological and radiological weapons. [27]

IV. *Allegations of use of chemical weapons in Indo-China*

In the 1960s, when the involvement of the United States in the war in Indo-China significantly increased, complaints were addressed to the United Nations that chemicals were being used by the US forces in military operations and against the civilian population.[9]

---

[9] For a general account of the use of chemical weapons in Indo-China, see Volume I.

In a cable dated 28 July 1964, addressed to the President of the Security Council, the Minister for Foreign Affairs of the Kingdom of Cambodia informed the Council that South Viet-Namese aircraft had violated Cambodian air space to a depth of several dozen kilometres on 13 and 20 June and on 9, 17, 20, 21, 22 and 23 July and had dumped toxic powder on the villages of Kes Chong, Pak Nhai, Pakpo, Lamak Chieng, Kamang Chong and Boloi, in the Srok of Dandaungpich, Province of Rattanakiri.[10] Seventy-six persons died as a result of this operation, "which is part of the chemical warfare which the United States—South Viet-Namese forces had hitherto waged only in South Viet-Nam". [28]

Replying to that accusation, the deputy US representative to the United Nations stated in a letter of 3 August 1964 that a careful investigation had been conducted, and it had been determined that no Republic of Viet-Nam nor US aircraft had conducted chemical operations of any character whatever, either adjacent to the Cambodian territory indicated, much less over Cambodian territory, on any of the dates or at any time in the period cited by the government of Cambodia. The operations conducted within South Viet-Nam by the government of Viet-Nam employed weed-killing chemicals of types used throughout the world, and extensively within the USA itself. These were spread in liquid form, left no powder residue, and were harmless to human beings. The US government would welcome an impartial investigation of the Cambodian charge. [29] Two days later, the government of South Viet-Nam also rejected the charge and requested that a scientist of repute be appointed or an international commission of observers and experts established to hold an investigation on the spot in order to clarify this matter, which, if it were true, could not fail to touch the conscience of mankind. The South Viet-Namese Foreign Minister informed the Security Council that since 1960 the armed forces of the Republic of Viet-Nam had been using a chemical product for defoliation purposes in certain areas of dense vegetation. "This new tactic has enabled our security forces to carry out counter-attacks, and thus by the end of 1963 Viet Cong sabotage acts had been reduced by 70 per cent in comparison with previous years." The chemical product in question—continued the South Viet-Namese communication—was not the one alluded to in the Cambodian cable which described it as a toxic powder. In fact it was a deep purple liquid which other countries had been using for a long time in gardening to destroy weeds and plant parasites. This chemical product, the formula and use of which were well known to the public, was easy to obtain in shops and had not so far caused fatalities among human beings or animals. It did not kill trees, nor did it contaminate

---

[10] The transcription of Cambodian geographical names is given as in the United Nations documents.

running water. Its only effect was to cause leaves to fall from their branches. The United States consumed 13 300 tons of this product annually. From 1948 to 1960 Malaysia had successfully used it in its so-called defoliation campaign in jungle areas "for the purpose of combating communism". [30]

In a document entitled "Clearing the undergrowth—What are the facts about defoliation in South Viet-Nam", the South Viet-Namese government further explained that chemical defoliants were being used in order to reduce the possibility of ambush or sabotage around military establishments, along roads, railways, canals, rivers and high tension power lines in certain areas of the country. The chemicals used were allegedly diluted 2,4,5-T (trichlorophenoxyacetic acid) and 2,4-D (dichlorophenoxyacetic acid). Chemical defoliants—contended the document—had nothing to do with toxic liquids or gases associated with chemical warfare; they bore no resemblance to such substances. [31]

On 13 August 1964, Cambodia submitted yet another complaint to the Security Council. It informed the Council that on the afternoon of 29 July 1964 two aircraft of the armed forces of the Republic of Viet-Nam had flown over the Kes-chong area at an altitude of about 400 meters spraying yellow and white powder on the Phums of Loeunhai and Loeu-Khuon, Srok of Andaung Pich, in the Province of Rattanakiri. On 30 July, at about 3 p.m., an aircraft of those forces had flown over the Chipou area, spraying chemical products on the Phum of Bos Krassaing, Khum of Chrak Motes, and on the Khum of Prey Koki, Srok of Svay Teap, in the Province of Svay Rieng. Following this spraying, about thirty victims among the inhabitants had been seized with itching, burning sensations in their bodies and sore throats. On 31 July, towards noon, two aircraft of the same forces had sprayed yellow and white powder on the Phum Loeunhai, 10 kilometres northeast of Bokeo, in the Province of Rattanakiri. On the same day, at about 6 o'clock, another aircraft had sprayed the same powder on the area about 300 metres to the south of Boloi, in the Province of Sattanakiri. Two men and a child had died and a woman had become seriously ill. [32]

The US representative again denied the Cambodian accusations. He reiterated the proposal for an impartial international enquiry, indicating that the World Health Organization or the International Committee of the Red Cross would seem well-qualified to carry out the investigation and to report the facts to the Security Council. [33]

In a note[11] of 26 March 1965 to the government of the United States (circulated as an official UN Security Council document), the Soviet government referred to reports concerning the use of poison gases against the

---

[11] The note was rejected by the United States Embassy in Moscow.

population of South Viet-Nam by the US armed forces. According to these reports—said the note—US helicopters and fighter-bombers carrying special equipment had been spraying South Viet-Namese population centres with chemical substances of a combat type, which affected the organs of sight, the respiratory tract and the gastro-intestinal system. The victims of the gas attacks were peaceful Viet-Namese villagers, old people, women and children. The Soviet government considered the use of chemical weapons as a violation of the accepted rules of international law and of the elementary principles of morality and humanity, and in particular of the Geneva Protocol of 1925. It asked the US government to take the necessary steps to put an end to the use of poisonous substances in South Viet-Nam. [34]

On 2 April 1965, the US representative addressed a letter to the President of the Security Council, in which he explained that the materials which were employed in Viet-Nam were used by police forces in riot control in many parts of the world and were commonly accepted as appropriate for such purposes. They were non-toxic and were not prohibited by the 1925 Geneva Protocol, or by any other understandings on the subject. Reference was made to the US Secretary of State's statement of 24 March 1965, claiming that tear gas was a minimum instrument; it was being employed in order to avoid the problem of whether to use artillery or aerial bombs that would inflict great damage upon innocent people. [35]

On 5 April 1965, at a meeting of the Committee established to study the question of defining aggression, the Soviet delegate repeated the charge that the United States had engaged in gas warfare. He said that there could be no doubt whatever that the Geneva Protocol prohibited the use of the chemical weapons which the US military had been employing in South Viet-Nam. The United States replied that poisonous gases had not been used in Viet-Nam and there was no intention of employing them. [36]

Three weeks later, in a statement to the UN Disarmament Commission, the Soviet representative condemned the use of chemical weapons in Viet-Nam as "a crime against humanity", "a flagrant act of lawlessness", "a crude violation of universally accepted standards of international law" and "a flouting of the elementary principles of morality". [37] The United States called those charges false and stressed again that the use of riot-control agents had nothing to do with gas warfare which was waged with deadly poisonous gas. [38]

Subsequent charges concerning the use of chemical weapons in Viet-Nam also met with US denials.[12]

---

[12] In May, June and August 1969, the permanent representative of Cambodia, in letters addressed to the President of the Security Council, submitted complaints con-

V. UN appeal for compliance with the 1925 Geneva Protocol

On 7 November 1966, at the twenty-first session of the UN General Assembly, the Hungarian delegation submitted a draft resolution whereby the Assembly, guided by the principles of the UN Charter and contemporary international law, considering that weapons of mass destruction constitute a danger for all mankind, and recalling that the Geneva Protocol of 17 June 1925 on the prohibition of the use of asphyxiating, poisonous and other gases and of bacteriological methods of warfare had been signed and adopted and was recognized by many states, would:

1. demand strict and absolute compliance by all states with the principles and norms established by the Protocol which prohibits the use of chemical and bacteriological weapons;

2. condemn any actions aimed at the use of chemical and bacteriological weapons;

3. declare that the use of those weapons for the purpose of destroying human beings and the means of their existence constitutes an international crime. [39]

cerning damages caused to rubber plantations, orchards and forests by defoliants dropped by US aircraft. [62]

The total damage inflicted on the national economy was claimed to have amounted to $8 684 810.

The Cambodian government was prepared to extend to the US government all necessary facilities to verify the damages in question. The amount given above did not take into account the long-term effects which could not as yet be determined. On this score, the Cambodian government reserved the right to claim further reparations.

On 10 November 1969, the US Department of State issued the following press release: [63]

"A group of United States experts visited Cambodia from June 30 to July 9, 1969, to study areas of Kompong Cham province damaged by herbicides, and to determine the cause, severity, origin and extent of damage. Members of the team were Dr Charles E. Minarik, Director, Plant Sciences Laboratories, Department of Defense; Dr Fred H. Tschirley, Assistant Chief, Crop Protection Research Branch, Agricultural Research Service; Dr Nader G. Vakill, Agronomy Adviser, AID Saigon, and Mr Jack B. Shumate, Chief, Forestry Branch, AID Saigon. The Cambodian Government cooperated fully in providing facilities for the study carried out by the experts.

"The team's report concluded that herbicide damage in the affected area was extensive due to a combination of two factors, (*a*) defoliation of fruit trees near the border as a result of drift from spray operations conducted in April and May 1969 in northern Tay Ninh province, Republic of Viet-Nam; and (*b*) defoliation of rubber, fruit and forest trees farther north probably by direct application of spray from the air on a north-south line running through two major rubber plantations.

"The team also concluded that few, if any, rubber or fruit trees have been killed. The degree and rate of recovery will depend on a number of factors, but should be well advanced by July/August 1970. The team recommended that a final assessment of damage be made then, based on the decline in latex and fruit production following the defoliation damage."

In presenting the draft resolution to the First Committee of the Assembly, the Hungarian representative referred to the use of chemical weapons in Viet-Nam, and quoted from a report of the "Committee for the Denunciation of War Crimes Perpetrated in South Viet-Nam by the United States of America", published by the South Viet-Nam National Liberation Front on 22 July 1966.

He stressed, however, that the problem was of a much wider scope, and of general concern for all the world. Anticipating objections to his proposal, he made the following points:

The prohibition of chemical and bacteriological weapons is valid not only for those who have acceded to the Geneva Protocol; in the opinion of outstanding international lawyers, the cumulative effect of customary law and of the existing instruments having binding force is to render such prohibition legally effective upon practically all states.

Practicing riot control and conducting warfare are two distinctly different problems. The first falls under the internal jurisdiction of every state and its excessive use might be considered under the terms of basic human rights, while the second is a matter between armed forces of different nations, and is, therefore, governed by international law.

Gas weapons are intended to break the backbone and the moral resistance of people; they are intended to destroy health, spread diseases and create starvation among wide strata of the population. The application of such weapons is on the fringe of genocide. Other chemical weapons, such as napalm which burns large surfaces of the human body, suffocating and disfiguring people, cause tremendous suffering to civilian populations. Toxic herbicides poison foods, drinking water and irrigation waters.

The Geneva Protocol clearly and unequivocally covers all kinds of weapons of chemical warfare. [40]

The Soviet representative, expressing support for the Hungarian draft resolution, drew attention to the opinion of scientists who, during the Pugwash conferences, had condemned the use of chemical and bacteriological weapons. During the twenty-first session of the General Assembly a large group of US scientists, including Nobel Prize winners, appealed to the US government, asking it to put an end to the use of chemical weapons in Viet-Nam. All this—in his opinion—testified to the urgency of the problem. [41]

The Bulgarian representative noted that the use of chemical and bacteriological weapons must be banned not only because of their direct harmful effects; it was a loophole for recourse to even more disastrous and powerful means for causing human suffering, annihilation and destruction. He was convinced that any weakening of the prohibition increased the menace to peace, security and human lives. [42]

The representative of Nepal felt that the use of chemical weapons in war was as dangerous and criminal as the use of nuclear weapons. [42]

The delegate of Byelorussia drew attention to the telegrams of 16 March and 28 April 1966, sent by the General Secretary of the Ministry of Information, Propaganda and Tourism of the Government of National Unity of Laos, and a member of the State Committee for the Implementation of the Geneva Agreements of 1962, to the Foreign Minister of the USSR, co-chairman of the Geneva Conference on Laos, accusing the United States of using chemical substances against the people of Laos. The cables stated that US aircraft had been dropping poisonous substances over large areas in the densely populated eastern parts of the province of Saravane, as a result of which large areas covered with forests, rice paddies and fruit orchards had been destroyed and many of the inhabitants had been seriously poisoned, some even died. [42]

The delegation of Tanzania considered the intentional use of chemical or disease agents or their toxic products as an employment of the forces of nature against people and their food supply. [43]

The US representative strongly opposed the proposal made by Hungary. He said that the Geneva Protocol of 1925, to which the USA was not party, was intended to prohibit the use in warfare of deadly gases such as mustard gas and phosgene; it did not apply to all gases, and it certainly did not prohibit the use of simple tear gas where necessary to avoid injury to innocent persons. The United States had never engaged in gas warfare since World War I, when it was necessary to use gas in retaliation. The US Secretary of State had said on 25 March 1965, with reference to Viet-Nam: "We are not engaged in gas warfare. It is against our policy to do so, as it is against the policies of most other Governments." What happened in Viet-Nam was the occasional use of tear gas by the forces of the United States and those of the Republic of Viet-Nam. Tear gas had nothing whatever to do with the Geneva Protocol. Evidence available for recent years showed that more than fifty countries had used tear gas for domestic riot-control purposes. When, for example, civil authorities must enforce law and order in the face of an unruly mob, they must often decide, when other means of persuasion have been exhausted, whether to use brute force and lethal weapons, and thus risk injury and death perhaps even to innocent bystanders, or to disperse the mob by recourse to riot-control agents, such as tear gas, which have no harmful after-effects. In Viet-Nam, when the Viet Cong takes refuge in a village and uses innocent civilians and prisoners as shields, would it be more humane to use rifle and machine-gun fire and explosive grenades to dislodge and destroy the Viet Cong and thus risk the lives of the innocent and wounded hostages?—asked the US representative.

As to the use of herbicides in Viet-Nam, he argued that these materials involved the same chemicals and had the same effects as those commonly used in the USA and a great many other countries to clear weeds and control vegetation; they were not bacteriological weapons, nor was their use contrary to international law.

Finally, the US representative said that on 10 November 1966, a US patrol in Viet-Nam had encountered gas which appeared to have the same effects as tear gas. During that operation 1 200 gas grenades had been captured, and they were of Chinese manufacture. [41]

The delegations of Kenya, Uganda and Tanzania proposed that the Hungarian draft resolution be amended by adding, in the preambular part, that weapons of mass destruction were incompatible with the accepted norms of civilization; inserting a new preambular paragraph which would affirm that strict observance of the rules of international law on the conduct of warfare was in the interest of maintaining these standards of civilization,[13] replacing operative paragraphs 2 and 3 by new formulations, so as to:

2. deplore the use of chemical and bacteriological weapons for the purpose of destroying human beings and the means of their existence; and

3. invite all states to accede to the Geneva Protocol of 17 June 1925. [44]

The amendments were sponsored by a few more African states [45], and their revised version was accepted by Hungary. [46]

Canada, Italy, the United Kingdom and the United States, however, considered the above changes insufficient and proposed to delete the word "contemporary" from the preambular part; add a new preambular paragraph noting that the Eighteen-Nation Committee on Disarmament has the task of seeking an agreement for a cessation of development and production of chemical and bacteriological weapons and other weapons of mass destruction and for the elimination of all such weapons from national arsenals as called for in the draft proposals for general and complete disarmament before the Committee; and replace operative paragraphs 1, 2 and 3 of the draft by a call for strict observance by all states of the principles and objectives of the Geneva Protocol and a condemnation of all actions contrary to those objectives. [47] In a subsequent document, following a suggestion made by Norway, replacement of operative paragraphs 1 and 2, only, was proposed. [48]

The Canadian representative gave the following motivation for the Western powers amendments. The term "contemporary", applied to international law, introduced unnecessary ambiguity. The statement contained in

---

[13] The first version of the amendments contained an affirmation that the maintenance of restrictions on the sovereignty of nations in the conduct of warfare was in the interest of maintaining these standards of civilization.

the draft resolution, that the Geneva Protocol prohibited the use of chemical weapons, was unacceptable without mentioning the specific qualifications of the actual text; the scope of the Protocol must not be interpreted—and extended—in the ambiguous manner which was found in operative paragraph 1 of the draft resolution. Even if every state ratified the Geneva Protocol, this would not finally dispose of the threat of the use of chemical and biological methods of warfare. To be sure that these weapons would not be used, measures must be taken in the general process of disarmament, as provided in the draft treaties submitted by the USSR and USA to the Eighteen-Nation Committee. [49]

The United States opinion was that whether or not, or by what procedure, states which had not yet done so should adhere to the Geneva Protocol was for each of them to decide in the light of constitutional and other considerations that may determine their adherence to any international instrument. What could be done, however, was to obtain from every country represented in the UN, whether or not party to the Geneva Protocol, a formal public expression of intent to observe strictly the objectives and the principles of the Protocol. [49]

The French delegation pointed out that a condemnation of chemical weapons in general, as contained in the Hungarian draft resolution, was not predicated upon the text of the Geneva Protocol. The British representative said that the Geneva Protocol did not refer to "chemical" weapons. It would be a dangerous precedent if the Assembly began to interpret this great international Protocol which was already the subject of considerable differences of interpretation. [49]

The representative of the Netherlands remarked that the Geneva Protocol was not devoid of vagueness and ambiguity, and felt that the moment had come to give serious consideration to a possible review of the Protocol by an appropriate body. In the light of scientific and technological developments of the postwar era, such a review was not only warranted but in fact overdue. [49] Belgium expressed similar views. [50] The delegate of Malta considered the Geneva Protocol outdated. He suggested that the Eighteen-Nation Committee on Disarmament should be invited to study the revision and updating of the Protocol, as well as the procedures necessary to give effective publicity through the United Nations to national production and to the stocks of chemical and bacteriological weapons in national arsenals. [49]

On 23 November 1966, the First Committee adopted the Western powers amendments. The draft resolution, as amended, passed in the Committee on the same day by 101 votes to none, with three abstentions (Cuba, France, Gabon) [50], and on 5 December 1966, in the General Assembly, by 91

votes to none, with four abstentions (Albania, Cuba, France, Gabon). [51]
The text was as follows:

The General Assembly,

Guided by the principles of the Charter of the United Nations and of international law,

Considering that weapons of mass destruction constitute a danger to all mankind and are incompatible with the accepted norms of civilization,

Affirming that the strict observance of the rules of international law on the conduct of warfare is in the interest of maintaining these standards of civilization,

Recalling that the Geneva Protocol for the Prohibition of the Use in War of Asphyxiating, Poisonous or Other Gases and of Bacteriological Methods of Warfare of 17 June 1925 has been signed and adopted and is recognized by many States,

Noting that the Conference of the Eighteen-Nation Committee on Disarmament has the task of seeking an agreement on the cessation of the development and production of chemical and bacteriological weapons and other weapons of mass destruction, and on the elimination of all such weapons from national arsenals, as called for in the draft proposals on general and complete disarmament now before the Conference,

1. Calls for strict observance by all States of the principles and objectives of the Protocol for the Prohibition of the Use in War of Asphyxiating, Poisonous or Other Gases, and of Bacteriological Methods of Warfare, signed at Geneva on 17 June 1925, and condemns all actions contrary to those objectives;

2. Invites all states to accede to the Geneva Protocol of 17 June 1925. [52]

The US representative, speaking in explanation of his vote at the plenary meeting of the Assembly, reiterated his government's policy with regard to the use of chemical and bacteriological weapons. He said that "it would be unreasonable to contend that any rule of international law prohibits the use in combat against an enemy, for humanitarian purposes, of agents that governments around the world commonly use to control riots by their own people", and added that, similarly, the Geneva Protocol did not apply to herbicides. [51]

At the same meeting, the Hungarian representative expressed his conviction that whatever gases were used in warfare, they were all poisonous in one degree or another. "Some of them kill instantly, some of them kill somewhat belatedly; some of them kill everybody and some of them kill only some people. So there is a difference only of degree in their danger." [51]

VI. *Allegations of use of chemical weapons in the Yemen*

During the 1963–67 civil war in the Yemen between the royalist regime, backed by Saudi Arabia, and the republican authorities, backed by the

UAR, allegations were made that lethal gas was used by the United Arab Republic (an estimated 50 000 Egyptian troops were then stationed in the Yemen).

In a cable of 20 February 1967, the permanent representative of Saudi Arabia drew the attention of the UN Secretary-General to the employment of lethal gas by Egyptian planes. Similar charges were contained in petitions sent to the United Nations and later circulated as Security Council documents.[14] In a letter of 22 March 1967 to the UN Secretary-General, the representative of Saudi Arabia complained about the use of gas in Kitaf,[15] Northern Yemen, on 5 January 1967. According to his communication, the gas had played havoc with the population, killing them by asphyxiation and rendering seriously ill those persons who had been even lightly contaminated. Members of the International Red Cross mission fled from the region and took refuge in Saudi Arabia. They used the telecommunication facilities in Najran to cable their organization to supply them with gas masks before they could return to the northern region of Yemen. (A copy of their cable was attached to the letter.)

Approximately two hundred patients of those who fled to Najran after having survived the lethal gas attack were admitted to the Najran hospital. Their medical examination revealed the following symptoms:

Difficult breathing accompanied by intense coughing.
Vomiting with foam mixed with blood issuing from the mouth.
Haemorrhage from the nose and mouth.
Congestion in the face and eyes.
Haemorrhage from the throat.
Low blood pressure.
Some of the patients unable to walk or move.
Some patients totally unconscious.
Some patients with swelling around the neck and the chest.
Some patients had blood in their urine.
Some patients suffered from subcutaneous haemorrhage.
Some patients had blood in their faeces.

Among these patients, twelve were in a very critical condition which necessitated their transfer to the military hospital in Taif in order to resume treatment. A medical report was signed by five physicians.

Beside human casualties, a number of animals were found dead as a result of the lethal gas attack, with apparent symptoms similar to those which were manifest in the persons killed by that gas. Even all vegetation,

---

[14] The allegations are dealt with in greater detail in Volumes I and V.
[15] The transcription of Yemeni and Saudi Arabian geographical names is given as in the United Nations documents.

including trees, contaminated by the gas, was affected to the extent that grass and foliage withered away.

The report of the Department of Legal Medicine released by the Saudi Arabian Ministry of Health indicated that the specimens of clothing and animal tissue as well as the blood analysis of certain patients had proven, by extensive chemical tests, that both compounds of chlorine and phosphorous gases had been employed in the attack on Kitaf. The gases were contained in thin encasements. (A copy of the report of the Department of Legal Medicine was enclosed.)

The representative of Saudi Arabia in his subsequent letters suggested that the Secretary-General should make a statement to decry the use of lethal gas in any region of the world. He also said that his government would welcome an investigation of what had happened in Kitaf. A neutral representative could interview many of the patients and each of the physicians that treated them in Najran. Such a representative could also visit the site of the attack to question the survivors who had escaped and eventually returned to Kitaf.

The Secretary-General replied that since the allegation had been firmly denied by the government in question,[16] the information submitted by Saudi Arabia might best have been presented to the competent deliberative organ of the United Nations. He did not believe that any useful purpose would be served by a statement of the nature suggested by Saudi Arabia.

The above exchange of communications was circulated on 6 April 1967 to the members of the UN Security Council. [53]

No action was taken by the United Nations on those and subsequent allegations of use of poisonous gas in the Yemen.

---

[16] On 1 February 1967, the UAR national guidance minister made the following statement.

"World news agencies have reported a statement made in the House of Commons this afternoon by the British Prime Minister, Mr Harold Wilson, who commented on the allegations disseminated by Saudi-Arabia and some propaganda elements co-operating with it, that the UAR used poison gas bombs against the village "Kataf" on Yemeni-Saudi border. The UAR deemed it wise hitherto to ignore these allegations which turned out to be untrue. But the remarks made by the British Premier in the House of Commons gave them certain colour. Although the British Premier was vague when he said that his Government had reason to believe that the allegations were true, his words might give a wrong impression.

"In the name of the UAR I have been entrusted to affirm once again and in a decisive manner that the UAR has not used poisonous gas at any time and did not resort to using such gas even when there were military operations in Yemen.

"I have also been entrusted with announcing officially that the UAR is ready to accept a fact-finding mission from the UN and is ready to make necessary arrangements for the mission to go to Yemen immediately. Yemen has agreed to give the mission all facilities to expose the anti-UAR propaganda and those who undertake it in London." [61]

One explanation for lack of UN action is given in the following documents.

On 27 July 1967, the British Foreign Office issued the following statement:

Mr. Duncan Sandys, M.P., Mr. Emanuel Shinwell, M.P. and Mr. Jeremy Thorpe, M.P. called on the Foreign Secretary on 26th July.

They drew Mr. Brown's attention to the House of Commons Motion, supported by over 200 M.P.s of all parties, in the following terms:

"That this House deplores the continued use of poison gas by the Egyptian forces in the Yemen and calls upon Her Majesty's Government to raise the matter urgently at the United Nations."

They strongly urged the Foreign Secretary to take steps, through the United Nations or otherwise, to try and prevent a continuance of these atrocities.

Mr. Brown made it clear that Her Majesty's Government deplored the use of poison gas by the U.A.R. armed forces, for which there was no possible justification. He said that Her Majesty's Government will be consulting with other Governments as to the best means of putting a stop to this clear breach of generally accepted rules of conduct.

In a letter of 15 September, 1967, to Mr Duncan Sandys, M.P., the Foreign Secretary, George Brown, wrote:

At our meeting on 26 July, I undertook to consult with like-minded Governments about ways of stopping the U.A.R.'s use of gas in the Yemen. We have now consulted a number of Governments to see if they are prepared to initiate action at the United Nations, but we have unfortunately been quite unsuccessful. While all the Governments we have approached deplore the U.A.R.'s use of poison gas, they all seem to feel they have compelling reasons of national interest for not publicly taking the lead in censuring the U.A.R.

The situation has now been very greatly altered by the agreement reached between King Faisal and President Nasser at Khartoum providing for the withdrawal of Egyptian forces from the Yemen and the cessation of all Saudi military aid. There are reliable reports that the withdrawal of Egyptian forces is to be completed within three months and preparations are already reported at Hodeida for their movement through that port. Despite past history, the chances of this agreement being carried out seem to be high.

In the light of this I must honestly say that I think the chances of getting any government to attack the U.A.R. in the United Nations for its use of poison gas in the Yemen are nil. Indeed to raise the matter now would almost certainly be counter-productive. I therefore consider, and I hope you will agree, that it is not appropriate for me to continue my efforts at this time to have this issue raised at the United Nations.

We shall have achieved our purpose by the departure of the U.A.R. forces from the Yemen.

On 19 September 1967, Mr Duncan Sandys replied:

The fact that Colonel Nasser has once again promised to withdraw his forces from the Yemen, and that you believe he will this time keep his word, is no reason to allow him to get away with this flagrant breach of the rules of war. As a warning to others in future conflicts, it is essential that the United Nations should be called upon to condemn these acts of barbarism. The Governments you approached may, as you say, have "reasons of national interest" for not wishing to censure Egypt. But if we gave the lead, others would surely feel obliged to support us; and even if they did not, should Britain be afraid to speak alone for humanity?

The official United States position was set out in the Seventh Annual Report, of 30 January 1968, of the US Arms Control and Disarmament Agency:

On June 2 the International Committee of the Red Cross at Geneva issued a statement confirming the use of poison gas in the Yemen. At that time the ICRC transmitted to authorities involved in the Yemen conflict a report on its investigation of the use of poison gas. The text of this report was published in the *New York Times* on July 28, 1967. The United States condemned the use of lethal gas as "clearly contrary to international law," and called on the authorities concerned in Yemen to heed the request of the ICRC not to resort in any circumstances whatsoever to their use.

VII. *Suggestion for a revision of the Geneva Protocol*

At the 1967 session of the Eighteen-Nation Committee on Disarmament, Sweden listed the cessation of the development and production of CB weapons among the topics which should be in the foreground of deliberations. [54] However, the Committee concentrated its major efforts on the elaboration of a non-proliferation treaty and was not able to devote sufficient time to the consideration of other matters.

On 7 December 1967, at the twenty-second session of the General Assembly, the delegation of Malta introduced a draft resolution which recommended that the Eighteen-Nation Disarmament Committee consider as a matter of urgency the problems relating to the use of chemical, biological and radiological weapons with a view to revision, updating or replacement of the Geneva Protocol for the Prohibition of Use in War of Asphyxiating, Poisonous or Other Gases and of Bacteriological Methods of Warfare; invited the Committee, if it believed it would facilitate such consideration, to establish a subcommittee for the above purpose; and requested the Secretary-General to prepare a concise report on the nature and probable effects of existing chemical, biological and radiological weapons and on the economic and health implications of the possible use of such weapons with particular reference to states that were not in a position to establish comprehensive methods of protection. [55]

In his statement to the First Committee of the General Assembly, on 12 December 1967, the Maltese representative pointed out that, as far as contemporary chemical warfare was concerned, the prohibition contained in the Geneva Protocol was scarcely more than marginally relevant. Toxic chemical agents, which might be used in modern warfare, were not necessarily either gases or liquids. Furthermore, the most dangerous were neither asphyxiating nor poisonous. The Protocol forbade bacteriological methods of warfare, but this covered only relatively few, and not the most dangerous, of the micro-organic agents that may be used in modern biological warfare. Finally, the prohibition in the Geneva Protocol extended only to the use in war of certain gases, analogous liquids and bacteria. Their use for hostile purposes in peacetime was not prohibited. This was a fatal omission in contemporary conditions, when wars were seldom declared and when some of the most dangerous chemical and biological weapons in the arsenals of states were eminently suited for use in circumstances in which no overt conflict existed.

The delegate of Malta also considered the Geneva Protocol excessively vague. He explained that the term "chemical weapons" in his draft resolution was used to signify "toxic chemical agents used for hostile purposes which produce their effects directly as a result of their chemical properties rather than as a result of blast, heat or other physical effects of a chemical reaction"; while the term "biological weapons" was used to signify "all micro-organisms including viruses, or their toxic products intentionally used for hostile purposes". He then quoted a paragraph in a petition of 14 February 1967 to the US President, signed by seventeen US Nobel Prize laureates in chemistry, biochemistry and physics, 127 members of the US Academy of Sciences and 5 000 other US scientists:

Chemical and biological weapons have the potential of inflicting, especially on civilians, enormous devastation and death which may be unpredictable in scope and intensity; they could become far cheaper and easier to produce than nuclear weapons, thereby placing great mass destructive power within the reach of nations not now possessing it; they lend themselves to use by leadership that may be desperate, irresponsible or unscrupulous.[17]

Speaking about the nature and capabilities of chemical weapons, the representative of Malta said that the meaning of the Geneva Protocol could not be stretched to include herbicides having low toxicity to man and animals, although their use could have seriously damaging effects on food supplies; nor were insecticides covered by the Protocol. Some insecticides belonged to the same class as nerve agents, had comparable toxicities and were extremely hazardous, as numerous accidental deaths to their users in many countries had demonstrated.

It was difficult, he continued, clearly to distinguish biological warfare from chemical warfare; many toxic chemical agents, such as toxins, were sometimes considered biological agents by the military.

His view was that chemical and biological weapons had been shrouded in official secrecy for too long; their nature, potential effects and the hazards involved in their possible use were not well known; technologically less advanced countries in particular were helpless even to detect the use of many of the more dangerous and insidious chemical and biological weapons in existence, hence the governments of those countries had no hope of protecting their populations. The need for publicity was vital. There was

---

[17] *Washington Post*, 15 February 1967.

also need for forging either a radically revised or a new international instrument which would establish an agreed standard of conduct for states in the field of chemical and biological weapons. [65]

The revised version of the Maltese draft resolution, of 13 December 1967, reaffirmed a UN resolution of 5 December 1966 calling for strict observance by all states of the principles and objectives of the Geneva Protocol; dropped the reference to radiological weapons; and included a request to the Eighteen-Nation Committee to consider the definition of chemical and biological weapons. [57]

On 11 December 1967, the delegation of Hungary tabled a draft resolution (subsequently co-sponsored by Madagascar and Mali) which, reaffirming the validity of the Geneva Protocol and regretting that there were states which still had not acceded to it, demanded strict and absolute compliance by all states with the principles and norms established by the Protocol; declared that the use of chemical and bacteriological weapons for the purpose of destroying human beings and the means of their existence constituted a crime against humanity; and appealed to those states which had not done so to accede to the Protocol. [58]

The Hungarian representative, in a statement of 12 December 1967, criticized the Maltese proposal. In his opinion, a revision of the Geneva Protocol was unwarranted. What was needed was not to update the Protocol, but to give a strong effect to its prohibitive clauses. It would be wrong, both politically and psychologically, to create, by a suggested revision, loopholes for those who wanted to avoid adhering to that instrument. The delegate of Hungary said that the effects of the use of chemical and bacteriological weapons were widely known. Chemical agents used in Viet-Nam upset the ecological balance of areas exposed to chemical raids. The rhythm of crop rotation became upset. The chemicals, washed into the streams, decimated or even killed off the entire fish population. By breaking the biological chain in plant life, they had a tremendous effect upon human existence in the areas concerned. Well-planted areas, under heavy cultivation, had been transformed into deserts. [59]

The Soviet representative also spoke about the use of gas and other poisonous substances in Viet-Nam. In his speech of 13 December 1967, he referred to the information furnished by the National Liberation Front of South Viet-Nam, according to which, in the course of 1965, those substances were the cause of 46 000 human casualties and of the loss of thousands of heads of cattle; 700 000 hectares of fields in Viet-Nam were poisoned in that year alone. He criticized the United States for its refusal to ratify the Geneva Protocol and argued that the Protocol had become a universally acknowledged norm of contemporary international law binding on all powers, including the USA, no matter what interpretation the US military might place upon it.

The USSR objected categorically to the proposal to revise the provisions of the Geneva Protocol; such a policy would lead only to the undermining of the legal standards which were a very important and fundamental restraint on chemical and bacteriological warfare. The Protocol condemned and banned the use of all forms of CB weapons, without any exception. [60]

Neither the Maltese nor the Hungarian draft was put to a vote.

VIII. *Summary and comment*

A number of restrictions imposed after the Second World War upon the defeated Germany, with a view to preventing the development of its military potential, had been eased by the mid-fifties when the Federal Republic was about to join the Western military and political organizations. However, before acceding to those organizations the West German government was invited to pledge that it would refrain from manufacturing in its territory chemical or biological weapons (in addition to nuclear weapons), and would submit to a supervision by competent authorities of the Western European Union.

The Protocol embodying a declaration to this effect is thus far the only instrument in existence containing undertakings not to produce atomic, biological and chemical weapons and providing for international control. The definitions are comprehensive and include in the category of chemical weapons irritant and growth-regulating substances.

The Treaty with Austria of 1955 also contained clauses prohibiting CB weapons.[18] But in the United Nations, measures specifically aimed at preventing chemical and biological warfare were then little discussed. They formed part of various proposals for an overall reduction and limitation of armaments and were mentioned in draft resolutions, memoranda and working papers often under a general heading of weapons of mass destruction. Some of the documents, submitted by France and Britain in the five-power subcommittee of the Disarmament Commission, suggested that under a disarmament convention information should be supplied by states on plants producing CB weapons, and that subsequently resident posts should be established or routine visits arranged to inspect the plants.

More detailed provisions were included in the Soviet and US plans for

---

[18] The peace treaties with Bulgaria, Finland, Hungary, Italy and Romania (signed on 10 February 1947), which imposed armaments limitations on those countries, included in the definition of "war material", as a separate category, asphyxiating, lethal, toxic or incapacitating substances intended for war purposes, or manufactured in excess of civilian requirements.

general and complete disarmament, but CBW was not given high priority in either of them. Main attention was devoted to nuclear weapons and the means of their delivery.

The cessation of production of CB weapons was envisaged at a second stage of disarmament, and the total destruction of stocks at a second (Soviet plan) or third stage (US plan). Verification of those measures (to be exercised by a special international organization) would constitute part of an overall system of supervision under a general disarmament treaty.

No progress was made in the discussion of general disarmament programmes. It was found necessary to adopt a step-by-step approach to a comprehensive agreement, by dealing with so-called partial or collateral measures of disarmament.

In the 1960s numerous complaints were made that chemicals were being used in Indo-China by the US forces in military operations and against the local population. In 1966, the matter was brought up in the United Nations by Hungary which, referring to the above charges, proposed that the use of chemical and bacteriological weapons for the purpose of destroying human beings and the means of their existence be declared an international crime.

The United States denied that it was engaged in chemical warfare. It argued that the substances used in Viet-Nam were not toxic and were not prohibited by the 1925 Geneva Protocol or any other international agreement: the herbicides employed involved chemicals commonly used to control vegetation, while tear gas was a riot-control agent which helped to reduce casualties. The USA thus reverted to the "humanitarian" argument which it had presented to the Preparatory Commission for the Disarmament Conference some thirty-six years before (see Chapter 3), and which it had later discarded (see Chapters 4 and 5).

Many members of the United Nations did not like the political undertones of the Hungarian proposal for CBW prohibition, but there was general support for a resolution calling for strict observance of the principles and objectives of the Geneva Protocol, condemning all actions contrary to those objectives, and inviting all states to accede to the Protocol.

In the light of the above resolution it may seem somewhat surprising that the United Nations took no action on the allegation of use of chemical weapons in the Yemen in 1967. The charge put forward by Saudi Arabia was that the UAR aircraft carried out attacks with lethal gases killing people as well as animals, and destroying some vegetation.

The United Arab Republic categorically refuted the accusations, but the UN's unwillingness even to discuss the matter was rather attributable to the crisis situation in the Middle East. In those circumstances many governments felt they had overriding reasons of national interest for not accusing the UAR. Besides, with both parties to the dispute being Arab, the need for solidarity may have prevailed in the face of war with Israel.

The attempt by Malta, made at the twenty-second UN General Assembly, to initiate a revision, updating or replacement of the Geneva Protocol by another instrument, met with little support.

Chapter 9. 1968-1969

I. *Proposal for separate consideration of biological methods of warfare*

In the course of the disarmament debate in 1968, reference was frequently made to chemical and biological warfare. The prohibition of CBW was generally considered one of the most urgent measures to be taken up following the conclusion of the Treaty on the Non-Proliferation of Nuclear Weapons.

The delegate of Ethiopia in the First Committee of the UN General Assembly said, on 6 May 1968, that, as a victim of the unrestricted use of gas on the eve of the Second World War, his country knew well the horrors of modern means of warfare, and had emerged from that experience stronger in its conviction that the use of all weapons of mass and indiscriminate destruction should be forever outlawed. [1]

On 1 July 1968, the USSR issued a memorandum concerning urgent measures to stop the arms race and achieve disarmament. Under the heading "Prohibition of the use of chemical and bacteriological weapons", the memorandum recalled the 1966 General Assembly resolution inviting all states to accede to the 1925 Geneva Protocol, and noted that that recommendation had not been implemented by certain countries, and above all by the USA. It further stated that the United States was "using chemical weapons in its aggressive war in Vietnam". The Soviet government proposed that the Eighteen-Nation Committee on Disarmament (ENDC) examine ways and means of ensuring that all states comply with the Geneva Protocol prohibiting the use of chemical and bacteriological weapons.[1] [2]

On 10 July 1968, the UN Secretary-General, speaking at a press conference in Geneva, said that it was unfortunate that the international com-

---

[1] Commenting on the Soviet memorandum, the US Deputy Secretary of Defense stated on 11 July 1968 that, although the United States was not a party to the Geneva Protocol of 1925, it had consistently supported the worthy objectives which the Protocol sought to achieve, and it believed that all states should do likewise. US General Wheeler said on the same day that the Soviet proposal was "obviously designed for other than serious negotiations". He based this observation on the fact that the proposal was accompanied by an accusation that the United States was involved in a war of aggression in Viet-Nam.
  The French government, in its reply of 19 August 1968 to the USSR memorandum, reasserted its interest in the universal implementation of the 1925 Protocol prohibiting the use of chemical and bacteriological weapons. It said it was prepared to study problems which might be raised by the prohibition not only of the use, but also of the manufacture, of such weapons; supervision was an essential condition for success.

253

munity was not aware of the dangers inherent in the development of materials for chemical and biological warfare. [3]

The Eighteen-Nation Disarmament Committee adopted a provisional agenda which, among non-nuclear measures, envisaged the discussion of the question of CB warfare. [32] On 16 July 1968, the British representative in the Committee said that he could not agree with those who claimed that nothing more was needed in the field of chemical and biological warfare than that all states should adhere to the 1925 Geneva Protocol and strictly observe its principles. There was a strong case for either revising the Geneva Protocol or trying to negotiate some additional instrument to clarify and strengthen its provisions while keeping the Protocol itself in being. The UK preference was for the latter course.

The problems involved in seeking to go beyond the Geneva Protocol seemed greater, and international opinion less clear, in the field of chemical weapons than in that of biological weapons. The former had already been used in war with terrible effect. The latter had never been used but they were generally regarded with even greater abhorrence.

It seemed, therefore, that one answer might be to make a distinction between chemical and biological weapons in the approach to the problems involved. It might be easier first to tackle agents of biological warfare and seek to conclude an instrument on biological warfare which would go beyond the Geneva Protocol and actually ban the production and possession of agents of biological warfare.

As far as chemical warfare was concerned, the British Representative thought that one must rest content for the moment with the Geneva Protocol. As an aid to further action, however, he took up a proposal contained in the draft resolution submitted by the Maltese delegation at the twenty-second session of the UN General Assembly and suggested that the Secretary-General should be requested to prepare a report on the nature and possible effects of chemical weapons and on the implications of their use, with a view to giving the ENDC an international scientific basis for future consideration of further measures for their limitation and control, as well as focusing public opinion on the issues involved. This would follow the precedent of the report on the effects of the possible use of nuclear weapons. [4]

Poland, on the other hand, suggested that the UN Secretariat should prepare a report on the effects of both chemical and bacteriological weapons. [5]

UK working paper on microbiological warfare

On 6 August 1968, the British delegation to the ENDC submitted the following working paper:

The United Kingdom Delegation consider that the 1925 Geneva Protocol is not an entirely satisfactory instrument for dealing with the question of chemical and microbiological warfare. The following points may be noted:

(i) Many states are not parties to the Protocol and of those that are parties many, including the United Kingdom, have reserved the right to use chemical and bacteriological weapons against non-parties, violators of the Protocol and their allies.

(ii) Jurists are not agreed whether the Protocol represents customary international law or whether it is of a purely contractual nature.

(iii) Even if all states were to accede to the Protocol there would still be a risk of large-scale use of the proscribed weapons as long as states have the right to manufacture such weapons and to use them against violators and their allies.

(iv) There is no consensus on the meaning of the term "gases" in the phrase "asphyxiating, poisonous or other gases and all analogous liquids, materials or devices". The French version of the Protocol renders "or other" as "ou similaires" and the discrepancy between "other" and "similaires" has led to disagreement on whether non-lethal gases are covered by the Protocol.

(v) The term "bacteriological" as used in the Protocol is not sufficiently comprehensive to include the whole range of microbiological agents that might be used in hostilities.

(vi) The prohibition in the Protocol applies to use "in war". There may therefore be doubt about its applicability in the case of hostilities which do not amount to war in its technical sense.

2. It is not to be expected that all these difficulties can be easily or speedily resolved. The United Kingdom Delegation suggest, however, that the problem might be made less intractable by considering chemical and microbiological methods of warfare separately. The Geneva Protocol puts them on an identical basis, but—

(i) As indicated in paragraph 1 (iv) above, there is disagreement on whether the ban covers all agents or only lethal ones. It would be extremely difficult to secure agreement on a new instrument banning the use of all agents of chemical warfare, particularly as some of those agents have legitimate peaceful uses for such purposes as riot control.

(ii) Chemical weapons have been used on a large scale in war in the past and are regarded by some states as a weapon they must be prepared to use if necessary in any future war, particularly as they fear they may be used against them. In any event, at the moment, they would be reluctant to give up the manufacture of chemical agents and the right to conduct research, etc., in this field.

3. The United Kingdom Delegation recognize that verification, in the sense in which the term is normally used in disarmament negotiations, is not possible in either the chemical or the microbiological field. The difficulty, as far as the microbiological field is concerned, is that the organisms which would be used are required for medical and veterinary uses and could be produced quickly, cheaply and without special facilities either in established laboratories or in makeshift facilities. As far as chemical agents are concerned it seems unlikely that states will be prepared to forego the right to produce and stockpile such agents for

possible use in war unless adequate verification procedures can be devised and applied and problems of definition, etc. resolved. However, the use of microbiological methods of warfare has never been established, and these are generally regarded with even greater abhorrence than chemical methods. The United Kingdom Delegation therefore consider that in this field the choice lies between going ahead with the formulation of new obligations and doing nothing at all—in which case the risks and the fears of eventual use of microbiological methods of warfare will continue and intensify indefinitely.

4. The United Kingdom Delegation therefore propose the early conclusion of a new Convention for the Prohibition of Microbiological Methods of Warfare, which would supplement but not supersede the 1925 Geneva Protocol. This Convention would proscribe the use for hostile purposes of microbiological agents causing death or disease by infection in man, other animals, or crops. Under it states would:—

(i) declare their belief that the use of microbiological methods of warfare of any kind and in any circumstances should be treated as contrary to international law and a crime against humanity;

(ii) undertake never to engage in such methods of warfare themselves in any circumstances.

5. The Convention should also include a ban on the production of microbiological agents which was so worded as to take account of the fact that most of the microbiological agents that could be used in hostilities are also needed for peaceful purposes. Thus the ban might be on production of microbiological agents on a scale which had no independent peaceful justification. Alternatively, the Convention might ban the production of microbiological agents for hostile purposes, or it might ban their production in quantities that would be incompatible with the obligation never to engage in microbiological methods of warfare in any circumstances.

6. Whatever the formulation might be, the ban would also need to cover ancillary equipment specifically designed to facilitate the use of microbiological agents in hostilities. In addition, the Convention would of course need to include an undertaking to destroy, within a short period after the Convention comes into force, any stocks of such microbiological agents or ancillary equipment which are already in the possession of the parties.

7. The Convention would also need to deal with research work. It should impose a ban on research work aimed at production of the kind prohibited above, as regards both microbiological agents and ancillary equipment. It should also provide for the appropriate civil medical or health authorities to have access to all research work which might give rise to allegations that the obligations imposed by the Convention were not being fulfilled. Such research work should be open to international investigation if so required and should also be open to public scrutiny to the maximum extent compatible with national security and the protection of industrial and commercial processes.

8. In the knowledge that strict processes of verification are not possible, it is suggested that consideration might be given *inter alia* to the possibility that a competent body of experts, established under the auspices of the United Nations, might investigate allegations made by a party to the Convention which appeared to establish a *prima facie* case that another party had acted in breach of the

obligations established in the Convention. The Convention would contain a provision by which parties would undertake to co-operate fully in any investigation and any failure to comply with this or any of the other obligations imposed by the Convention would be reported to the Security Council.

9. As regards entry into force of the Convention, the appropriate international body might be invited to draw up a list of states (say 10–12) that it considers most advanced in microbiological research work. The Convention might come into force when ratified by all those states and a suitably large number of other states.

10. Consideration should be given to the possibility of including in the Convention an article under which the parties would undertake to support appropriate action in accordance with the United Nations Charter to counter the use, or threatened use, of microbiological methods of warfare. If such an article were included it might be endorsed by the Security Council in rather the same way as the Council welcomed and endorsed the declarations made by the United States, the Soviet Union and the United Kingdom in connection with the Non-Proliferation Treaty. [6]

Discussion of the UK working paper

The British representative explained that he had used the term "microbiological" rather than "biological", because man himself may be regarded as a "biological agent". He stressed that the 1925 Geneva Protocol should remain in force, but contested the opinion that the Protocol had prevented the use of CB weapons in the past, notably in World War II, and that by implication the Protocol could be relied upon to prevent the use of these weapons in the future. He knew of no evidence to support the view that Hitler had not resorted to the use of gas because of respect for the Geneva Protocol, and recalled Hitler's lack of concern for the Geneva Convention concerning prisoners of war. In his opinion a more likely explanation of Germany's restraint was fear of retaliation. Chemical warfare had been used in the 1930s and had been used again since then. The most eloquent evidence of the fear of the use of these weapons, and of lack of faith in the Protocol's power to prevent their use, lay in the fact that the armed forces of all the major powers were trained and equipped to defend themselves at any rate against chemical methods of warfare, and that those countries were engaged in expensive research programmes to produce counter-measures against attack by microbiological agents. The British delegation was sure that neither of these precautions would be abandoned even if the Protocol was ratified by all states.

The representative of the UK further recognized that the greatest difficulty to be faced in connection with a possible convention was that of verification, since parties to any arms control or disarmament agreement were entitled to be reasonably satisfied to the greatest practicable extent that other parties were carrying out their obligations under the agreement. This principle was well illustrated by the safeguards requirements of the Non-Proliferation Treaty. However, no comparable system was possible for microbiological or chemical weapons. Any such system would be so intrusive as to be quite unacceptable, and even then could not be fully effective. Almost all the material and equipment to be dealt with had legitimate peaceful purposes; it would be wrong to inhibit work of real value to humanity in combating disease, for example, and impracticable to inspect every laboratory in every country. Failing a fully effective system of verification, and the British delegation believed that it was beyond the wit of man to devise one, arrangements could be provided which should satisfy states that they would not be exposing themselves to unacceptable risks. [7]

The Soviet delegate considered that the British proposal meant the reopening of issues which had long been solved. The Geneva Protocol, a useful and important international document, might be destroyed without being replaced by a better or indeed by any other instrument that would provide for the prohibition of the use of CB weapons. The need was rather for ensuring that the Protocol should be strictly observed. Having achieved this, the next measure could be the cessation of the manufacture of CB weapons and their destruction. [8]

The USSR supported the idea advanced by the delegation of Poland concerning the preparation of a report on the consequences of the possible use of chemical and bacteriological weapons. [9]

India could not agree with the view that the Geneva Protocol, being a declaratory prohibition and without a system of international control—which in this case it would be extremely difficult to provide—was of no use. [10]

The representative of Sweden considered the Geneva Protocol as part and parcel of established international law. The Protocol should be strengthened through accession by all states and the abolition of reservations made to it. Nothing should be done that would damage or undermine the ban already expressed in it. Moreover, it would be both desirable and natural explicitly to give the Protocol a broad interpretation precisely in order to retain it without amendment, and to consider all existing biological and chemical weapons as belonging in one set and the prohibition to use any of them as valid without exceptions. Some joint collective statement in the General Assembly or elsewhere might be useful which, without regard to the various positions and practices of the past as to the extent of the existing ban, would enable states to register adherence to a ban on all biological and chemical means of warfare, comprehensively interpreted. Sweden did not share the British view that the treatment of biological weapons should be separated from that of chemical weapons. Beyond the confirmation of and extended adherence to the Geneva Protocol, problems must be tackled related to the possibility of prohibiting also the dissemination, production and stockpiling of biological and chemical means of warfare, including the means of their delivery, as well as concerning their elimination. The difficulties, which were often presented as stumbling blocks, resided in the question of control. Perfect control over the production and possession of biological and chemical weapons was not possible. However the following measures might be suggested:

Universal openness about activities in this field should be attained in order gradually to create confidence. When an international ban on production of biological and chemical means of warfare comes into being, coupled with openness about laboratories and factories, considerable protection against violations should be obtained already through an element of what one might call "control through public morale"—or even "control through public shame".

An international agency, for instance the World Health Organization, might undertake a key role in collecting, systematizing and disseminating all information pertaining to CBW available from national and scientific sources. Such a continuous survey could serve the control function of monitoring any suspicious build-up of capabilities for biological and chemical warfare.

A system of periodic reporting could be worked out under which states would transmit information about resources, stocks and research in factories, stores and laboratories, about personnel employed, future plans, and so on. Needs for peaceful purposes should then be indicated. The activities in the sphere of science which should be made the subject of reporting would have to be defined. Lists might be drawn up and periodically revised by agreement.

More active steps in such a gradually expandable verification system would imply efforts to check against possible lacunae in the flow of information of suspicious trends, to press for further information, to question the appropriateness of certain research or stockpiling. That would constitute the beginning of a process of "verification-by-challenge".

Finally, thought would have to be given to the acceptability of some system of inspection *in loco,* voluntary, by mutual visits to laboratories by scientific experts, or prescribed in a treaty. The task would not be too difficult, particularly in regard to biological means of warfare. Inspectors could visit laboratories and factories of possible interest from the viewpoint of warfare capacity. The scientific documentation, plus the periodic reporting and the systematized compilation mentioned above, would constitute a preliminary control system and could then serve as a point of departure for possible further investigations by inspectors. The whole sequence might be made fairly similar to the control system provided by the International Atomic Energy Agency safeguards for controlling the Non-Proliferation Treaty. [11]

The US representative felt that a serious problem was posed by the need to verify a ban on the production and possession of microbiological agents. If the British proposal for such a ban had wide support in principle, he would recommend that a working group be formed under the auspices of the Eighteen-Nation Disarmament Committee to study this problem as well as other problems relating to such a ban, and that this group report at a later date. He said he could support a study on the nature and possible effects of chemical and bacteriological weapons, either together or separately. To support such a study was not to imply that the Geneva Protocol should be revised, superseded or supplemented. The US government had made no decision on whether revision was required. [8]

On 28 August 1968, the ENDC agreed to recommend to the UN General Assembly that the Secretary-General appoint a group of experts to study the effects of the possible use of chemical and bacteriological means of warfare. [12]

II. *United Nations decision to study the effects of CBW*

Secretary-General's suggestions

The UN Secretary-General's annual *Report on the Work of the Organization,* published in September 1968, contained the following passages concerning CBW:

> While progress is being made in the field of nuclear disarmament, there is another aspect of the disarmament problem to which I feel too little attention has been devoted in recent years. The question of chemical and biological weapons has been overshadowed by the question of nuclear weapons, which have a destructive power several orders of magnitude greater than that of chemical and biological weapons. Nevertheless, these too are weapons of mass destruction regarded with universal horror. In some respects they may be even more dangerous than nuclear weapons because they do not require the enormous expenditure of financial and scientific resources that are required for nuclear weapons. Almost all countries, including small ones and developing ones, may have access to these weapons, which can be manufactured quite cheaply, quickly and secretly in small laboratories or factories. This fact in itself makes the problem of control

and inspection much more difficult. Moreover, since the adoption, on 17 June 1925, of the Geneva Protocol for the Prohibition of the Use in War of Asphyxiating, Poisonous or Other Gases and of Bacteriological Methods of Warfare, there have been many scientific and technical developments and numerous improvements, if that is the right word, in chemical and biological weapons, which have created new situations and new problems. On the one hand, there has been a great increase in the capability of these weapons to inflict unimaginable suffering, disease and death to ever larger numbers of human beings; on the other hand, there has been a growing tendency to use some chemical agents for civilian riot control and a dangerous trend to accept their use in some form in conventional warfare.

Two years ago, by resolution 2162 B (XXI), the General Assembly called for the strict observance by all States of the principles and objectives of the Geneva Protocol of 1925, condemned all actions contrary to those objectives and invited all States to accede to the Protocol. Once again I would like to add my voice to those of others in urging the early and complete implementation of this resolution. However, in my opinion, much more is needed.

During the twenty-three years of the existence of the United Nations, there has never been a thorough discussion in any United Nations organ of the problems posed by chemical and biological weapons, nor has there been a detailed study of them. Recently the matter has been receiving more attention and it is felt that the time has come to deal with it more fully. I therefore welcome the recommendation of the Conference of the Eighteen-Nation Committee on Disarmament to the General Assembly that the Secretary-General appoint a group of experts to study the effects of the possible use of chemical and bacteriological means of warfare. I believe that such a study, which would explore and weigh the dangers of chemical and biological weapons, would prove to be a most useful undertaking at the present time. It could attract attention to an area of multiplying dangers and of diminishing public appreciation of them. It could also serve to clarify the issues in an area which has become increasingly complex. Certainly a wider and deeper understanding of the dangers posed by these weapons could be an important element in knowing how best to deal with them. [13]

Consideration of the proposal for a CBW study

In the autumn of 1968 the problem of chemical and biological warfare was discussed at the twenty-third session of the UN General Assembly.

On 18 November 1968, the delegations of Canada, Hungary, India, Mexico, Poland, Sweden and the United Arab Republic submitted to the First Committee of the Assembly a draft resolution, which (1) requested the Secretary-General to prepare a concise report in accordance with the proposal contained in the introduction to his annual report and the recommendation of the ENDC; (2) recommended that the report be based on accessible material and prepared with the assistance of qualified consultant experts appointed by the Secretary-General; (3) called upon governments, national and international scientific institutions and organizations to cooperate in the preparation of the report; (4) requested that the report be transmitted to the ENDC, the UN Security Council and the General Assembly, possibly by 1 July 1969, and to the governments of member states; (5) recommended that governments should give the report wide distribution, so as to acquaint public opinion with its contents; (6) reiterated the call for strict observance by all states of the principles and objectives of the 1925 Geneva Protocol and invited all states to accede to that Protocol. [14]

The delegation of Malta, later joined by that of Trinidad and Tobago, formally proposed that the Secretary-General's report should include an indication of the nature, means of delivery, and effects, including economic and health implications, of the possible use of chemical, bacteriological and other biological means of warfare, with particular reference to the position of states that are unable to establish adequate measures of detection and protection against the possible use of these means of warfare. [15]

The delegate of Poland submitted that the purpose of the Secretary-General's report should be to inform fully and authoritatively all governments and world public opinion about the consequences of the possible use of chemical and bacteriological means of warfare; to be instrumental in strengthening the prohibition contained in the Geneva Protocol; to facilitate further examination of CBW problems and, above all, speed up the solution of the question of strict and universal observance of the banning of those weapons. He doubted whether it would be practicable to include in the resolution a detailed list of problems to be studied. He felt that the Secretary-General and the experts should be given a broad mandate. [16]

The representative of Malta explained that his main concern was that the proposed study should take into account the position in which the majority of UN members found themselves, namely that of not possessing adequate means of detection and of being unable to take measures of protection against these weapons in the event of their use. [17]

The representative of Hungary criticized the Maltese proposal, saying that there was nothing the proponents of chemical and bacteriological warfare would accept more willingly than a specification of weapons. That would make things easier for them—he argued—since the weapons kept secret would not be included in the list, and they would have a free hand to use them. It is "precisely the fact that the prohibition was formulated in the Geneva Protocol in such a comprehensive way, permitting an interpretation *per analogiam,* that makes every attempt to 'improve' that international instrument suspicious". The Hungarian delegate quoted the conclusions reached by lawyers participating in the International Conference on Human Rights, organized by the United Nations in Teheran in the spring of

1968: (a) that the Geneva Protocol of 1925 is validly included in the present law of warfare; and (b) that all states which have not yet done so should become parties to the Protocol. He also named instances of use of chemical or bacteriological warfare: by Italy against the Ethiopians; by Japan against China and against US troops; by the USA in Korea and in Viet-Nam. He said that such formidable weapons were used against the developing countries and the liberation movements. [18]

The Mongolian representative insisted that the question of respect for the provisions of the 1925 Geneva Protocol was an urgent matter "since the United States continues to use poison gas and other chemical means of warfare in South Vietnam", and since "the colonialists in southern Africa use poisonous means of warfare against the patriotic forces of the African people". [19]

The Soviet delegation expressed conviction that the adoption of the draft resolution would be conducive to strict respect for the provisions of the 1925 Geneva Protocol by all states, and would encourage adherence to the Protocol. It categorically opposed the amendments proposed by Malta as irrelevant and as leading to a revision or weakening of the Geneva Protocol. [20]

Ireland hoped that the report of the experts would enable conclusions to be drawn on the adequacy of the Geneva Protocol of 1925 in the present conditions. [21]

The Belgian delegate drew attention to the existence of some gases which were neither asphyxiating nor poisonous and were sometimes used as means of internal repression. He asked that the recommended study should establish a clear-cut distinction between chemical weapons falling under the Geneva Protocol and tear gas and other gases which were in police arsenals. [22]

The representative of Australia said that the study should not cover agents such as defoliants, herbicides and riot-control agents, "to which the 1925 Protocol clearly does not apply". [23]

The French representative stated that his government, which was the depositary of the Geneva Protocol, attached great importance to that international instrument and deemed it necessary to keep it in force. The French delegation had no objections to entrusting the UN Secretary-General with the task of investigating the effects of the possible uses of chemical and bacteriological weapons, it being understood that the experts would be able to deal with all biological weapons. France would like to see a later study on the possibility of prohibiting, under effective control, the manufacture of weapons dealt with in the draft resolution. [17]

The US representative suggested that the study to be prepared by the UN Secretary-General should deal equally and individually with the effects of chemical and biological weapons. In his view, the scientific and technological differences between the two systems, as well as differences which obtained in their operational applications, warranted such a particular approach to each category of weapons. While the language in the ENDC's recommendation specifically referred to chemical and bacteriological means of warfare, the latter should embrace the types of weapons also referred to as biological. This form of warfare was also at times referred to as microbial, bacterial, microbiological, or germ warfare. It should be understood that it meant disease-causing living micro-organisms, "be they bacteria, or viruses or whatever they might be, used as deliberate weapons of war". The US representative stressed that his delegation regarded operative paragraph 6 of the draft resolution as not intended to prejudge for political purposes the results of the study to be undertaken. [24]

Italy formulated some criteria which—in its view—might underlie future negotiations:

The Geneva Protocol of 1925 seemed to be inadequate, primarily because the definitions contained therein did not cover the entire range of modern chemical weapons. At the same time, the Protocol had a certain use, because any new agreement must complete it and not replace it or weaken it.

In order to be able to prohibit the manufacture of chemical and bacteriological weapons, some system of control had to be devised, and any such system must be set up on a non-discriminatory basis.

Any new agreement must also be imbued with the principle of universality rather than that of reciprocity; this presupposed the adherence of a number of countries, sufficiently broad and qualified, to make it worth while. [25]

The draft resolution, sponsored by twenty-one powers, was revised so as to recommend that the Secretary-General's report also take into account the views expressed and the suggestions made during the discussion of this item at the twenty-third session of the General Assembly. [17]

Following this amendment, Malta and Trinidad and Tobago withdrew their proposal.

The draft was adopted in the First Committee on 10 December 1968, by a vote of 112 to none, with one abstention; [17] and on 20 December 1968, by the General Assembly, by a vote of 107 to none, with two abstentions [26].

The adopted text was as follows:

The General Assembly,

Reaffirming the recommendations contained in its resolution 2162 B (XXI) of 5 December 1966 calling for strict observance by all States of the principles and objectives of the Protocol for the Prohibition of the Use in War of Asphyxiating, Poisonous or Other Gases, and of Bacteriological Methods of Warfare signed at Geneva on 17 June 1925, condemning all actions contrary to those objectives and inviting all States to accede to that Protocol,

Considering that the possibility of the use of chemical and bacteriological weapons constitutes a serious threat to mankind,

Believing that the people of the world should be made aware of the consequences of the use of chemical and bacteriological weapons,

Having considered the report of the Conference of the Eighteen-Nation Committee on Disarmament which recommended that the Secretary-General should appoint a group of experts to study the effects of the possible use of such weapons,

Noting the interest in a report on various aspects of the problem of chemical, bacteriological and other biological weapons which has been expressed by many Governments and the welcome given to the recommendation of the Conference of the Eighteen-Nation Committee on Disarmament by the Secretary-General in the introduction to his annual report on the work of the Organization submitted to the General Assembly at its twenty-third session,

Believing that such a study would provide a valuable contribution to the consideration by the Conference of the Eighteen-Nation Committee on Disarmament of the problems connected with chemical and bacteriological weapons,

Recalling the value of the report of the Secretary-General on the effects of the possible use of nuclear weapons,

1. Requests the Secretary-General to prepare a concise report in accordance with the proposal contained in paragraph 32 of the introduction to his annual report on the work of the Organization submitted to the General Assembly at its twenty-third session and in accordance with the recommendation of the Conference of the Eighteen-Nation Committee on Disarmament contained in paragraph 26 of its report,

2. Recommends that the report should be based on accessible material and prepared with the assistance of qualified consultant experts appointed by the Secretary-General, taking into account the views expressed and the suggestions made during the discussion of this item at the twenty-third session of the General Assembly;

3. Calls upon Governments, national and international scientific institutions and organizations to co-operate with the Secretary-General in the preparation of the report;

4. Requests that the report be transmitted to the Conference of the Eighteen-Nation Committee on Disarmament, the Security Council and the General Assembly at an early date, if possible by 1 July 1969, and to the Governments of Member States in time to permit its consideration at the twenty-fourth session of the General Assembly;

5. Recommends that Governments should give the report wide distribution in their respective languages, through various media of communication, so as to acquaint public opinion with its contents;

6. Reiterates its call for strict observance by all States of the principles and objectives of the Protocol for the Prohibition of the Use in War of Asphyxiating, Poisonous or Other Gases, and of Bacteriological Methods of Warfare signed at Geneva on 17 June 1925, and invites all States to accede to that Protocol. [27]

III. Report of the UN Secretary-General on CB weapons

During the ENDC session in spring 1969, the United Kingdom gave notice of its intention to submit shortly a draft convention prohibiting biological weapons. The convention—indicated the British delegate—would go further than the 1925 Geneva Protocol in covering all weapons now understood by the word "biological", and in prohibiting absolutely not only their use but also their production and possession. [28]

Sweden again expressed the belief that an authoritatively agreed clarification as to the widest possible interpretation of the prohibitions contained in the Geneva Protocol should be obtained in regard to biological and chemical means of warfare. [29]

Most Committee members, however, preferred to defer action until the Secretary-General's report on the effects of the possible use of CB weapons had been submitted.

In accordance with the UN resolution of 20 December 1968, the UN Secretary-General appointed a fourteen member group of consultative experts[2] to assist him in the preparation of the report. Their report was published on 1 July 1969. [30] It described the basic characteristics of CB weapons; their probable effects on military and civilian personnel; environmental factors affecting the use of CB weapons; possible long-term effects on human health and ecology; economic and security implications of the

---

[2] They were the following,

Dr Tibor Bakacs, Professor of Hygiene, Director-General of the National Institute of Public Health, Budapest; Dr. Hotse C. Bartlema, Head of the Microbiological Department of the Medical-Biological Laboratory, National Defence Research Organization TNO, Rijswijk, Netherlands; Dr Ivan L. Bennett, Director of the New York University Medical Center and Vice-President for Medical Affairs, New York University; Dr S. Bhagavantam, Scientific Adviser to the Minister of Defence, New Delhi; Dr Jiri Franek, Director of the Military Institute for Hygiene, Epidemiology and Microbiology, Prague; Dr Yosio Kawakita, President of University of Chiba, Professor of Bacteriology, Chiba City, Japan; M. Victor Moulin. *Ingénieur en chef de l'armement, Chef du Bureau Défense chimique et biologique, Direction technique des armements terrestres,* Saint Cloud, France; Dr M. K. McPhail, Director of Chemical and Biological Defence, Defence Chemical, Biological and Radiation Laboratories, Defence Research Board, Ottawa; Academician O. A. Reutov, Professor of Chemistry at the Moscow State University, Moscow; Dr Guillermo Soberón, Director, *Instituto de Investigaciones Biomédicas, Universidad Nacional Autónoma de México,* Mexico City; Dr Lars-Erik Tammelin, Chief of Department for Medicine and Chemistry, Research Institute for National Defence, Stockholm; Dr Berhane Teoume-Lessane, Medical Co-Director and Head of Department of Viruses and Rickettsiae, Imperial Central Laboratory and Research Institute, Addis Ababa; Colonel Zbigniew Zoltowski, Professor of Medicine, Epidemiologist and Scientific Adviser to the Ministry of National Defence, Warsaw; Sir Solly Zuckerman, Chief Scientific Adviser to the Government of the United Kingdom, Professor Emeritus, University of Birmingham. William Epstein, Director of the Disarmament Affairs Division of the United Nations Secretariat, served as Chairman of the group of experts, and Alessandro Corradini as Secretary of the group.

development, acquisition and possible use of CB weapons and systems of their delivery. It is briefly summarized here:

Introduction

No form of warfare has been more condemned than has the use of CB weapons.

Most of the knowledge concerning the use of chemical weapons is based upon the experience of the First World War. The agents used in that war were much less toxic than those which could be used today, and they were dispersed by means of relatively primitive equipment as compared with what is now available and in accordance with battlefield concepts of a relatively unsophisticated kind.

Since the Second World War, bacteriological (biological) weapons have also become an increasing possibility. But because there is no clear evidence that these agents have ever been used as modern military weapons, discussions of their characteristics and potential threat have to draw heavily upon experimental field and laboratory data and on studies of naturally occurring outbreaks and epidemics of infectious disease, rather than on direct battlefield experience.

The outstanding characteristic of these weapons, particularly the biological, is the variability—amounting under some circumstances to unpredictability—of their effects. Depending on environmental and meteorological conditions, and on the agent used, the effects might be devastating or negligible. They could be localized or widespread. They might bear not only on those attacked but also on the side that initiated their use, whether or not the attacked military forces retaliated in kind. Civilians would be even more vulnerable than the military.

To appreciate the risks of biological warfare, one has only to remember how a natural epidemic may persist unpredictably and spread far beyond the initial area of incidence, even when the most up-to-date medical resources are used to suppress the outbreak. The difficulties would be considerably increased were deliberate efforts made to propagate pathogenic organisms. Mass disease following an attack, especially of civilian populations, could be expected not only because of the lack of timely warning, but also because effective measures of protection or treatment simply do not exist or cannot be provided on an adequate scale.

Chemical and biological warfare agents

Chemical agents of warfare are taken to be chemical substances—whether gaseous, liquid, or solid—which might be employed because of their direct toxic effects on man, animals and plants. Bacteriological agents of warfare

are living organisms, whatever their nature, or infective material derived from them, which are intended to cause disease or death in man, animals, or plants, and which depend for their effects on their ability to multiply in the person, animal, or plant attacked.

All biological processes depend upon chemical or physico-chemical reactions, and what may be regarded today as a biological agent could, tomorrow, as knowledge advances, be treated as chemical. Because they themselves do not multiply, toxins, which are produced by living organisms, are treated in the report as chemical substances. As a class, chemical agents produce their injurious effects more rapidly than do bacteriological agents. The time between exposure and significant effect may be minutes, or even seconds, for highly toxic gases or irritating vapours. Blister agents take a few hours to produce injury. Most chemicals used against crops elicit no noticeable effect until a few days have elapsed. On the other hand, a bacteriological agent must multiply in the body of the victim before disease (or injury) supervenes. This period is rarely as short as one or two days, and may be as long as a few weeks or even longer. For both chemical and bacteriological agents the speed of action is affected by the quantity absorbed, but this secondary factor does not obscure the basic difference between the two classes of agents in the time they take to manifest their effects.

The effects of most chemical agents that do not kill quickly do not last long, except in the case of some agents such as phosgene and mustard, where they might continue for some weeks, months, or longer. On the other hand, bacteriological agents that are not quickly lethal cause illness lasting days or even weeks, and on occasion involve periods of prolonged convalescence. The effects of agents which act against plants and trees could last for weeks or months and, depending on the agent and the species of vegetation attacked, could result in death.

Because they infect living organisms, some bacteriological agents can be carried by travellers, migratory birds, or animals to localities far from the area originally attacked. The possibility of this kind of spread does not apply to chemical agents. But control of contamination by persistent chemical agents could be very difficult. Should large quantities of chemical agents penetrate the soil and reach underground waters, or should they contaminate reservoirs, they might spread hundreds of kilometers from the area of attack, affecting people remote from the zone of military operations.

In circumstances that favour their persistence, herbicides, defoliants, and perhaps some other chemical agents might linger for months, stunting the growth of surviving or subsequent plant life, and even changing the floral pattern through selection. Following repeated use, certain chemical agents could even influence soil structure. The risk of residual effects with some bacteriological agents is potentially greater, mainly because they could lead to disease that might become epidemic if man-to-man transmission occurred readily. Bacteriological agents also might find unintended hosts in the animals and plants of an area, or be transported by infected individuals over great distances to new environments.

The experts described some of the properties of chemical agents affecting man and animals: nerve agents; blister agents (vesicants); choking agents; blood agents; toxins; tear and harassing gases; psychochemicals; as well as agents affecting plants—herbicides (defoliants). They also gave the characteristics of the methods of delivery of those agents.

The following calculations were presented of possible effects of a nerve gas attack on a city:

The population density in a modern city may be 5 000 people per square kilometre. A heavy surprise attack with non-volatile nerve gas by bombs exploding on impact in a wholly unprepared town would, especially at rush hours, cause heavy losses. Half of the population might become casualties, half of them fatal, if about 1 ton of agent were disseminated per square kilometre. If such a city were prepared for attack, and if the preparations included a civil defence organization with adequately equipped shelters and protective masks for the population, the losses might be reduced to one half of those which would be anticipated in conditions of total surprise.

Given a town with a total population of 80 000, a surprise attack with nerve gas could thus cause 40 000 casualties, half of them fatal, whereas under ideal circumstances for the defence, fatalities might number no more than 2 000. It is inconceivable, however, that the ideal would ever be attained.

Bacteriological agents could be used with the intention of killing people or of incapacitating them for short or long periods. The agents, however, cannot be defined rigidly as either lethal or incapacitating, since their effects are dependent upon many factors relating not only to themselves but also to the individuals they attack. Any disease-producing agent intended to incapacitate may, under certain conditions, bring about a fatal disease. Similarly, attacks that might be intended to provoke lethal effects might fail to do so.

A number of natural diseases of man and domestic animals are caused by mixed infections (e.g., swine influenza, hog cholera). The possible use of two or more different organisms in combination in bacteriological warfare needs to be regarded seriously because the resulting diseases might be aggravated or prolonged.

Victims of an attack by bacteriological weapons would, in effect, have contracted an infectious disease. The diseases would probably be known, but their symptoms might be clinically modified.

The experts described some of the properties of selected viruses, rickettsiae, bacteria, fungi, protozoa, in relation to their effects on man, animals and plants; as well as the methods of their possible delivery.

As in the case of chemical agents, estimates were given of possible consequences of a biological attack, and the conclusion reached was that large-scale bacteriological attacks could have a serious impact on the entire target country. Depending on the type of agent used, the disease might well spread to neighbouring countries.

Whatever might be done to try to save human beings, nothing significant could be done to protect crops, livestock, fodder, and foodstuffs from a chemical and bacteriological weapons attack. Persistent chemical agents could constitute a particular danger to livestock. Water in open reservoirs could be polluted as a result of deliberate attack, or perhaps accidentally, with chemical or bacteriological weapons. The water supply of large towns could become unusable, and rivers, lakes and streams might be temporarily contaminated.

Comparative estimates were made of disabling effects of hypothetical attacks on totally unprotected populations using a nuclear, chemical or bacteriological (biological) weapon that could be carried by a single strategic bomber. According to those estimates the area affected would be: in the case of a nuclear weapon (1 megaton)—up to 300 sq. km; in the case of a chemical weapon (15 tons of nerve agent)—up to 60 sq. km; in the case of a biological weapon (10 tons)—up to 100 000 sq. km.

Production of weapons

Today a large number of industrialized countries have the potential to produce a variety of chemical agents. Many of the intermediates required in their manufacture, and in some cases even the agents themselves, are widely used in peacetime. Such substances include, for example, phosgene, which some highly developed countries produce at the rate of more than 100 000 tons a year; ethylene-oxide, which is used in the manufacture of mustard gases, is also produced on a large scale in various countries; mustard gas and nitrogen mustard gases can be produced from ethylene-oxide by a relatively simple process. Similar remarks were made with regard to biological agents. The development of sophisticated and comprehensive weapons systems for chemical or bacteriological warfare would require a very costly additional effort. Nonetheless, the possibility that a peacetime industry could be converted to work for military purposes increases the responsibility of governments that are concerned about preventing chemical and biological war from ever breaking out.

Despite the fact that the development and acquisition of a sophisticated armoury of chemical and bacteriological weapons systems would prove very costly in resources and would be dependent on a sound industrial base and a body of well-trained scientists, any developing country could in fact acquire, in one way or another, a limited capability in this type of warfare—either a rudimentary capability that it developed itself, or a more sophisticated one that it acquired from another country.

Conclusion

All weapons of war are destructive to human life, but chemical and bacteriological weapons stand in a class of their own as armaments that exercise their effects solely on living matter. The idea that bacteriological weapons could deliberately be used to spread disease generates a sense of horror. The fact that certain chemical and bacteriological agents are potentially unconfined in their effects, both in space and time, and that their large-scale use could conceivably have deleterious and irreversible effects on the balance of nature, adds to the sense of insecurity and tension. Considerations such as these set them into a category of their own in relation to the continuing arms race.

The present inquiry has shown that the potential for developing an armoury of chemical and bacteriological weapons has grown considerably in recent years, not only in terms of the number of agents, but also in their toxicity and in the diversity of their effects. At one extreme, chemical agents exist and are being developed for use in the control of civil disorders; others have been developed in order to increase the productivity of agriculture. But even though these substances may be less toxic than most other chemical agents, their ill-considered civil use, or use for military purposes, could turn out to be highly dangerous. At the other extreme, some potential chemical agents that could be used in weapons are among the most lethal poisons known. In certain circumstances the area over which some of them might exercise their effects could be strictly confined geographically. In other conditions some chemical and bacteriological weapons might spread their effects well beyond the target zone. No one could predict how long the effects of certain agents, particularly bacteriological weapons, might endure and spread and what changes they could generate.

Moreover, chemical and bacteriological weapons are not a cheap substitute for other kinds of weapons. They represent an additional drain on the national resources of those countries by which they are developed, produced, and stockpiled. The cost cannot be estimated with precision; this would depend on the potential of a country's industry. To some the cost might be tolerable; to others it would be crippling, particularly when account is taken of the resources that would have to be diverted to the development of testing and delivery systems. No system of defence, even for the richest countries of the world, and whatever its cost, could be completely secure.

Because chemical and bacteriological weapons are unpredictable, in varying degree, either in the scale or duration of their effects, and because no certain defence can be planned against them, their universal elimination would not detract from any nation's security. Once any chemical or bacteriological weapon had been used in warfare, there would be a serious risk of escalation, both in the use of more dangerous weapons belonging to the same class and of other weapons of mass destruction. In short, the development of a chemical or bacteriological armoury and a defence implies an economic burden without necessarily imparting any proportionate compensatory advantage to security. At the same time it imposes a new and continuing threat to future international security.

Were these weapons ever to be used on a large scale in war, no one could predict how enduring the effects would be, and how they would affect the structure of society and the environment in which we live. This overriding danger would apply as much to the country that initiated the use of these weapons as to the one which had been attacked, regardless of what protective measures it might have taken in parallel with its development of an offensive capability. A particular danger also derives from the fact that any country could develop or acquire, in one way or another, a capability in this type of warfare, despite the fact that this could prove costly. The danger of the proliferation of this class of weapons applies as much to the developing as it does to developed countries.

The momentum of the arms race would clearly decrease if the production of these weapons were effectively and unconditionally banned. Their use, which could cause an enormous loss of human life, has already been condemned and prohibited by international agreements, in particular the Geneva Protocol of 1925, and, more recently, in resolutions of the General Assembly of the United Nations. The prospects for general and complete disarmament under effective international control and hence for peace throughout the world, would brighten significantly if the development, production, and stockpiling of chemical and bacteriological agents intended for purposes of war were to end and if they were eliminated from all military arsenals.

If this were to happen, there would be a general lessening of international fear and tension. It is the hope of the authors that this report will contribute to public awareness of the profoundly dangerous results if these weapons were ever used, and that an aroused public will demand and receive assurances that governments are working for the earliest effective elimination of chemical and bacteriological weapons.

The Secretary-General's proposals

In the foreword the Secretary-General said that he had decided to accept the unanimous report of the consultant experts in its entirety. He urged that the following measures be undertaken:

1. To renew the appeal to all states to accede to the Geneva Protocol of 1925;

2. To make a clear affirmation that the prohibition contained in the Geneva Protocol applies to the use in war of all chemical, bacteriological and biological agents (including tear gas and other harassing agents), which now exist or which may be developed in the future;

3. To call upon all countries to reach agreement to halt the development, production and stockpiling of all chemical and bacteriological (biological) agents for purposes of war and to achieve their effective elimination from the arsenal of weapons.

Consideration of the Secretary-General's report

The UN Secretary-General's report was welcomed in the CCD[3] and in the UN General Assembly as a contribution toward increased knowledge about CBW and as a useful basis for consideration of CBW problems.

Some criticism was voiced by the representative of Brazil who felt that the report fell short of its purpose: except for a few historical examples, not a single country was mentioned; no expenditure figures were given; the numbers of laboratories and of people working for military purposes in this field were not specified. Ecuador held similar views. Australia made a reservation about the Secretary-General's suggestion that the UN should interpret the Geneva Protocol to include such agents as tear gas, herbicides and defoliants. The United Kingdom thought that the right course would be to start by banning the use as well as the production and possession of

[3] In summer 1969, the membership of the Eighteen-Nation Committee on Disarmament was enlarged. Japan and Mongolia joined the Committee on 3 July; Argentina, Hungary, Morocco, the Netherlands, Pakistan and Yugoslavia joined on 7 August 1969.

On 26 August 1969, it was decided that the new name of the Committee would be "The Committee on Disarmament" and that the new name of the Conference would be "The Conference of the Committee on Disarmament" (CCD).

The twenty-fourth UN General Assembly endorsed the agreement on the title and on the composition of the Committee.

biological weapons. The report was, nonetheless, commended in a UN resolution of 16 December 1969 [33] and helped to move CB disarmament to the forefront of disarmament negotiations.

IV. *WHO report on CBW hazards to health*

On 28 November 1969, a World Health Organization group of consultants submitted to the UN Secretary-General a report[4] on *Health Aspects of Chemical and Biological Weapons.* [31]

They differed somewhat from the UN experts with respect to the emphasis and the assessment of possible effects of CBW on public health. They arrived, however, at essentially the same technical conclusions.

The summary of the WHO report read as follows:

A. Qualitative considerations

... The rapid action of the lethal *chemical agents* would preclude any large reduction of mortality by specific treatment. Possible protection by gas masks or shelters requires a highly disciplined and prepared population, a condition that is not fulfilled in most countries today, and it would pose serious economic and psychosocial problems if such a defence programme were to be implemented.

The outstanding characteristics of *biological weapons* for potential use in warfare are the following:

(*a*) The large variety of biological agents and the possible combinations available for such purposes.

(*b*) The possibilities for manipulating currently circulating strains of microorganisms for warfare purposes, by producing antigenically modified or antibiotic-resistant types (tularaemia, plague, anthrax, influenza) that would by-pass available prophylactic or therapeutic procedures.

(*c*) The unpredictability of the direct effects. A biological attack intended to be highly lethal might prove relatively ineffective, whereas an attack intended to be merely incapacitating might kill an unexpectedly large proportion of the target population. Also, certain agents (anthrax, coccidioidomycosis) could persist for long periods in a resistant spore form, which could be spread over very large distances by wind carriage in the course of time.

(*d*) The unpredictability of secondary effects such as the likelihood of contagion and the danger that epidemics might be initiated. There is the additional danger that epidemics might occur unintentionally through escape of virulent strains being purposely sought in laboratories.

(*e*) Although biological agents themselves are easy to produce, complex production and delivery systems are needed if even minimal reliance is to be placed on the outcome of an attack, except perhaps where the intention is simply to produce social disruption by a limited sabotage effort (e.g., the introduction of smallpox).

[4] This was an expansion of the material that WHO had earlier submitted to the Secretary-General's group of experts, at the latter's request.

Of the above characteristics, (*a*) and (*b*) would favour the attacker, whereas (*c*) and (*d*) would reduce the value of biological weapons from a military point of view.

B. Quantitative estimates

1. Assessments have been made of the primary effects of possible small-scale airborne attacks on cities of 0.5–5 million population in industrially developed and developing countries. The postulated mode of attack consisted of one or a few bombers dispersing specific chemical or biological agents along a 2-km line perpendicular to the direction of the wind. On the basis of the particular assumptions employed, the following conclusions have been reached:

(*a*) Of the known chemical warfare agents, only the nerve gases, and possibly botulinal toxin, have a casualty-producing potential comparable to that of biological agents.

(*b*) Under atmospheric conditions favourable to the attacker, an efficiently executed attack on a city with 4 tons of sarin (requiring some 15–20 tons of weapons) could cause tens of thousands of deaths in an area of about 2 km^2. Even in unfavourable conditions there could be thousands of deaths. If 4 tons of VX were used in such an attack, the casualties would not be appreciably greater in unfavourable meteorological conditions, but in favourable conditions this small attack would affect an area of about 6 km^2 and could cause anywhere between 50 000 and 180 000 deaths.

(*c*) If a suitably stabilized botulinal toxin or a fine aerosol of VX (particles of 5μ diameter) were developed and 4 tons were employed, several hundreds of thousands of deaths could result because of the greater coverage possible with such agents—12 km^2 for botulinal toxin and 40 km^2 for monodispersed VX aerosol. A larger total weight of weapons, perhaps 2–3 times that needed for the agents in (*b*) above, would have to be used to deliver these forms of botulinal toxin and VX.

(*d*) If a biological agent such as anthrax were used, an attack on a city by even a single bomber disseminating 50 kg of the dried agent in a suitable aerosol form would affect an area far in excess of 20 km^2, with tens to hundreds of thousands of deaths. A similar attack with any one of a number of other more labile biological agents could affect from 1 km^2 to more than 20 km^2, depending upon agent used, with tens to hundreds of thousands of casualties and many thousands of deaths.

2. Limited sabotage of a communal water supply with the typhoid fever bacillus, LSD, or a stable botulinal toxin, could cause considerable disruption and deaths in a large city, affecting tens of thousands of people.

3. Sabotage-induced or open attacks, causing the secondary spread of epidemics of yellow fever, pneumonic plague, smallpox or influenza, might under certain conditions ultimately result in many millions of illnesses and deaths.

4. The numbers of potential casualties and deaths recorded in this report represent the possibilities arising out of a very small and limited attack already well within the capabilities of a number of nations, with the possibility that an ever-increasing number of countries will acquire similar capabilities. With tech-

nologically advanced weapons and a larger scale of attack, achievable without too much difficulty by militarily advanced powers, the magnitude of destructiveness attendant upon the use of chemical and biological weapons would be considerably increased.

The main conclusions of the WHO report were:

1. Chemical and biological weapons pose a special threat to civilians. This is because of the often indiscriminate nature of such weapons, and because the high concentrations in which they would be used in military operations could lead to significant unintended involvement of the civilian population within the target area and for considerable distances downwind.

2. The large-scale or, with some agents, even limited use of chemical and biological weapons could cause illness to a degree that would overwhelm existing health resources and facilities.

3. Large-scale use of chemical and biological weapons could also cause lasting changes of an unpredictable nature in man's environment.

4. The possible effects of chemical and biological weapons are subject to a high degree of uncertainty and unpredictability, owing to the involvement of complex and extremely variable meteorological, physiological, epidemiological, ecological, and other factors.

5. Although advanced weapons systems would be required for the employment of chemical and biological agents on a militarily significant scale against large civilian targets, isolated and sabotage attacks not requiring highly sophisticated weapons systems could be effective against such targets in certain circumstances with some of these agents.

V. *Summary and comment*

At the request of the UN General Assembly an international group of experts prepared a report on chemical and bacteriological (biological) weapons and the effects of their possible use. The overall assessment was that certain chemical and biological agents are potentially unconfined in their effects, both in space and time. Their large-scale use could conceivably have deleterious and irreversible effects on the balance of nature. The danger would apply as much to the country that initiated the use of these weapons as to the one which had been attacked. No system of defence, whatever its cost, could be completely secure. CB weapons are not a cheap substitute for other kinds of weapons. Their elimination would not detract from any nation's security.

The report was welcomed by many countries as a contribution toward increased knowledge about CBW. It was later supplemented by a study of World Health Organization consultants on health aspects of CBW.

In presenting the report, the UN Secretary-General recommended universal accession to the Geneva Protocol, an affirmation that the prohibition applied to the use in war of all chemical and biological agents (including tear gas and other harassing agents), and a cessation of the development, production and stockpiling of those agents, as well as their elimination.

The comparison between the UN report and that of 1924 (see Chapter 1) reveals the tremendous progress made in forty-five years in the field of chemistry and biology. The significant result of this progress is the greater variety and potency of injurious effects which the newly developed agents can induce on human beings, animals and plants, in spite of considerable improvements in means of defence. Especially striking is the difference in the estimation of the bacteriological weapon. While in 1924 it was considered a not particularly formidable arm, the latest findings suggest that large-scale bacteriological attacks could have a terrible impact upon the population and the economy of the target country.

The authors of the two reports agreed, however, that CB weapons pose a threat to humanity.

In the discussion in the United Nations and the Disarmament Committee, the prevailing feeling was that no new treaty instrument was necessary to prohibit the use of CB weapons: the existing constraints under the Geneva Protocol should be affirmed and strengthened. As to the prohibition of the development, production and stockpiling of CB weapons, there was a consensus that a new instrument or instruments should be elaborated.

The United Kingdom suggested to deal first with BW and conclude an agreement which would ban the use, production and possession of agents of biological warfare. Thus, for the first time a split was recommended in the treatment of the two categories of weaponry prohibited together in the 1925 Protocol.

The main reasons for the split, put forward by the UK, were that biological weapons, regarded with particular abhorrence, had never been employed, while chemical weapons had been used in the past and some states would be reluctant at the moment to give them up; in addition, the latter weapons posed problems of definition and presented difficulties in verification.

The British arguments were found unconvincing and met with opposition. In 1968, even the United Kingdom's allies, notably the USA, did not seem to like the suggested approach. The Socialist and non-aligned countries insisted on a joint treatment of chemical and biological weapons.

Part II. Developments in 1969–1970

Unilateral renunciations of CBW

In the latter part of 1969 and in 1970, the international debate about chemical and biological weapons was particularly lively and specific. It was stimulated by more public information about the nature [1] and dangers [2] of CB weapons, by some accidents which occurred with chemical weapons, as well as by the continued use of chemicals in military operations in Viet-Nam. The protests of scientists, especially microbiologists, against the corruption of science for military purposes, also increased.

The United Nations appealed to all states to renounce the use of CB weapons and to abolish them. The World Health Organization as well as some of the non-governmental organizations, in particular the International Committee of the Red Cross, the Special Committee of Non-Governmental Organizations on Disarmament[1] and the International Association of Microbiological Societies, made similar requests.

Under these circumstances, some governments found it opportune to define and make public their policies on CBW.

On 25 November 1969, the US President announced that the United States was unilaterally renouncing BW and intended to ratify the 1925 Geneva Protocol prohibiting the use in war of asphyxiating, poisonous or other gases, and of bacteriological methods of warfare. His statement with regard to chemical weapons reaffirmed the renunciation of the first use of lethal chemical weapons, extended the renunciation to the first use of incapacitating[2] chemicals, and promised submission of the 1925 Geneva Protocol to the Senate for its advice and consent to ratification. No reference

---

[1] The Committee was set up pursuant to a resolution adopted by the eleventh Conference of non-governmental organizations in consultative status with the UN Economic and Social Council.

[2] The WHO report divides biological and chemical agents into three types: "A lethal agent is one intended to cause death when man is exposed to concentrations well within the capability of delivery for military purposes. An incapacitating agent is one intended to cause temporary disease or to induce temporary mental or physical disability, the duration of which greatly exceeds the period of exposure. A harassing agent (or short term incapacitant) is one capable of causing rapid disablement that lasts for little longer than the period of exposure." The report adds: "The above classifications are not toxicological categories, for the effects of a chemical warfare agent depend as much on the way it is used as on its toxicological properties. If too much of an agent intended for harassment is used, it may kill or severely injure. Likewise, if a low concentration of a lethal agent is disseminated, its effects may be only incapacitating or harassing."

was made to harassing[2] chemicals, such as tear gas, or to anti-plant chemicals; later it was made plain that these were not included.

With regard to biological weapons, the statement proclaimed renunciation of the use of lethal biological agents and weapons, and all other methods of biological warfare; confinement of biological research to defensive measures such as immunization and safety measures; and disposal of existing stocks of bacteriological weapons. [3]

It was not clear whether the renunciation of BW embraced toxins, an important category of possible warfare agents,[3] but on 14 February 1970 another announcement was made: the USA decided to renounce offensive preparations for and the use of toxins as a method of warfare; to confine the military programmes for toxins, whether produced by bacteriological or any other biological method or by chemical synthesis, to research for defensive purposes only; to destroy all existing toxin weapons and all existing stocks of toxins which are not required for a research programme for defensive purposes. [4] Plans for the destruction of biological agents and toxin stockpiles were made public by the US Defense Department in December 1970.

The decision to dismantle the US resources for biological warfare was characterized by Sweden as the only true disarmament measure that had been taken during the whole post-war period, and as the only decision involving any measure of military sacrifice on the part of superpowers. [56]

The United States appeared to hope that other states would also unilaterally renounce biological weapons. It urged, nevertheless, that such decisions be converted into an international obligation through a convention.

Shortly after the US statement on CBW, the United Kingdom declared that it had never had any biological weapons, did not have any now, and had no intention of acquiring any. [5] A similar statement was made in March 1970 by Canada, which added that it did not possess any chemical weapons either, and did not intend to develop, produce, acquire, stockpile or use such weapons at any time in the future unless they were used against the military forces or the civil population of Canada or its allies. [6]

In April 1970, the Swedish government, in a foreign policy message to the parliament, called attention to the fact that Sweden neither possessed, nor intended to manufacture, any biological or chemical means of warfare [7]. In September 1970, the Yugoslav government stated that it did not

---

[3] Toxins are poisonous substances produced by living organisms including plants, animals and bacteria. In contrast to the organisms that produce them, toxins are not capable of reproduction.

possess, and did not intend in the future to develop, produce, acquire, stockpile or in any other manner possess, biological means of warfare, and that the research work in this area would in the future also be limited exclusively to the necessary measures of protection in case Yugoslavia were exposed to an attack with such weapons. [56]

The Netherlands recalled that in 1930, when it ratified the Geneva Protocol, it was among the first countries to renounce unconditionally the use of biological weapons. [8]

Mexico suggested, as an intermediate step pending a comprehensive ban on CB weapons, that all states make declarations renouncing unilaterally the use in war of biological weapons, their manufacture and stockpiling; the renunciation would acquire a contractual character when overall agreement was achieved. [9] Several delegations to the Disarmament Conference and the twenty-fifth UN General Assembly emphasized that unilateral decisions can be no substitute for international multilateral instruments binding on governments and their successors.

Throughout the discussion in the CCD and the UN General Assembly in 1969–70, a distinction was drawn between measures to prohibit the use of, and measures designed to abolish, CB weapons.

Prohibition of use of CB weapons

Adherence to the Geneva Protocol

For years there has been pressure from many quarters to bring about universal adherence to the 1925 Geneva Protocol. In 1969, the United Nations invited all states which had not yet done so to accede to or ratify the Geneva Protocol in commemoration of the forty-fifth anniversary of its signing and the twenty-fifth anniversary of the UN. [10] The appeal was reiterated in 1970. [53]

From the outset, the country principally aimed at was the United States, the only big power not yet party (though a signatory) to the Geneva Protocol. (China, France, the UK and the USSR have been parties since the late 1920s.) Furthermore, the United States was known to have a very large programme for CBW research, production and stockpiling, both at home and abroad, and to have made extensive use of some types of chemicals in the hostilities in Indo-China.

The United States has taken the position that it respects the principles and objectives of the Geneva Protocol, but for years it resisted formal commitments in this field. The US President's statement on CBW of November 1969 marked a major change in this policy. In renouncing production and stockpiling and any use of biological weapons, including toxins, as well as first use of lethal and incapacitating chemicals against any country, whether or not bound by similar commitments, the United States went farther than the ban under the Geneva Protocol.

However, the non-inclusion in the renunciation concerning chemical warfare, of harassing and anti-plant agents the use of which in Indo-China set in motion the recent international drive for the ratification of the Geneva Protocol, raised a serious problem in view of the US declaration that it would ratify the Protocol. The majority of states consider the ban under the Geneva Protocol as all-inclusive (see next section) and no party has entered reservations limiting the types of weapons to which it applies.

The procedure of ratification of the Protocol by the United States was considerably delayed. Apparently there was some debate within the US administration as to whether the Protocol should be approved with or without exceptions. Eventually, on 19 August 1970, the US President asked the Senate to approve[4] the Protocol; he transmitted to the Senate a report by the Secretary of State, setting forth the understanding of the Protocol by the United States, namely that it does not prohibit the use in war of riot-control agents and chemical herbicides, and that smoke, flame and napalm[5] are also not covered by the Protocol.

The US administration proposed to ratify the Protocol with a reservation permitting the use of chemical weapons and agents if an enemy or its allies were to employ them first. Renunciation of use of biological weapons would be unconditional. [11]

Proposals were put forward in the UN for the withdrawal of a reservation, previously made to the Geneva Protocol, which exempts parties from prohibition of use of CB weapons against non-parties. Italy, supported by some non-aligned countries, stressed, both in the UN General Assembly and at the Disarmament Conference, that renunciation of the right to use arms banned by the Protocol against non-signatory states (when these had not themselves violated the ban) would extend the geographical scope of the Protocol, make it more universal and enhance its efficacy. [12]

If the reservation is formally withdrawn, it could be argued that the text of the Protocol itself requires clarification since, as it stands now, the parties "agree to be bound as between themselves".

There is less support for dropping another reservation which states that the obligations would cease to be binding in regard to all enemy states whose armed forces or whose allies failed to respect the Geneva Protocol.[6] The Soviet Union said that the reservations had played an important role in preventing a widespread use of CB methods of warfare and had served as the basis for the warning issued by the Allies to the German government concerning the possible use of chemical weapons by the latter during World War II. It further contended that the adoption of a convention aimed at completely eliminating CB weapons from military arsenals, with the participation of a wide range of states, would make the question of reservations to the Protocol pointless. [36]

Canada stated, with respect to chemical weapons, that it must reserve the right to retaliate (if these weapons are used against the civil population

---

[4] The 1925 Geneva Protocol was before the US Senate until 8 April 1947. It was then withdrawn by the US President along with a number of other pending treaties, "with a view to placing the treaty calendar on a current basis". [51]

[5] The UN Secretary-General's report on *Respect for Human Rights in Armed Conflicts,* of 18 September 1970, contained a suggestion that a study be made of napalm weapons and the effects of their use, so as to facilitate action by the United Nations with a view to curtailing or abolishing such uses of the weapons in question as might be established as inhumane. [65]

A proposal to ban the use of napalm was presented by Sweden, in June 1971, at an International Red Cross conference in Geneva.

[6] The question whether the prohibition of use of CB weapons should be absolute, or whether states should retain the right of retaliation, was widely discussed at the Disarment Conference in 1932. (See, e. g., pp. 122 and 133.)

or armed forces of Canada or of its allies) "until a verifiable agreement can be concluded". [55]

Indeed, the retention of the right to use CB weapons, even in extraordinary situations, would seem incompatible with a ban on possession of those weapons.

But some countries, like Ethiopia [63], Nepal [60] or Ghana [60], asked for the removal of all reservations to the Geneva Protocol without reference to a future agreement prohibiting the production and stockpiling of CB weapons. The UN Secretary-General also appealed to all parties who had signed the Protocol with reservations to renounce them.[7] [52]

Japan deposited the instrument of ratification of the Geneva Protocol on 21 May 1970, without attaching any reservation. It proposed that each state should undertake never, under any circumstances, to engage in CB warfare. [13]

Since the United Nations call of 1969 for universal adherence to the Geneva Protocol, the number of parties to the Protocol has considerably increased. By the end of 1970 it reached 82.

Tear gas and herbicides

During the debate about the Geneva Protocol, sharp controversy arose over the scope of the prohibition, the point at issue being whether the use of tear gas and anti-plant chemicals in war is banned under international law.[8]

In the discussion on the subject in the United Nations, the United States and Australia maintained that tear gas and anti-plant chemicals were not prohibited: tear gas was employed by many countries for domestic riot-control purposes, and anti-plant agents involved the same chemicals and had the same effect as the materials commonly used in many countries to control vegetation. They contended that in some cases the use of tear gas, herbicides and defoliants in warfare may be more humane than the use of conventional weapons.

The United States also argued that chemical anti-plant agents were not intended to be covered by the Geneva Protocol, while the issue of riot-control agents had always been controversial.[9] If states wished to seek

---

[7] The legal implications of the reservations to the Geneva Protocol are discussed in Volume III.
[8] The legal and political problems involved are discussed extensively in Volumes III and V, respectively.
[9] In 1930, the United States objected to the prohibition of use of tear (lachrymatory) gas in war, but in 1932–33, during the Disarmament Conference, it agreed to such a prohibition and was prepared to accept some restrictions even on tear gas for internal police operations. (See pages 134 and 172.)

agreement upon a uniform interpretation, they would have to enter special negotiations.

The majority of UN members, however, has taken the position that the existing rule of international law prohibiting chemical warfare already covers all chemical weapons, and supported the UN Secretary-General's recommendation to this effect (see page 273).

The main arguments of those who favoured a comprehensive ban were that no sharp demarcation line can be drawn between harassing and other anti-personnel chemical agents; military applications of tear gas are different from civil applications and in war there is also a risk that tear gas may provoke the use of even more harmful agents; military employment of anti-plant chemicals may cause serious injury to people as well as long-term damage to the environment; the negotiating history of the Geneva Protocol supports the view that it was meant to be comprehensive, and if some kinds of chemical warfare were condoned, the Protocol would be undermined.

In December 1969, the United Nations discussed a resolution, initiated by Sweden, which stated that the 1925 Geneva Protocol embodied the generally recognized rules of international law prohibiting the use in international armed conflicts of all biological and chemical methods of warfare, regardless of any technical developments. It declared as contrary to those rules the use in international armed conflicts of:

Any chemical agents of warfare—chemical substances, whether gaseous liquid or solid—which might be employed because of their direct toxic effects on man, animals or plants;

Any biological agents of warfare—living organisms, whatever their nature, or infective material derived from them—which are intended to cause disease or death in man, animals or plants, and which depend for their effects on their ability to multiply in the person, animal or plant attacked. [16]

The aim of the resolution, as explained by its sponsors, was to ensure that no acquiescence was taken to have occurred in a restrictive interpretation of the prohibition of CB means of warfare, since it was neither possible nor desirable to meet the military requirements of any state by making an exception to the comprehensive ban.

The resolution was adopted by eighty positive votes. It was opposed by three countries: the USA which has used tear gas and herbicides in Viet-Nam; Australia which is fighting in Viet-Nam with the United States; and Portugal which has been accused in the United Nations of using CB weapons against the local population of its African territories. [62]

Thirty-six states (mostly NATO members and other US allies) abstained. Many of them did not seem to approve the US stand [58] and took no official position on the substance of the issue. Their reservations were mainly of a legal nature, both constitutional and procedural, in particular with regard to the competence of the UN General Assembly to interpret existing international instruments through resolutions. Some abstaining countries were not parties to the Geneva Protocol. The United States pointed out that UN General Assembly resolutions are only recommendatory, and it was up to the parties themselves to decide the scope of the Protocol's coverage.

The UN resolution on CBW did not have an immediate effect upon government policies.

The British Secretary of State for Foreign and Commonwealth Affairs announced on 2 February 1970 that, while tear gases and shells producing poisonous fumes are prohibited under the Geneva Protocol, CS[10] and other such gases not significantly harmful to man in other than wholly exceptional circumstances are not. [14] It was later explained that the use of the substance in question in very high concentrations in enclosed spaces over long periods would be an exceptional circumstance. [15] The UK government thus modified its interpretation of the Protocol, which a previous government had enunciated in 1930 and which subsequent governments had upheld. (See pages 102 and 167.)

The Canadian statement of March 1970 (see page 281) did not include tear gas and other crowd and riot-control agents in the commitment not to develop, produce, acquire, stockpile or use chemical weapons; it explained that "their use or the prohibition of their use in war presents practical problems in relation to the use of the same agents by police and armed forces for law enforcement purposes which require detailed study and resolution". [6] In August 1970, the US government reiterated that it did not consider riot-control agents to be covered by the Geneva Protocol and proposed to ratify the Protocol with that understanding.

However, a few months later, on 1 October 1970, the Netherlands, which had abstained from voting on the December 1969 resolution, officially expressed readiness—in the framework of international negotiations—to take account of a majority opinion in the United Nations, namely that "the use of all biological and chemical agents of warfare—including tear gases—should be prohibited".[11] [64]

---

[10] CS is a chemical irritant (C and S being the initials of the two discoverers), nowadays widely used as a tear gas by police forces, and as a harassing agent by military forces in Viet-Nam.
[11] The decision followed a report on the desirability of a ban on the use of tear gases in armed conflicts, presented to the Minister for Foreign Affairs of the Netherlands by the Netherlands Advisory Committee on questions of disarmament and international security and peace. The majority of the Committee expressed the view that to be effective a ban on chemical weapons would have to be all-embracing;

Norway, another NATO country which in 1969 had been among the abstaining states, stated in November 1970 that current international negotiations should aim at achieving an effective ban on the use of CB weapons and on their development, production and stockpiling, including a ban on the use in warfare of tear gases. [59]

Japan, while being of the view that the Geneva Protocol could not be interpreted as prohibiting the use of riot-control agents, said that this position should not be construed as indicating that the government of Japan was not in favour of prohibiting the use in war of such agents.[12]

A somewhat different situation developed with regard to anti-plant agents due in the first instance to warnings by US scientists that certain chemical defoliants posed dangers of birth defects to the population living in the areas being sprayed. The United States imposed restrictions on the domestic use of some such chemicals as health hazards. In April 1970, the US Department of Defense suspended the use of the herbicide *Orange* in Indo-China, but operations using other agents continued, including the destruction of crops.

On 26 August 1970, the US Senate rejected an amendment to the military procurement bill, providing that no funds authorized under the bill would be used to procure, maintain or use herbicides. The opponents of the amendment argued that the herbicides were used for the protection and safety of the US troops and that this immediate benefit outweighed any adverse economic effects and possible long-term ecological consequences. [17]

Nevertheless, as a result of sharp criticism, voiced in particular by the American scientific community,[13] the US President directed in December 1970 that there should be strict conformance in Viet-Nam with policies governing the use of herbicides in the United States; that a program be initiated for an orderly yet rapid phase-out of the herbicide operations; that during the phase-out the use of herbicides in Viet-Nam be restricted to the perimeter of fire-bases and US installations or remote unpopulated areas; and that the ban on herbicide *Orange* remain in effect. [61]

With the exception of Australia, which has troops in Viet-Nam, no state has explicitly supported the military use of anti-plant agents. The Nether-

---

a ban on tear gases should apply to armed conflicts in which armed forces are engaged in hostilities; such a ban would leave unimpaired the use of tear gases for policing purposes (including such use by military personnel) in cases of internal disturbances or riots in occupied territories—situations where a proper and fitting use of tear gas could be guaranteed and where the danger of escalation would be minimal.

[12] Letter of 24 May 1971 from the Embassy of Japan in Stockholm.

[13] A report of the Herbicide Assessment Commission of the American Association for the Advancement of Science, presented in December 1970, described the harm caused to the land and to the people of Viet-Nam by the US herbicide spraying programme.

lands government expressed the view that it was necessary to establish a clear rule for the future, which would exclude the use of herbicides and defoliants for warlike purposes. [8, 64] Norway considered that a ban on the use of CB weapons should include a ban on the use of herbicides in warfare. [59]

The World Health Assembly, in a special resolution adopted on 21 May 1970, declared that the use not only of CB weapons but also of any CB agents for the purposes of war might lead to a disturbance of ecological processes which in its turn would menace the existence of modern civilization. [50]

Finland suggested that the International Court of Justice be requested to provide an advisory opinion as to the scope of the rules and prohibitions of the Geneva Protocol. [60]

Prohibition of production and possession of CB weapons

Separate or joint treatment of chemical and biological weapons

The CCD and the United Nations extensively discussed whether, in dealing with the prohibition of development, production and stockpiling, chemical and biological weapons should be treated jointly or separately.

The following main arguments were advanced by the United States and the United Kingdom in favour of a separate treatment according priority to biological weapons:

1. The two categories of weapons differ as regards the potential toxicity, speed of action, duration of effect, specificity, controllability and residual effects; the biological weapon is the only self-propagating weapon in existence and is the most odious of all weapons.

2. Weight for weight, biological agents are of potentially much greater contaminating power, much more difficult to control in action and more unpredictable in effect than chemical agents. While chemical weapons affect smaller areas and can be used with a certain amount of precision, biological weapons are totally indiscriminate and are likely to affect vast areas and large civilian populations far removed from the scene of their use.[13] Biological weapons may therefore be thought of principally as strategic weapons, chemical weapons primarily as tactical ones. If one renounces one of these categories, one cannot make up for that by an increased production of the other, because each serves a different function.

3. While chemical weapons have been used in warfare and a number of countries have a CW capability or are conducting research in this field, biological weapons have never been used and few nations appear to have engaged in substantial efforts to develop them.

4. A ban on biological weapons poses a less difficult verification problem than that on chemical weapons.

---

[13] In this connection reference was made to the UN report which contained comparative estimates of disabling effects of hypothetical attacks using a nuclear, chemical or bacteriological (biological) weapon. (See page 270.)

5. If biological weapons were eliminated, it would be less likely that they would ever be utilized, and thus the Geneva Protocol would be strengthened to that extent.

The United Kingdom warned that in view of rapid scientific development there was a danger in delaying an agreement on biological weapons: the conditions of production, stockpiling and use of these weapons might be soon completely modified.

The proponents of a joint treatment of chemical and biological weapons— the Socialist and a number of non-aligned countries—argued that:

1. Both categories of weapons are classified as weapons of mass destruction,[14] whether destined for strategic or tactical use.

2. Both have been dealt with together in a number of international agreements and documents; a separate treatment would lead to the weakening of the 1925 Geneva Protocol.

3. All biological processes depend upon chemical or physico-chemical reactions, and what may be regarded today as a biological agent could, tomorrow, as knowledge advances, be treated as chemical.[15]

4. The fact that a certain quantity of a chemical agent will produce a lethal effect in an area smaller than that affected by the same quantity of a biological agent appears insignificant in view of the enormous stockpiles of agents which have already been accumulated.

5. A combination of biological and chemical weapons can be used with a view to obtaining greater effectiveness or to making their detection more difficult.

6. The means of delivery of chemical and biological agents are similar and in the armed forces of many countries the same services deal with CB means of warfare and protection.

7. If biological weapons, which have never been used and which are considered to be of little military effectiveness, were to be dealt with now and chemical weapons, which have already been used with disastrous effects, were left for later examination, the chemical arms race may intensify and may even seem legitimized. If anything, chemical warfare should have priority.

---

[14] The USA did not share this opinion. It stated that, while biological weapons are unquestionably weapons of mass destruction, it would be inaccurate to give the same label to the whole class of chemical weapons; an incapacitating chemical agent is not, in its view, a weapon of mass destrucion. [18]

[15] This was the opinion of experts, included in the UN Secretary-General's report on CB weapons.

Developments in 1969–1970

During the 1970 spring session of the CCD, an attempt was made by Sweden to analyse the question from a substantive point of view so as to determine how far it was feasible to treat chemical and biological weapons together, or to what extent it was necessary to give them separate treatment.

It appeared from the analysis that *research* with regard to both chemical and biological agents would have to be exempted from prohibition as well as from obligatory verification. *Development* of warfare agents and of devices for their dissemination, including preparation of instructions and manuals, as well as *training,* could be prohibited unconditionally and the prohibition might be dealt with in one comprehensive treaty; only with regard to the verification aspect might such differences exist that would call for separate treatment. It would seem to be possible to prohibit simultaneously the *testing* of chemical and biological warfare agents; for the purpose of verification, the surveillance of the sites and the security arrangements for testing areas might provide some leads, but in order to provide more conclusive evidence different techniques for various chemical and biological means of warfare might have to be foreseen.

As far as *production* was concerned, the Swedish view was that biological agents lent themselves almost entirely to unconditional prohibition, with some exceptions for quantities needed for laboratory work and for developing protective substances. Unconditional prohibition was also possible for single purpose chemical agents such as nerve gases and for toxins. However, to establish boundary lines between production of chemical agents having a legitimate use in peaceful activities, and production for direct warfare purposes, one would have to resort to conditional prohibition or prohibition with partial restraints. Technically the problem might be dealt with either in one comprehensive treaty with specified exemptions, or in a separate treaty or protocol. Such agents as were generally excluded from civilian use could be automatically included in a treaty of unconditional international prohibition. For all agents under unconditional prohibition the most effective means of verification should be sought; for other chemical agents it might suffice to prescribe a procedure of obligatory reporting to some international agency on their production, stockpiling and use for civilian purposes. (Sweden subsequently suggested a tentative list of chemical warfare agents which could be subject to conditional and unconditional prohibition, respectively, with regard to production and stockpiling.) *Transfers* between countries of all biological agents of warfare and of an increasing number of chemical agents would have to be prohibited unconditionally. *Destruction or decontamination* of CB weapons may be prescribed under a general prohibitory rule, but the technically separate types of treatment required would call for different modalities if the destruction was to be verified. [19]

Morocco believed that a legal instrument prohibiting the development, production and stockpiling of chemical and biological weapons (with a provision for their destruction) could include definitive verification procedures relating only to biological weapons; the total elimination of such weapons could be effective upon the entry into force of that instrument. In view of the technical difficulties connected with the verification regarding chemical weapons, the instrument should provide for subsequent examination of the problem in order to arrive, within a prescribed period of time, at a supplementary document laying down verification procedures for chemical weapons; the latter document would put into effect the total and definitive implementation of the provisions prohibiting such weapons. A working paper containing this proposal was submitted in July 1970. [20]

Nepal suggested the conclusion of an agreement banning biological weapons, along the lines proposed by the United Kingdom, but coupled with a moratorium with regard to chemical weapons and providing for "verification by challenge". [60]

On 25 August 1970, the group of twelve non-aligned members of the CCD stated in a joint memorandum that it was essential that both chemical and biological weapons should continue to be dealt with together in taking steps towards the prohibition of their development, production and stockpiling and their effective elimination from the arsenals of all states. It expressed the conviction that an effective solution to the problem should be sought on this basis. [21] The basic points of the memorandum were subsequently incorporated in a resolution of the twenty-fifth UN General Assembly. [53]

The value of the memorandum was weakened by conflicting interpretations. The Soviet Union interpreted it as supporting the view that there should be a single international instrument, like the one put forward by the Socialist countries, which would cover the prohibition of both biological and chemical weapons. The United States understood it to mean that a solution may be reached through a series of actions, all of them representing steps towards the total prohibition of CBW, and that it would be consistent with this approach to reach agreement banning biological agents and toxins, along the lines of the UK draft convention, while continuing work on the prohibition of other agents. Argentina, one of the authors of the memorandum, explained that the memorandum was not meant to support either of the two opposing views. [22, 54]

The "joint versus separate treatment" dispute may appear academic. It would seem immaterial whether there are one, two or several international instruments covering the prohibition of production and possession of CB weapons as long as there is confidence that the whole range of the weapons in question would be quickly banned. But such confidence was lacking. A good number of UN members felt that a convention limited to biological weapons—and it was with such a convention that the advocates of a separate treatment wanted to start—might not be followed by a similar agreement on chemical weapons. The representative of France, for example, pointed out in the UN that if chemical weapons were not dealt with together with biological weapons, there would be reason for fear that "any solution concerning them would be postponed indefinitely". [54] Some countries, in criticizing a partial approach to CB disarmament, referred to the Nuclear Test-Ban Treaty of 1963: despite the determination, expressed in the preamble to the treaty, to seek to achieve the discontinuance of all test explosions of nuclear weapons and to continue negotiations to this end, the treaty remained partial and underground tests continue unabated.

The US representative in the CCD, explaining his government's position on the military usefulness and roles of each of the two categories of weaponry, indicated the motives underlying the US decision to renounce biological means of warfare. He said that the US government considered that biological weapons have no value as a deterrent against use by others because retaliation in kind would not be an acceptable or rational response to a biological attack. They have no value as a means of redressing military balance either, because few, if any, military situations can be imagined in which a state would try to redress a military imbalance by retaliating with weapons whose effects would not show up for days. For these reasons, even the known retention of biological weapons by one state should not affect another state's decision to give them up; inspection was not necessary. On the other hand, chemical weapons have, in the US view, obvious usefulness in certain military situations, primarily as battlefield weapons. They are more predictable than biological weapons and, unlike the latter, they can produce immediate effects, which is an important quality for use in combat: hence the belief that CW capability is important for national security. The inability of an attacked nation to retaliate with chemicals could give a military advantage to any government which might resort to using them. In particular, the one-sided possession of nerve agents could offer unacceptable advantages to the country possessing them. Anyone suggesting retaliation with nuclear weapons in the event of a chemical attack would, in the US opinion, abrogate his responsibility to find meaningful arms-control solutions to the problems of chemical weapons. [8, 23] The United States said that the possibility of eliminating chemical capabilities depended upon developing appropriate verification and that there was a long and difficult road ahead. In a special working paper it drew attention to the magnitude and complexity of the problem of control. [24]

The divergent approaches were reflected in draft conventions submitted to the CCD and the UN General Assembly.

Draft convention on BW

On 10 July 1969, at the Conference of the Committee on Disarmament, the United Kingdom, developing its earlier proposals (see page 254), tabled a draft convention providing for undertakings: never, in any circumstances, by making use for hostile purposes of microbial or other biological agents causing death or disease by infection or infestation in man, other animals, or crops, to engage in biological methods of warfare (Article I); not to produce or otherwise acquire, or assist in or permit the production or acquisition of microbial or other biological agents of types and in quantities that have no independent peaceful justification for prophylactic or other purposes, as well as of ancillary equipment or vectors the purpose of which is to facilitate the use of such agents for hostile purposes; not to conduct, assist or permit research aimed at production of the kind prohibited above; to destroy, or divert to peaceful purposes, within three months after the convention comes into force for a given party, any stocks of such agents or ancillary equipment or vectors as have been produced or otherwise acquired for hostile purposes (Article II).

Any party believing that biological methods of warfare have been used against it would be entitled to lodge a complaint with the UN Secretary-General, submitting all evidence at its disposal, and request that the complaint be investigated and that a report on the result of the investigation be submitted to the Security Council; any party believing that another party has acted in breach of other undertakings under the convention would be entitled to lodge a complaint with the Security Council and request that the complaint be investigated (Article III).

Each party would affirm its intention to provide or support appropriate assistance to any other party, if the Security Council concludes that biological methods of warfare have been used against that party (Article IV).

Each party would have the right to withdraw from the convention, if it decided that extraordinary events, related to the subject matter of the convention, had jeopardized the supreme interests of its country (Article IX). (A similar provision appears in the Test-Ban and Non-Proliferation Treaties.)

The preamble reaffirmed the validity of the 1925 Geneva Protocol and Article VI stated that nothing contained in the convention shall be construed as in any way limiting or derogating from obligations assumed under the Protocol.

Special provision was made for negotiations on effective measures to strengthen the existing constraints on the use of chemical methods of warfare (Article V).

The United Kingdom stressed that the convention would not prohibit the development of a passive defensive capability against biological warfare. It did not make it clear, however, whether it was permitted under such justification to develop new biological warfare agents.

The UK delegation submitted, as a document complementary to the draft convention, a draft Security Council resolution by which the UN Secretary-General would be requested to take measures enabling him to investigate without delay complaints lodged with him, as well as complaints with the Security Council, if so requested by the Council; and the Security Council would declare its readiness to give urgent consideration to complaints lodged with it, and to any report that the Secretary-General may submit on the result of his investigation of a complaint, and to consider urgently what action should be taken or recommended in accordance with the UN Charter, if it concluded that the complaint was well-founded. [25]

On 26 August 1969, taking account of some of the critical remarks made by different delegations, the United Kingdom revised the text of its draft by introducing the following amendments:

The undertaking by a party not to engage in biological methods of warfare (Article I) was now qualified by the clause: "insofar as it may not already be committed in that respect under Treaties or other instruments in force prohibiting the use of chemical and biological methods of warfare". The purpose of the amendment was to make it clear that existing commitments under the Geneva Protocol and other agreements were not affected by the draft convention; some countries in becoming parties to the convention would undertake additional commitments under Article I, others would not. The ban was extended to cover microbial or other biological agents causing damage in addition to those causing death or disease (Article 1).

To emphasize the right to develop defence measures, which would include in particular vaccines for protection against possible biological attack, the exception to the prohibition of production or acquisition was modified to read: "independent justification for prophylactic or other peaceful purposes" (Article II).

The complaints lodged with the Security Council would have to be supported by all evidence at the disposal of the complaining party, as in the case of complaints lodged with the UN Secretary-General (Article III).

To avoid the impression that negotiations on chemical weapons would aim at a convention more limited in scope than the draft convention on biological weapons, the words "the use of" were dropped in Aricle V, to read "effective measures to strengthen the existing constraints on chemical methods of warfare."

The related draft Security Council resolution was also changed by adding a preambular paragraph which reaffirmed the right of individual and collective self-defence recognized in Article 51 of the UN Charter. [26]

In 1970, during the spring session of the CCD, the Netherlands suggested that the undertaking not to produce should apply to biological agents "that are not exclusively required for prophylactic or protective purposes" and to leave out of the corresponding part of Article II of the draft the word "independent" which could lead to confusion, and also the term "peaceful" which may give rise to different interpretations. [8] The United Kingdom agreed to delete the word "independent", but felt that the substitution of "protective purposes" for "other peaceful purposes" would place too restrictive an interpretation on the legitimate peaceful uses which would be exempt from the prohibitions.

The United States, whose policy on toxins was by then identical to its policy on biological weapons, proposed to include toxins in the UK draft convention because the production of bacteriological toxins in any significant quantity would require facilities similar to those needed for the production of biological agents. Though toxins of the type useful for military purposes could conceivably be produced by chemical synthesis in the future, the end products would be the same in the effects of their use and those effects would be indistinguishable from toxins produced by bacteriological or other biological processes. [27] The United States also suggested the deletion in Article I of the phrase "by infection or infestation" in order to put the emphasis of the prohibition on the agents themselves rather than on the manner in which a disease is introduced.

Article I, as proposed by the USA, would provide for an undertaking never, in any circumstances, by making use for hostile purposes of microbial or other biological agents or toxins causing death, damage or disease to man, other animals or crops, to engage in biological methods of warfare.

Article II would also undergo a modification, so as to include toxins in the convention's prohibitions and requirements concerning production, acquisition, research and destruction. [28]

The UK considered that the formulation of its draft already covered the prohibition of production and acquisition of toxins but agreed to making a specific mention to that effect and accepted the US amendments. [29] (For the text of the revised UK draft convention of 18 August 1970, see page 322.)

The UK draft was criticized by the Socialist and many non-aligned countries chiefly for not dealing with chemical weapons. The amendment concerning toxins was found insufficient by the critics who also argued that by including toxins in a BW draft treaty, the UK and USA confirmed the possibility of dealing with biological and chemical weapons in one document. (Toxins were classified in the UN Secretary-General's report as biologically-produced *chemical* substances.) A mere assurance that negotiations on measures to strengthen the existing constraints on chemical methods of warfare would be pursued was considered inadequate.

The complete prohibition of use of biological weapons, that is even in self-defence or retaliation, was obviously a step forward, when compared to the Geneva Protocol; some countries however thought that its inclusion in a convention dealing with production was unnecessary. In Sweden's opinion, such a clause, because it was confined to biological weapons, could even be a risky undertaking. [19]

Sweden and the United Arab Republic stressed the need for some system of verification to ensure abidance by the commitment not to produce biological weapons, irrespective of the complaints procedure. (The problem of verification is dealt with in more detail in a later section.)

Draft convention on CBW

On 19 September 1969, Bulgaria, Byelorussia, Czechoslovakia, Hungary, Mongolia, Poland, Romania, Ukraine and the USSR submitted to the twenty-fourth UN General Assembly a draft convention prohibiting both chemical and biological weapons.

The undertakings provided for were: not to develop, produce, stockpile or otherwise acquire CB weapons (Article 1); to destroy within a specified period or to divert to peaceful uses all previously accumulated CB weapons (Article 2); not to assist, encourage or induce any particular state, group of states or international organizations to develop, produce or otherwise acquire and stockpile CB weapons (Article 3. Each party shall be internationally responsible for compliance with the provisions of the convention by legal and physical persons exercising their activities in its territory, and also by its legal and physical persons outside its territory (Article 4); it would take the necessary legislative and administrative measures to prohibit the development, production and stockpiling of CB weapons and to destroy such weapons (Article 5). The parties would consult one another and cooperate in solving any problems which may arise in the application of the provisions of the convention. The convention would enter into force after the deposit of a specified number of instruments of ratification, including those of the permanent members of the UN Security Council. [30]

In response to criticism by Western delegations concerning the inadequacy of the verification system, Hungary, Mongolia and Poland suggested, on 14 April 1970, the inclusion of a new article by which the parties to the convention would be entitled to lodge complaints of possible violations of its prohibitions with the UN Security Council. Such a complaint should include all possible evidence confirming its validity as well as a request for its consideration. Each party would undertake to cooperate in carrying out any investigations which the Security Council may undertake on the basis of the complaint received. A special Security Council resolution declaring the readiness of the Council to consider any such complaints, to take all the necessary measures for their investigation and to inform the parties of the result of the investigation, was also proposed. [31]

Referring to doubts whether in questions relating to the application of measures of disarmament one could rely on the Security Council in view of the veto power of its permanent members, Poland noted that no better system of security than the one provided in the UN Charter had as yet been elaborated and there would probably be no changes in this respect in the foreseeable future; the present system seemed to be fully sufficient for the purpose of a CBW convention.

The fact that the proposed convention dealt with both chemical and biological weapons was widely welcomed. Many nations, however, found the draft deficient with regard to verification and control. The inclusion of a complaints procedure, patterned after the corresponding clause of the UK draft, muted the criticism to some extent, but failed to remove it. The United States said that it would have no way of knowing, if the draft convention of the Socialist countries were adopted, whether the chemical weapons possessed by the Soviet Union had been destroyed pursuant to the convention or whether the Soviet Union was continuing to produce chemical munitions or was retaining a capability to produce such munitions. The method of consultation between the parties, as proposed in the draft, was considered lacking in precision. The Western delegations pointed out that a state cannot be held responsible for acts committed by unauthorized individuals outside its territorial limits and that Article 4 of the draft was therefore unenforceable. The lack of a provision for amendments was also criticized.

The requirement for the convention to be ratified by all the permanent members of the UN Security Council could delay indefinitely the entry into force of the CBW prohibition; none of the previously concluded arm-control agreements contained such a requirement.

Important objections were raised by the United Kingdom and the United States with regard to the very object of the prohibition. The draft was limited to banning "weapons". This could mean that the development, production and stockpiling of agents or their intermediates would be permitted as long as they were not "weaponized", that is placed in munitions, and that the components of weapons would not be abolished. The United States raised the question of chemicals which were used in industry but could also be used directly to inflict casualties on the battlefield by being released from ordinary industrial containers. Under such circumstances, the parties would preserve the capability for quick, if not immediate, retaliation and also the right to do so, since the draft made no provision for banning the use of CB weapons, and the Geneva Protocol did not provide for absolute prohibition of use either.

The USA stated that the draft convention of the Socialist countries could not be a basis for negotiation, mainly because it did not cope with the problems inherent in the task of controlling chemical weapons. [18]

On 23 October 1970, further changes were introduced in the Socialist draft convention [49]: the obligations not to develop, produce, stockpile or otherwise acquire, as well as to destroy or divert to peaceful uses the accumulated stocks, were extended to include equipment and vectors[16] specially designed for the use of CB weapons as means of warfare—in addition to CB weapons themselves.

Mongolia—one of the sponsors of the draft—described this change as a significant addition covering an "important ingredient of the CB weapon system". It drew attention to the fact that specially designed equipment and carriers were as indispensable for the purpose of waging chemical and germ warfare as the agents themselves, and that without them CB agents tended to turn into a burdensome stock, the maintenance and storage of which were fraught with risks and danger. [57] No definition of "equipment" was offered. Neither was it clear what would be the status of installations which, though not specifically designed for CB warfare, could nevertheless be used for the purpose, without, or with only slight, adaptations.

An article was added providing for an undertaking to facilitate international cooperation in the field of peaceful chemical and bacteriological (biological) activities, including the exchange of CB agents and equipment

---

[16] The word "vector" was used in the English version. In the Russian text, the term employed was "sredstva dostavki", that is "means of delivery".

for the processing, use or production of these agents for peaceful purposes (Article VIII).

According to Hungary—another sponsor of the draft—the aim of the new article was to create a new basis for states to cooperate in the propagation of scientific and technological information, first of all in the interest of the developing nations; it was meant to ensure that—as stated in the preamble to the draft convention—the scientific discoveries in the field of chemistry and biology should be used only for peaceful purposes. [54]

An amendment clause was included, as well as a provision for a review conference to be held five years after the entry into force of the convention, with a view to assuring that the purposes of the preamble and the provisions of the convention are being realized (Articles IX and X). (The text of the revised draft convention is on page 326.)

Verification[17]

In the view of the United Kingdom, verification in the sense in which the term is normally used in disarmament negotiations is not possible in the field of BW. A provision therefore was made in the UK draft convention for a complaints procedure to deter would-be violators. Quick and automatic investigation—contended the UK—should be possible where a party alleged that biological methods of warfare had been used against it because, in that case, the complainant would provide all the facilities for carrying out an investigation. In other cases, facilities for conducting an inquiry would have to be provided by the accused party. The investigating body would establish the types and quantities that were in production and report the justification for that production offered by the state concerned. The UN Security Council and individual parties would then have to decide whether the justification was adequate and to act accordingly.

Canada remarked that efforts to devise verification mechanisms other than those involved in the investigation of complaints concerning use, development, production or stockpiling of biological weapons seemed technically futile because of the high risk of undetected evasion of any other procedures that might be promulgated. A political decision by governments accepting the risks inherent in a complaints procedure would therefore appear to be the most logical solution. [32]

In any event, as explained by the United Kingdom, the draft convention tabled by it did not depend for its efficacy on verification but on the nature of the biological weapon as it exists today. [56]

[17] A detailed analysis of the verification problem can be found in Volume V.

With regard to an agreement covering chemical weapons, the United Kingdom thought that verification would need to be extremely reliable. It saw the main difficulty in reducing the risk of entering into such an agreement to an acceptable level: verification measures involving intrusiveness were unacceptable to a number of states, while the likelihood of detecting violations through external means, such as observation satellites and remote sensors, was low. [33] Also, Canada found that remote (extraterritorial) sampling for the verification of an adherence to a chemical disarmament agreement was not feasible. [67]

The United States attached paramount importance to controlling a ban on chemical weapons. It pointed out that the capacity for producing CW agents grew out of, and was linked to, the commercial chemical industry. It quoted data showing that the raw materials for various CW agents, and even some agents themselves, were produced in vast amounts in a great many locations throughout the world. [24] It maintained that the production of chemical nerve agents involved chemical processing which utilized production facilities and equipment similar to the equipment and processes used by a major segment of the world chemical industry. The problem of identification of nerve-agent production facilities could not therefore be solved by off-site observation. [34] As to economic data monitoring, the United States considered that under optimum conditions such monitoring could be of ancillary use, but alone would not provide an answer to the verification problem and could not serve as a substitute for direct on-site inspection. [35] The USA drew the conclusion that reliable international verification arrangements, involving inspection, were needed so as to have confidence that whatever bans are placed on chemical weapons were being observed. The United States admitted that it was unable to define the measures required; the problem had to be studied further.

The position of the Soviet Union and other Socialist states on the question of ascertaining whether or not CB weapons are being produced was that any system of international verification would be impractical in view of the specific features of chemical and bacteriological substances: the process of manufacturing such substances for peaceful purposes was essentially no different from that of their production for military purposes. They asserted that international control would be tantamount to the intrusion of foreign personnel in chemical and biological enterprises: "There would have to be a controller in every pharmacy, drug store, garage or any place where chemical and bacteriological (biological) weapons might be produced." Their conclusion was that such a procedure was impossible and that it would be more appropriate to leave control to the national governments which would see to it that no firm, juridical or physical person, would produce chemical and biological weapons; any problems which may arise in the application of the provisions of the convention could be solved by the parties through consultation and cooperation. [36]

The Socialist states envisaged the possibility of on-site inspection, if and when the UN Security Council decided to conduct such inspection under the complaints procedure.

A number of proposals were submitted regarding *administrative measures* to be taken by states as safeguards at the national level.

Yugoslavia suggested enactment of a law prohibiting research for weapons purposes and the development, production or stockpiling of agents for CB weapons; enactment of a law for compulsory publication of the names of institutions and facilities engaged in or which, by their nature, could engage in the activities prohibited under the treaty, as well as data concerning the production of materials or agents which could be used for the production of CB weapons; promulgation of a decision to eliminate existing stockpiles and to abolish proving grounds for the testing of the prohibited weapons, and all installations related exclusively to such weapons; cessation of the training of troops in the use of CB weapons and deletion from army manuals of all relevant instructions except for sections dealing with protection.

In Yugoslavia's opinion all such national legislative measures should be preceded by establishing civilian administration or control of all institutions now engaged in the research, development or production of chemical and biological weapons. In enacting the laws, an exception could be made, in line with the provisions of a treaty on the complete prohibition of chemical and biological weapons, for types and quantities of agents used for riot-control purposes within the country. [37]

Czechoslovakia thought that self-supervision could be carried out by national bodies having an international reputation, such as academies of science. [38]

Mongolia put forward an idea of establishing special government agencies, on the pattern prescribed in the Single Convention on Narcotic Drugs, to ensure compliance with a CBW convention; the agencies might include representatives of important research institutes, medical and veterinary services, departments responsible for chemical industries, etc. Mongolia suggested, in addition to a national system of compulsory registration of CB agents, control of the import and export of agents which could be converted into weapons, as well as control of the manufacture, import and export of equipment and apparatus that could be used for the production of CB weapons. [39]

Poland advised including in textbooks dealing with chemistry and biology an indication that the use of CB agents for any warlike purposes constitutes a violation of international law and is liable to prosecution. [39]

Mexico felt that individuals could be active participants in the denunciation of treaty violations and thus become agents of disarmament and champions of the interests of the international community. [9]

In the view of Mongolia, a review conference held on a regular basis could, in the light of new developments in science and technology, recommend to the parties other appropriate measures to be applied in order to secure further the implementation and operation of the convention. [39] Sweden thought that particular attention at such a conference should be paid to changes in the recognition of the application of chemical and biological agents for warfare or for peaceful purposes, respectively. [7]

The UAR said that a provision on withdrawal, incorporated in the treaty, would ensure respect for the obligations assumed. [40]

The *complaints procedure* provided for in both draft conventions was generally considered an important part of the verification system.

Japan believed that on the basis of present technical knowledge and experience a violation of the prohibition of the use of CB weapons could be verified with a considerable degree of certainty, provided that the UN Secretary-General could act without delay with the cooperation of competent international experts. In the case of suspicion that the prohibition of development, production and stockpiling of these weapons was violated, verification under a complaints procedure, especially with regard to chemical weapons, would be more difficult, but still to a considerable extent effective. [41] Statistics of certain chemical substances concerning the amount of their production, preferably on a factory basis, exportation and importation as well as consumption for different purposes, might be used as part of the data forming the evidence for a possible complaint. In *ad hoc* inspections based on the complaints procedure, gaschromatography could be applied to microanalyse substances from the chemical plant concerned existing in liquid wastes, the soil and dust in and around the premises, on the production devices or on the workers' clothes. [42]

Japan also suggested that a roster of experts available for investigations should be provided for and kept by the UN Secretary-General. [41]

The complaints formula which was expected to function mainly as a restraint needed, in the view of a number of non-aligned countries, clarification as to the way it would operate in practice and as to the consequences that would ensue if the Security Council were to be convinced of the accuracy of an alleged breach of the obligations. Some of these countries favoured a graduated approach beginning with complaints lodged with either the UN Secretary-General or a specially established international organ, preferably upon consultation between the states concerned. A system of "verification by challenge", outlined by Sweden, would permit a party suspected of having violated its engagements to free itself from that suspicion through the supply of relevant information, not excluding invitation to inspection. [43] The resort to the Security Council could then be an ultimate step, although the right of any party to address itself directly to the Security Council, at any stage, would remain unaffected.

Attitudes to the adequacy of administrative undertakings plus a complaints procedure were varied and rather tentative, but most nations, apart from the Socialist group, considered them to be insufficient for CW disarmament which—they alleged—posed problems of a different dimension. It was asked how a suspicion that a violation had been committed was to be established to justify a complaint. The requirement for *international control arrangements* was repeatedly emphasized.

Some proposals to this effect were:

1. International open exchange of information on pertinent peaceful, scientific, technical and other activities (Sweden). [7, 43]

2. Compulsory periodic reporting on chemical and biological agents, applying to both qualitative and quantitative factors, to an international organ having the duty of receiving, storing, analyzing and distributing the information contained in the reports.

As to the selection of such an organ, Sweden thought that for biological agents and for some chemical agents the World Health Organization seemed to be the proper body, as it already had the essential know-how, while for some other chemical agents, particularly those going through industrial production for civilian uses, the Food and Agriculture Organization might be a possibility. [43]

Japan suggested that a certain level of lethal dose by hypodermic injection be employed as a criterion, so as to limit the scope of the chemical items to be accounted for. It prepared a tentative list of substances to be reported on. [42]

Yugoslavia suggested that an international organ mentioned above should be able, at the request of states, to carry out preliminary investigations in order to ascertain whether a violation of the treaty had occurred. [44]

3. Appropriately regulated access, in accordance with the concept of verification by challenge, to institutions which prior to the ban were engaged in research, development, production and testing of CB weapons, as well as to institutions which by their nature could be engaged in such activities;

lists of these facilities would have to be declared by governments (Yugoslavia). [45]

4. Control from the air by satellites and other devices for remote detection (Yugoslavia). [45]

5. Tracing in each state the flow of materials which may be used for production of the most dangerous agents. This could deter the use of highly poisonous organophosphorus agricultural chemicals as CW agents (Japan). [46]

The non-aligned countries represented in the CCD expressed the opinion that verification should be based on a combination of appropriate national[18] and international measures, which would complement and supplement each other, thereby providing an acceptable system which would ensure effective implementation of the prohibition. [21]

To facilitate consideration of the question of verification, Sweden, in a working paper submitted to the CCD, analysed the ways in which the problem had been dealt with in various arms-control treaties and proposals. It appeared that in most cases in the past a combination of several methods had been deemed necessary. Sweden attached particular importance to a gradual system of successive steps or measures of increasing severity, where the initial step or steps were mainly of a fact-finding nature. Only when the fact-finding machinery led to the suspicion or certainty that circumvention of an obligation had taken place, would it seem necessary to resort to more far-reaching steps. [66]

Japan favoured an international meeting of experts to study technical aspects of verification relating to the prohibition of chemical and biological weapons. [41] The idea was supported by a number of Western and non-aligned countries.

Italy submitted specific suggestions as to how a group of experts convened for the purpose of studying the problems of control over chemical weapons should function. [47] It also introduced a working paper enumerating technical questions to be examined. [48] Canada invited governments to reply to a number of questions concerning national policies and controls, the production and stockpiling of chemicals, and research and development. [32] The Soviet Union, however, thought that consideration of the CBW prohibition should not be channeled into a discussion of technical details, since the problem was essentially political. [36]

---

[18] When the same opinion was stated in the United Nations, Malta observed that the "national measures" that had been suggested were related more to self-restraint than to verification. [55]

Summary and comment

In 1969–70 there has been a new wave of pressure to bring about general adherence to the Geneva Protocol prohibiting the use of CB weapons, as well as to stop the development and production of those weapons and to eliminate their stockpiles. The international discussion on the subject was more specific than at any time since World War II. The UN Secretary-General's report on chemical and biological weapons and the effects of their possible use, and a WHO report on health aspects of CBW, stimulated and helped the debate along. The United Nations and some important non-governmental organizations called upon states to take urgently appropriate measures.

In these circumstances, the unilateral renunciation by the United States of biological weapons, including toxins, and a decision to dispose of existing stockpiles, was welcomed as an important disarmament event. A few other Western and non-aligned countries declared that they had no intention of manufacturing or otherwise acquiring CB weapons, or only biological weapons. There was some expectation that the remaining powers would follow suit, but the consensus was that unilateral decisions should not be a substitute for multilateral binding agreements.

The drive for universal prohibition of use of CB weapons was reinforced by new ratifications of the Geneva Protocol. By the end of 1970, the number of parties reached 82.

The United States, the only big power not yet party to the Geneva Protocol, renounced unconditionally the use of biological weapons and the first use of lethal and incapacitating chemicals, and declared the intention to ratify the Protocol with an understanding, however, that the use in war of riot-control agents and chemical herbicides is not prohibited.

No party to the Geneva Protocol has entered reservations limiting the types of weapons to which it applies and the weight of international opinion has been that the existing rule of law prohibiting chemical weapons comprehensively covers all chemical agents. The UN resolution to this effect, adopted by a majority of member states in December 1969, was a formal expression of that opinion.

Although the UN resolution did not have an immediate effect upon the policies of the opponents of a tear-gas prohibition (shortly after its adoption the UK government made a statement claiming the legality of use of such gas in war), a few months later some of the US allies which had previously abstained from expressing an opinion on the subject admitted the desirability of and the need for a ban on the use of tear gas in warfare.

On the question of herbicides the United States got even less support than on tear gas. With the exception of Australia, there had been no explicit support for the military use of anti-plant agents. The US government itself, pressed by the American scientific community, decided first to restrict and then to phase out the herbicide operations in Viet-Nam.

A qualified ratification of the Geneva Protocol, such as has been proposed by the United States, may raise serious problems, because it is precisely the use of harassing and anti-plant chemicals in actual hostilities that has provoked the recent animated debate about CBW.

Proposals were made in the United Nations that to enhance the efficacy of the Geneva Protocol the reservations entered into by a number of states, limiting the applicability of the prohibition to parties, and to first use only, should be withdrawn so as to make the ban universal and absolute. While nobody denies that the reservation *intra partes* has become obsolete, there is some unwillingness, especially on the part of the big powers, to renounce the retaliatory use, most particularly of chemical weapons, as long as the possession of those weapons has not been internationally prohibited. (The unilateral renunciation of BW by the United States is unconditional.) Indeed, full effectiveness of the prohibition of use of CB weapons may be achieved only when the weapons in question are abolished.

A controversy arose as to whether, in dealing with the prohibition of development, production and stockpiling, chemical and biological weapons should be treated jointly or separately, in one or more international legal instruments. A good number of UN members felt that a convention limited to biological weapons—and it was with such a convention that the proponents of a separate treatment wanted to start—may not be followed by a chemical disarmament.

This feeling was reinforced by the assessment made by the United States of the military usefulness of the two categories of weaponry: while biological weapons appeared to have no value as a deterrent and to be of little effectiveness in any military situation, chemical weapons appeared less so, and were judged as important for "national security". The insistence of the United States on separate treatment of the two categories was therefore interpreted by some people as evidence of reluctance to get rid of chemical weapons.

The divergent approaches were reflected in draft conventions submitted to the Disarmament Conference and the UN General Assembly. None of them received general acceptance.

Summary and comment

The UK draft prohibited the production and acquisition only of biological weapons. The draft supported by the USA and some Western European and British Commonwealth countries was criticized for its partial approach. The inclusion of toxins in the prohibition, following an amendment proposed by the USA, was found insufficient. A mere assurance that negotiations on measures to strengthen the existing constraints on CW would be pursued was considered inadequate. No specific objections however were raised with regard to the provisions as far as they went.

The draft banning both biological and chemical weapons, which was submitted by a group of Socialist countries, was found deficient chiefly because it did not make fuller provision for verification and also because it did not explicitly prohibit the production of biological and chemical warfare agents, in addition to weapons.

The Socialist countries proposed that the convention, before coming into force, must be ratified by all the states that are permanent members of the UN Security Council. The implication of the proposal is that the Socialist countries are not ready to give up CB weapons, especially chemical, as long as all the other big powers, including those not participating in the negotiations—China and France—have not done so, too. This requirement did not figure in any previous arms limitation agreement, and seems likely to derive from strategic considerations. The fulfilment of the requirement involves the question of Chinese membership in the UN Security Council as well as the willingness of the Chinese People's Republic to accede to the convention. The authors must have been aware that this clause could delay or prevent the entry into force of the convention.

The non-aligned members of the CCD in a joint memorandum (the basic points of which were later included in a UN resolution) stated that chemical and biological weapons should continue to be dealt with together in taking steps towards the prohibition of their development, production and stockpiling as well as their elimination.

Control, especially in the field of CW prohibition, presents a serious difficulty in view of the fact that some warfare agents or intermediates are also used extensively for peaceful purposes. The United States asserted that reliable international verification arrangements, involving inspection, were needed, but it failed to define the measures required. The United Kingdom argued that it had not been possible to envisage any control machinery for CW prohibition that would not be either too inquisitorial or too weak.

The Soviet Union considered the intrusion of foreign personnel in industrial enterprises as unacceptable and suggested that control should be exercised by national governments.

The non-aligned countries' opinion that an effective verification system should be based on a combination of appropriate national and international undertakings was generally accepted, but the argument continued as to the extent of the latter.

A procedure under which the parties would be authorized to lodge with the UN Security Council complaints of possible violations of the convention was devised to serve as a restraint on would-be violators. Sweden suggested to complement it with "verification by challenge" which would permit a party under suspicion of having violated its obligations to free itself from that suspicion through the supply of relevant information or invitation to inspection.

The "complaints procedure" was believed to be an important element of the verification system, but many nations felt that with respect to CW prohibition such a deterrent alone would not be sufficiently effective.

Little attention was given to the methods of destroying CB weapons. The experience of the United States had shown that serious problems may arise when it comes to safe disposal of redundant stocks.

Demands were made for a thorough expert study of the verification problem. The Soviet Union argued that the pursuit of technicalities might become a convenient way of avoiding the political decision concerning the ban.

At the summer 1970 session of the CCD, the United States rejected the Socialist countries' draft convention as a basis for negotiation and declared that to insist on a single agreement covering both biological and chemical weapons would be to resign oneself to no concrete advance for a considerable period of time. The United Kingdom warned that in view of rapid scientific development there was a danger in delaying an agreement on biological weapons. The changes subsequently introduced in the Socialist draft did not alter the position of the Western powers. A proposition to combine a ban on biological weapons with that on single-purpose chemical weapons (nerve agents) was alluded to but not discussed.

A compromise solution, suggested by Morocco, aroused some interest in both the East and the West: CBW would be prohibited in one legal instrument, but while the provisions concerning biological prohibition would enter into force at once, those relating to chemical prohibition would become effective after a specified period of time, during which a supplementary document on chemical verification would be worked out.[1] The approach was consistent with the requirement, put forward by the non-

---

[1] Subsequently, Morocco proposed that no time-limits be set, but that negotiations should take place without delay and, if possible, immediately after the drafting of the principal text.

aligned countries, that biological and chemical prohibition should be dealt with together.

The twenty-fifth UN General Assembly, having considered the report of the Disarmament Conference, asked the latter to study further the problem of chemical and biological methods of warfare. CBW prohibition became one of the most important items on the disarmament negotiations agenda in 1971.

Postscript

In the spring of 1971, there was a major turning point in CB disarmament negotiations.

The USSR and its allies, which for years had been insisting on a joint treatment of chemical and biological weapons, and had considered their prohibition an indivisible entity, suddenly reversed their position. They accepted a partial (or gradual) approach advocated by the United Kingdom and the USA, and on 30 March 1971 produced at the Conference of the Committee on Disarmament a draft convention for biological disarmament only. Up to that moment, the majority of the CCD had upheld the view that such a minimum solution would not be satisfactory.

Before analyzing the draft of the Socialist countries, a brief summary is given of the discussion which preceded its submission.

Non-aligned states' proposals for CB disarmament

In a working paper tabled on 16 March 1971 [1], Sweden proposed a model for a comprehensive agreement concerning both chemical and biological means of warfare. The paper defined the scope of the prohibition and outlined verification procedures (For the details, see the attached "Guide to CB disarmament proposals", page 373.)

The most rigorous methods of control envisaged by Sweden would be those dealing with chemicals more toxic than 1 mg per kg body weight, with toxins and with biological agents without any recognized peaceful use.

The production of these compounds would in principle be prohibited. Any deviation from this general rule would have to be reported to an international agency, the report giving the reasons for the production (scientific use, protective measures, etc.). In case of any large-scale production (e.g., over 1 kg) or in case of suspected undeclared production, the international agency might be entitled to conduct an on-site inspection, either on the invitation of the producing or suspected party, or obligatory.

If verification methods ready for immediate application were found wanting, Sweden suggested accepting the idea of total, comprehensive agreement with the inclusion of an article which would set one or several deadlines for a more detailed elaboration of verification procedures. [2]

The Netherlands worked out a chemical formula to help in specifying organophosphorus compounds likely to have nerve-gas properties. The formula would be handled in connection with a toxicity level of 0.5 mg/kg determined subcutaneously. [3]

Yugoslavia asked that a convention should carefully define chemical weapons and warfare agents. [16] None of the drafts submitted thus far contained such a definition.

The United Kingdom noted that control suggested by Sweden for the most toxic chemicals would depend on the information provided from unverified national reporting, and thought that there was no value in allowing for possible on-site inspection, when these procedures only arose if the reporting state itself confessed to exceeding the prescribed limit [4].

The United States, too, maintained the view that failure to comply with a chemical warfare treaty would offer few chances of detection by any currently developed national means, i.e., using national resources to detect possible violations by others. To illustrate the difficulty of assuring compliance, it pointed out that diversion of only 1 per cent of annual production of elemental phosphorus in the USA could serve to produce 10 000 tons of nerve agents in a year, that is, enough to fill 3 million artillery rounds.[1] [5]

Japan doubted whether in the absence of precise knowledge of the nature and the amount of stockpiles or the capabilities of production of chemical weapons, a meaningful discussion could be conducted on the prohibition of those weapons and in particular on the verification of the prohibition. It suggested therefore promoting international communication, to the extent possible, with regard to the present state of affairs concerning chemical weapons. [6]

For the first time in the disarmament debate the methods of destruction of CB weapons were examined.

Sweden reviewed the possible technical means of destroying stockpiles [7]. (See page 386.) Its conclusion was that the safe methods existed, but because of the high toxicity and infectiousness of the agents, destruction costs could be high due to the need for special technical facilities.

Sweden favoured some form of international surveillance of the destruction of CB weapons and agents. In the case of diversion to peaceful purposes of stocks of warfare agents, an international involvement should not be excluded either. This could take the form, for instance, of transfers for research or health protection for the benefit of developing countries. [8]

The USA reiterated that it saw no practical prospect for early progress

---

[1] As stated by the US representative to the CCD, the USA had not been producing nerve agents since mid-1968.

Postscript

on the basis of a comprehensive approach to chemical and biological weapons [2] and assured that in any biological weapons convention it would support an unambiguous commitment engaging all parties to undertake further negotiations regarding limitations on chemical weapons [9]. It also announced that in accordance with its statement of renunciation of BW, of 1969, the entire US arsenal of biological weapons would be eliminated in about a year. The destruction of stocks would be done through a complicated process involving boiling the agents and then plowing them into the soil. There would be no attempt to dump any of the agents into the ocean. The estimated cost was just over $12 million. [14] After the destruction of stocks, the biological facilities would be open for public inspection and international visitors. [20]

Socialist countries' draft convention on BW

The seven Socialist members of the CCD submitted, on behalf of Bulgaria, the Byelorussian SSR, Czechoslovakia, Hungary, Mongolia, Poland, Romania, the Ukrainian SSR and the USSR, a draft convention on the prohibition of the development, production and stockpiling of bacteriological (biological) weapons and toxins and on their destruction [10].

The draft, which was slightly revised on 15 April 1971, took up, in essence, the UK proposal of 18 August 1970 (see p. 322). It provided for an undertaking not to develop, produce, stockpile or otherwise acquire microbiological or other biological agents or toxins of such types and in such quantities as are not designed for the prevention of disease or for other peaceful purposes; as well as auxiliary equipment or means of delivery designed to facilitate the use of such agents or toxins for hostile purposes (Art. I). The parties would also undertake to destroy within a period of three months after the entry into force of the convention—observing all the necessary precautions—or to divert to peaceful uses all previously accumulated weapons as well as the equipment and means of delivery (Art. II). It was explained by the sponsors of the draft that the terms "weapons" or "means of warfare", as used in the text, covered all bacteriological agents and toxins which can be used for purposes of war. [11, 16]

Each state would be internationally responsible for compliance with the provisions of the convention by legal or physical persons of that state (Art. IV).

The enforcement of the prohibition would be assured through national legislative and administrative measures, consultations among the parties and a complaints procedure (Art. V, VI and VII), in the same way as had been stipulated in the earlier nine powers draft banning both chemical and biological weapons. Also the provisions for international cooperation in the field of peaceful bacteriological activities (Art. X), were based on the previous draft.

The significance of the 1925 Geneva Protocol was emphasized; it was pointed out that the Protocol embodied generally recognized rules of international law (Art. VIII) and prohibited the use of any chemical and biological means of warfare (preamble).

There was a commitment to negotiate chemical disarmament (preamble and Art. IX); in particular, a review conference to be held five years after the entry into force of the convention would have to assure that the provisions concerning negotiations on chemical weapons were being realized (Art. XII). Poland even suggested setting a time-limit for the conclusion of those negotiations. [13]

The convention would enter into force after its ratification by a certain number of governments, including those designated as depositaries (Art. XIII). The latter were not specified, but it was assumed that following the precedent set by other arms control treaties, the UK, US and Soviet governments would be designated depositary governments. The requirement for the convention to be ratified by all the permanent members of the UN Security Council, as had been envisaged in the Socialist draft on CBW prohibiton, was now left out; it could thus enter into force without the participation of China or France. (For the full text of the draft convention see p. 331.)

While the British draft characterized the biological methods of warfare as those causing death, damage or disease to man, animals or crops, no description of the prohibited weapons was given in the Socialist draft. Neither draft defined toxins, though both explicitly included them with biologicals.

The undertaking not to produce agents "of such types and in such quantities as are not designed for the prevention of disease or for other peaceful purposes", as provided for in the Socialist draft, or "of types and quantities that have no justification for prophylactic or other peaceful purposes", as provided for in the British draft, permitted some production to go on. The provision, while necessary, meant that there would be a risk of marginal infringements and of allegations of infringements, but this might be reduced if secrecy surrounding biological research were reduced.[2]

No verification of the destruction of stocks was provided for in either of the drafts, nor were the parties required to announce that their stocks, if any, had been destroyed. Thus no country would be put in a position where it was forced to indicate whether it had possessed biological weapons.

---

[2] The problem of defensive research is discussed in Volume V.

Unlike the British draft, the Socialist draft did not contain a ban on the use of biological methods of warfare. The USSR and its allies considered that the matter was already "clearly and unequivocally" settled by the Geneva Protocol and that a provision banning only biological warfare would detract from the Protocol which prohibited both biological and chemical warfare. (This point of view was shared, among others, by Brazil, India, Morocco, Sweden, Yugoslavia and the UAR.) It was also argued that it would be pointless to prohibit the use of weapons in a convention providing for their destruction (Poland), and that the reservations to the Geneva Protocol concerning the right to use bacteriological weapons against non-parties or in retaliation would be rendered purposeless by such a convention (Morocco, UAR). No one, however, indicated whether those reservations would be formally withdrawn or allowed legally to subsist.

The Socialist draft, unlike the British, did not provide for a prohibition of research aimed at production of agents and equipment; and it did not include a withdrawal clause.

The reasons given by the USSR and its allies for submitting a draft convention prohibiting only biological weapons were: reluctance of the USA and other Western powers to renounce chemical means of warfare, and lack of prospect for a speedy achievement of a comprehensive agreement in which the militarily important states, capable of producing and stockpiling CB weapons, would participate; the desire to break the deadlock in a spirit of compromise and to ensure that the next UN General Assembly session mark a further step in disarmament; the wish to improve the international situation and strengthen peace and security. [8, 13]

Other arguments put forward by the Socialist countries in favour of a separate biological disarmament treaty resembled those which had been advanced by the USA and UK and previously found unacceptable by the Soviet Union. It was claimed, in particular, that such a treaty would be very important in that it would eliminate a whole category of weapons of mass destruction; that it would strengthen the Geneva Protocol; and that far from legalizing chemical weapons it would facilitate chemical disarmament. [16, 17]

In reporting and commenting on the new initiative, some authoritative Soviet journals and newspapers[3] took the view that common sense favoured a stage-by-stage approach; no mention was made of the UK proposal for biological disarmament, which had been on the table since 1968 in the form of a working paper, and since 1969 as a draft convention.

---

[3] See *New Times* No. 14, 1971; *Izvestia* 7 April 1971; and *Moscow News*, 10 April 1971.

The Soviet move, though expected by some people after the experience with the sea-bed treaty,[4] aroused consternation among those non-aligned powers which had been consistently arguing that chemical and biological disarmament should not be dealt with separately. They contended that a biological disarmament treaty would not meet the aspirations of the international community, as reflected in UN resolutions. Sweden qualified it as a very marginal measure of disarmament, restricted as it was to weapons which were judged to be militarily insignificant, compared to the banning of the militarily more relevant chemical weapons, and urged that a comprehensive solution be given priority. It remarked that thus far the method of concluding so-called partial treaties had led to acquiescence in the *status quo* for important armaments sectors, and wondered whether the result of a treaty prohibiting only biological weapons would not be to legitimize proliferation in regard to chemical weapons. [12]

Mexico could not comprehend how an artificial division of the subject of CBW prohibition and an isolated treatment of biological disarmament, which until recently had been regarded by the Soviet Union as intrinsically prejudicial, undesirable and unacceptable, could suddenly become a beneficial solution. [15]

Pakistan thought that a separate agreement on biological weapons would inevitably create an impression that an agreement on chemical weapons was not within reach in the foreseeable future. [22]

Yugoslavia felt that undertakings alone, in a partial treaty, to continue negotiations on unresolved issues have a limited value. [16]

On the other hand, Argentina and Brazil considered the submission of the new draft a laudable effort [14, 18], Nigeria appreciated the circumstances that had motivated the action of the Socialist states [25], and the United Arab Republic said that a difficult obstacle had been overcome, so that some positive though modest results could be achieved. [19]

The USA and other Western powers welcomed the Socialist countries' step and the United Kingdom expressed appreciation for it. The British critical remarks concerned mainly the non-inclusion of the ban on use of biological weapons and of the procedure for automatic investigation of complaints of use through the submission of such complaints to the UN Secretary-General. The argument was that the right to employ biological weapons under certain circumstances, as reserved by a number of countries under the Geneva Protocol, should be completely ruled out, and that the

---

[4] In 1969, in spite of the overwhelming support for the Soviet proposal to prevent both nuclear and conventional arms race on the sea-bed, the USSR unexpectedly decided to accept the US position and agreed to mere denuclearization of that environment.

fact-finding stage of the complaints procedure should be separated from and precede the stage of political consideration by the Security Council.

The UK regretted the omission of the ban on relevant research, which was contained in its own draft, as well as of the provision concerning appropriate assistance to a party against which biological weapons had been used; the proposed assistance was meant primarily as action of medical or relief nature, taken at the request of the victim [31]. The UK also suggested that the commitment to further negotiations on chemical methods of warfare should be rephrased so as to make it realistic [12, 13].

The USA and the United Kingdom opposed the contention that the Geneva Protocol embodied universally recognized rules of international law. They objected to the references (in the preamble) to the UN Secretary-General's report recommending, among others, a ban on tear gas and other harassing agents, and to the UN resolution prohibiting the use of *any* CB means of warfare, as an interpretation of the Geneva Protocol which was open to challenge. In addition, the United States called for the employment of consistent language on weapons and agents; considered that a three months period was not sufficient to destroy the stockpiles of biological weapons and toxins; and was in favour of a withdrawal clause similar to that contained in some other arms control agreements.

Canada thought that depositary governments should be notified that the destruction of stocks or their diversion to peaceful uses had been carried out by nations possessing biological and toxin weapons by the end of the period allotted for that purpose. [31]

The opponents of a separate convention on biological disarmament could either continue to urge a comprehensive prohibition of CB weapons and dissociate themselves from any partial arrangement, or try to amend the tabled draft as far as possible, in particular with a view to establishing a strong commitment to chemical disarmament. They chose the latter course.

Morocco submitted that a possibility be envisaged of a moratorium on chemical weapons pending the conclusion of an agreement on their elimination. [18] It suggested that the reservations to the Geneva Protocol should be declared null and void with regard to the prohibition of the use of bacteriological and toxin weapons. [33, 34]

Sweden asked that all toxins that have potential use as warfare agents, whatever their origin or mode of preparation (i.e., both natural and synthetically produced) should be covered by the convention. [24, 25] It also proposed that the principle of verification agreed between the parties be included in the convention along with verification through appropriate international procedures within the framework of the United Nations. [25]

Pakistan referred to the Polish suggestion to set a time-limit for negotiations and asked for a commitment that a convention on chemical methods of warfare should be concluded within three years of the entry into force of the convention on biological weapons. [22]

Yugoslavia asked that a conference to review the progress in the field of chemical disarmament be held sooner than provided for in the Socialist countries' draft. [22]

The UAR proposed that the undertaking under the Socialist draft be reinforced by the term "never in any circumstances", so as to make it virtually impossible for a party to formulate a reservation to the biological convention. [19] It also suggested that the provision containing an undertaking of the parties to consult one another and cooperate, should take into account the fact that in certain instances relations between states do not allow its implementation. [21] Pakistan and Nigeria shared this view.

Argentina said that a state could be responsible only for acts committed on its own territory or other areas under its jurisdiction, whatever the nationality of the person violating the convention, and suggested to amend Art. IV of the draft accordingly. [17]

The provision regarding the exchange and cooperation related to the use of biological substances for peaceful purposes, was generally found constructive.

The Socialist draft became the basis of the Disarmament Committee's work. A revised version of this draft submitted on 5 August 1971 [26] took into account the critical remarks made by the United States. The USA recorded its agreement by tabling simultaneously a parallel and identical draft under its own sponsorship. [27] The agreed text[5] also incorporated a few suggestions made by the non-aligned countries, but their most important postulates remained unsatisfied. (For the text see page 336.)

Subsequently, Hungary, Mongolia and Poland produced a draft UN Security Council resolution in which the Council would declare its readiness to consider complaints of breaches of the obligations contained in the convention and take measures for their investigation. [30]

On 17 August 1971, a group of eleven non-aligned members of the Disarmament Committee (Brazil, Burma, Ethiopia, India, Morocco, Mexico, Nigeria, Pakistan, Sweden, UAR and Yugoslavia) tabled amendments to the draft convention [32], aimed mainly at establishing a firm link between the prohibition of biological and chemical weapons. The language used would make it clear that the biological convention represented only a first step towards a comprehensive prohibition of CB weapons. The parties would have to accept the principle of complete prohibition of chemical weapons and

---

[5] It was re-issued on 12 August 1971 for technical reasons.

undertake to continue negotiations in good faith with a view to reaching early agreement on prohibition of their development, production and stockpiling and on their destruction, and on appropriate measures concerning equipment and means of delivery specifically designed for the production or use of chemical agents for weapons purposes.

In addition, Mexico proposed that, pending the above agreement, the parties should undertake to refrain from any further development, production or stockpiling of those chemical agents for weapons purposes, which because of their degree of toxicity have the highest lethal effects. The agents in question would be listed in a protocol annexed to the convention. [35]

The USSR and the USA were willing to negotiate changes in order to develop a widely acceptable draft biological convention for submission to the twenty-sixth UN General Assembly [28]. A new document was presented on 28 September 1971, under the sponsorship of Bulgaria, Canada, Czechoslovakia, Hungary, Italy, Mongolia, the Netherlands, Poland, Romania, the USSR, the United Kingdom and the USA. [36] (For the text see page 403.) The non-aligned members of the Disarmament Committee declined to endorse it, because not all their proposals had been accepted.

The official position of the great powers not participating in the negotiations was not known. France had all along preferred a comprehensive approach to CBW prohibition, but was preparing a law forbidding the manufacture, stockpiling, acquisition or transfer of biological weapons. [37] China's close ally, Albania, qualified the draft biological convention as a document deceiving the world public opinion by giving the impression that something was being done in the field of disarmament. [29]

The discussion on chemical disarmament was expected to continue, but the concern expressed by Mexico [15] seemed to be widely shared, namely that the superpowers were not yet ready to give up chemical weapons.

The arguments related to the verifiability of obligations, or to the involvement of all the great powers, did not prove convincing to many countries. Indeed, the very same problems were easily solved or just put aside once it was decided to do away with biological means of warfare. As in other disarmament negotiations, it is rather the usefulness of the weapons in question, as perceived by the military, in real or imaginary situations, which matters to the powers concerned.

Ratification of the Geneva Protocol

In March 1971, the US Senate Foreign Relations Committee held hearings on the question of the ratification of the Geneva Protocol.

While most members of the Committee supported the objectives of the Protocol, the main controversy was whether to accept the US government's interpretation, submitted as an informal "understanding" of its scope, that the use of riot-control agents such as tear gas or chemical herbicides was not prohibited.

A number of senators expressed the opinion that a ratification of the Geneva Protocol with the proposed "understanding" would erode its effectiveness, would be challenged by other nations and would complicate the chances of reaching a comprehensive treaty on CB disarmament; that the military value of the chemicals in question was low and their use would run counter to the US security interest in preventing the proliferation of CB weapons.

Those who opposed the above position represented the view of the US Department of Defense that tear gas and anti-plant agents were legitimate, humane weapons of war; that they caused less suffering than napalm and flame-throwers which would have to be used if tear gases were abandoned; and that local commanders should have the option to use chemicals to "save lives".

The chairman of the committee, in his letter to the US President of 15 April 1971, regretted that the Protocol had come before the Senate at a time when the USA was at war and employing chemical weapons which most nations considered to be prohibited. [23]

The majority of the Foreign Relations Committee being in favour of US ratification without any exemptions, the administration was asked to reconsider its "understanding" with regard to tear gas and herbicides. The US government undertook a detailed study of the political and military aspects of the use of those chemicals.

While the US use of the controversial chemicals in Viet-Nam apparently diminished in 1971, Portugal was condemned in April 1971, by the UN Decolonization Committee, for employing herbicides and defoliants as methods of warfare in its African territories.

By July 1971 the number of parties to the 1925 Geneva Protocol reached eighty-eight. (For the list of parties see page 342.)

Appendix 1.

Revised UK draft convention for the prohibition of biological methods of warfare and accompanying draft Security Council resolution, of 18 August 1970[1]

The States concluding this Convention, hereinafter referred to as the "Parties to the Convention",

Recalling that many States have become Parties to the Protocol for the Prohibition of the Use in War of Asphyxiating, Poisonous or other Gases, and of Bacteriological Methods of Warfare, signed at Geneva on 17 June 1925,

Recognizing the contribution that the said Protocol has already made, and continues to make, to mitigating the horrors of war,

Recalling further United Nations General Assembly Resolutions 2162B (XXI) of 5 December 1966, and 2454A (XXIII) of 20 December 1968, which called for strict observance by all States of the principles and objectives of the Geneva Protocol and invited all States to accede to it,

Believing that chemical and biological discoveries should be used only for the betterment of human life,

Recognizing nevertheless that the development of scientific knowledge throughout the world will increase the risk of eventual use of biological methods of warfare,

Convinced that such use would be repugnant to the conscience of mankind and that no effort should be spared to minimize this risk,

Desiring therefore to reinforce the Geneva Protocol by the conclusion of a Convention making special provision in this field,

Declaring their belief that, in particular, provision should be made for the prohibition of recourse to biological methods of warfare in any circumstances.

Have agreed as follows:

ARTICLE I

Each of the Parties to the Convention undertakes, insofar as it may not already be committed in that respect under Treaties or other instruments in

[1] Disarmament Conference document, CCD/225/Rev. 2.

UK draft convention

force prohibiting the use of chemical and biological methods of warfare, never in any circumstances, by making use for hostile purposes of microbial or other biological agents or toxins causing death, damage or disease to man, other animals, or crops, to engage in biological methods of warfare.

ARTICLE II

Each of the Parties to the Convention undertakes:

(a) not to produce or otherwise acquire, or assist in or permit the production or acquisition of:
  (i) microbial or other biological agents or toxins of types and in quantities that have no justification for prophylactic or other peaceful purposes;
  (ii) ancillary equipment or vectors the purpose of which is to facilitate the use of such agents or toxins for hostile purposes;
(b) not to conduct, assist or permit research aimed at production of the kind prohibited in sub-paragraph (a) of this Article; and
(c) to destroy, or divert to peaceful purposes, within three months after the Convention comes into force for that Party, any stocks in its possession of such agents or toxins or anciliary equipment or vectors as have been produced or otherwise acquired for hostile purposes.

ARTICLE III

1. Any Party to the Convention which believes that biological methods of warfare have been used against it may lodge a complaint with the Secretary-General of the United Nations, submitting all evidence at its disposal in support of the complaint, and request that the complaint be investigated and that a report on the result of the investigation be submitted to the Security Council.

2. Any Party to the Convention which believes that another Party is in breach of any of its undertakings under Articles I and II of the Convention, but which is not entitled to lodge a complaint under Paragraph 1 of this Article, may lodge a complaint with the Security Council, submitting all evidence at its disposal, and request that the complaint be investigated.

3. Each of the Parties to the Convention undertakes to co-operate fully with the Secretary-General and his authorized representatives in any investigation he may carry out, as a result of a complaint, in accordance with Security Council Resolution No . . .

ARTICLE IV

Each of the Parties to the Convention affirms its intention to provide or support appropriate assistance, in accordance with the United Nations

Charter, to any Party to the Convention, if the Security Council concludes that biological methods of warfare have been used against that Party.

ARTICLE V

Each of the Parties to the Convention undertakes to pursue negotiations in good faith on effective measures to strengthen the existing constraints on chemical methods of warfare.

ARTICLE VI

Nothing contained in the present Convention shall be construed as in any way limiting or derogating from obligations assumed by any State under the Protocol for the Prohibition of the Use in War of Asphyxiating, Poisonous or other Gases, and of Bacteriological Methods of Warfare, signed at Geneva on 17 June 1925.

ARTICLE VII

[Provisions for amendments.]

ARTICLE VIII

[Provisions for Signature, Ratification, Entry into Force, etc.]

ARTICLE IX

1. This Convention shall be of unlimited duration.
2. Each Party shall in exercising its national sovereignty have the right to withdraw from the Convention, if it decides that extraordinary events, related to the subject matter of this Convention, have jeopardized the supreme interests of its country. It shall give notice of such withdrawal to all other Parties to the Convention and to the United Nations Security Council three months in advance. Such notice shall include a statement of the extraordinary events it regards as having jeopardized its supreme interests.

ARTICLE X

[Provisions on languages of texts, etc.]

Revised draft Security Council resolution

The Security Council,

Welcoming the desire of a large number of States to subscribe to the Convention for the Prohibition of Biological Methods of Warfare, and thereby undertake never to engage in such methods of warfare; to prohibit the production and research aimed at the production of biological weapons; and to destroy, or divert to peaceful purposes, such weapons as may already be in their possession,

Noting that under Article III of the Convention, Parties will have the right to lodge complaints and to request that the complaints be investigated,

Recognizing the need, if confidence in the Convention is to be established, for appropriate arrangements to be made in advance for the investigation of any such complaints, and the particular need for urgency in the investigation of complaints of the use of biological methods of warfare,

Noting further the declared intention of Parties to the Convention to provide or support appropriate assistance, in accordance with the Charter, to any other Party to the Convention, if the Security Council concludes that biological methods of warfare have been used against that Party,

Reaffirming in particular the inherent right, recognized under Article 51 of the Charter, of individual and collective self-defence if an armed attack occurs against a Member of the United Nations, until the Security Council has taken measures necessary to maintain international peace and security.

1. Requests the Secretary-General
    (a) to take such measures as will enable him
        (i) to investigate without delay any complaints lodged with him in accordance with Article III.1 of the Convention;
        (ii) if so requested by the Security Council, to investigate any complaint made in accordance with Article III.2 of the Convention; and
    (b) to report to the Security Council on the result of any such investigation.
2. Declares its readiness to give urgent consideration
    (a) to any complaint that may be lodged with it under Article III.2 of the Convention; and
    (b) to any report that the Secretary-General may submit in accordance with operative paragraph 1 of this Resolution on the result of his investigation of a complaint; and if it concludes that the complaint is well-founded, to consider urgently what action it should take or recommend in accordance with the Charter.
3. Calls upon Member States and upon Specialized Agencies of the United Nations to co-operate as appropriate with the Secretary-General for the fulfilment of the purposes of this Resolution.

Appendix 2.

Revised draft convention on the prohibition of the development, production and stockpiling of chemical and bacteriological (biological) weapons and on the destruction of such weapons, submitted by Bulgaria, Byelorussian SSR, Czechoslovakia, Hungary, Mongolia, Poland, Romania, Ukrainian SSR and the USSR on 23 October 1970[1]

The States Parties to this Convention,

Convinced of the immense importance and urgent necessity of eliminating from the arsenals of States such dangerous weapons of mass destruction as chemical and bacteriological (biological) weapons,

Guided by the desire to facilitate progress in the achievement of the objectives of general and complete disarmament,

Desiring to contribute to the strengthening of confidence between peoples and the general improvement of the international atmosphere,

Believing that scientific discoveries in the field of chemistry and bacteriology (biology) must in the interests of all mankind be used solely for peaceful purposes,

Recognizing nevertheless that the development of scientific knowledge throughout the world will increase the risk of the use of chemical and bacteriological (biological) methods of warfare,

Convinced that such use would be repugnant to the conscience of mankind and that no effort should be spared to minimize this risk,

Recognizing the important significance of the Geneva Protocol of 17 June 1925 for the Prohibition of the Use in War of Asphyxiating, Poisonous or Other Gases, and of Bacteriological Methods of Warfare, an instrument which embodies generally recognized rules of international law and conscious also of the contribution which the said Protocol has already made, and continues to make, to mitigating the horrors of war,

Reaffirming their adherence to the purposes and principles of that Protocol and calling upon all States to comply strictly with them,

[1] UN document, A/8136.

Recalling United Nations General Assembly resolutions 2162B (XXI) of 5 December 1966 and 2454A (XXIII) of 20 December 1968 which condemned all actions contrary to the Geneva Protocol of 17 June 1925, and also resolutions 2603A and B (XXIV) of 16 December 1969 which, *inter alia,* confirmed once again the generally recognized character of the rules of international law embodied in the Geneva Protocol of 17 June 1925,

Noting the conclusions contained in the report submitted to the United Nations General Assembly and the Disarmament Committee on the grave consequences for mankind that might result from the use of chemical and bacteriological (biological) weapons,

Expressing their desire to contribute to the implementation of the Purposes and Principles of the Charter of the United Nations,

Have agreed as follows:

ARTICLE I

Each State Party to this Convention undertakes not to develop, produce, stockpile or otherwise acquire chemical and bacteriological (biological) weapons, or equipment or vectors specially designed for the use of chemical and bacteriological (biological) weapons as means of warfare.

ARTICLE II

Each State Party to this Convention undertakes to destroy within a period of ...—observing all the necessary precautions—or to divert to peaceful uses all previously accumulated chemical and bacteriological (biological) weapons in its possession, as well as equipment and vectors specially designed for the use of chemical and bacteriological (biological) weapons as means of warfare.

ARTICLE III

Each State Party to the Convention undertakes not to assist, encourage or induce any individual State, group of States or international organizations to develop, produce or otherwise acquire and stockpile chemical and bacteriological (biological) weapons.

ARTICLE IV

Each State Party to the Convention shall be internationally responsible for compliance with its provisions by legal and physical persons exercising their activities in its territory, and also by its legal and physical persons outside its territory.

ARTICLE V

Each State Party to the Convention undertakes to take as soon as possible, in accordance with its constitutional procedures, the necessary legislative and administrative measures to prohibit the development, production and stockpiling of chemical and bacteriological (biological) weapons and to destroy such weapons.

ARTICLE VI

The States Parties to the Convention undertake to consult one another and to co-operate in solving any problems which may arise in the application of the provisions of this Convention.

ARTICLE VII

1. Each State Party to the Convention which finds that actions of any other State Party constitute a breach of the obligations assumed under articles I and II of the Convention may lodge a complaint with the Security Council of the United Nations. Such a complaint should include all possible evidence confirming its validity, as well as a request for its consideration by the Security Council. The Security Council shall inform the States Parties to the Convention of the result of the investigation.

2. Each State Party to the Convention undertakes to co-operate in carrying out any investigations which the Security Council may undertake, in accordance with the provisions of the United Nations Charter, on the basis of the complaint received by the Council.

ARTICLE VIII

1. The States Parties to the Convention undertake to facilitate, and have the right to participate in, the fullest possible exchange of equipment, materials and scientific and technological information for the peaceful uses of chemical and bacteriological (biological) agents.

2. This Convention shall be implemented in a manner designed to avoid hampering the economic or technological development of States Parties to the Convention or international co-operation in the field of peaceful chemical and bacteriological (biological) activities, including the international exchange of chemical and bacteriological (biological) agents and equipment for the processing, use or production of chemical and bacteriological (biological) agents for peaceful purposes in accordance with the provisions of this Convention.

Socialist draft convention

ARTICLE IX

Any State Party may propose amendments to this Convention. Amendments shall enter into force for each State Party accepting the amendments upon their acceptance by a majority of the States Parties to the Convention and thereafter for each remaining State Party on the date of acceptance by it.

ARTICLE X

Five years after the entry into force of this Convention, a conference of States Parties to the Convention shall be held at Geneva, Switzerland, in order to review the operation of this Convention with a view to assuring that the purposes of the preamble and the provisions of the Convention are being realized. Such review shall take into account any new scientific and technological developments relevant to this Convention.

ARTICLE XI

1. This Convention shall be open to all States for signature. Any State which does not sign the Convention before its entry into force in accordance with paragraph 3 of this article may accede to it at any time.

2. This Convention shall be subject to ratification by signatory States. Instruments of ratification and instruments of accession shall be deposited with the Governments of ... which are hereby designated the Depositary Governments.

3. This Convention shall enter into force after the deposit of the ... instrument of ratification by Governments, including the instruments of ratification of the Governments of States which are permanent members of the United Nations Security Council and of other Governments designated as Depositaries of the Convention.

4. For States whose instruments of ratification or accession are deposited subsequent to the entry into force of this Convention, it shall enter into force on the date of the deposit of their instruments of ratification or accession.

5. The Depositary Governments shall promptly inform all signatory and acceding States of the date of each signature, the date of deposit of each instrument of ratification or of accession and the date of the entry into force of this Convention, and shall transmit other notices to them.

6. This Convention shall be registered by the Depositary Governments pursuant to Article 102 of the Charter of the United Nations.

ARTICLE XII

This Convention, of which the Chinese, English, French, Russian and Spanish texts are equally authentic, shall be deposited in the archives of the

Depositary Governments. Duly certified copies of this Convention shall be transmitted by the Depositary Governments to the Governments of the signatory and acceding States.

In witness whereof the undersigned, duly authorized, have signed this Convention.

DONE in . . . copies at . . ., this . . . day of . . . . . ., . . .

Appendix 3.

Draft convention on the prohibition of the development, production and stockpiling of bacteriological (biological) weapons and toxins and on their destruction, submitted by Bulgaria, Byelorussian SSR, Czechoslovakia, Hungary, Mongolia, Poland, Romania, Ukrainian SSR, and the USSR on 15 April 1971[1]

The States Parties to this Convention,

Determined to act with a view to achieving effective progress towards general and complete disarmament and, above all, with a view to prohibiting and eliminating nuclear, chemical, bacteriological (biological) and all other types of weapons of mass destruction,

Convinced that the prohibition of the development, production and stockpiling of bacteriological (biological) weapons and toxins and their elimination will facilitate the achievement of general and complete disarmament,

Convinced of the immense importance and urgent necessity of eliminating from the arsenals of States such dangerous weapons of mass destruction as weapons using bacteriological (biological) agents and toxins,

Desiring to contribute to the strengthening of confidence between peoples and the general improvement of the international atmosphere,

Believing that scientific discoveries in the field of bacteriology (biology) must in the interests of all mankind be used solely for peaceful purposes,

Recognizing nevertheless that in the absence of appropriate prohibitions the development of scientific knowledge throughout the world would increase the risk of the use of bacteriological (biological) methods of warfare,

Convinced that such use would be repugnant to the conscience of mankind and that no effort should be spared to minimize this risk,

Recognizing the important significance of the Geneva Protocol of 17 June 1925 for the Prohibition of the Use in War of Asphyxiating, Poisonous or Other Gases, and of Bacteriological Methods of Warfare, and conscious also of the contribution which the said Protocol has already made, and continues to make, to mitigating the horrors of war,

[1] Disarmament Conference document CCD/325/Rev. 1.

Reaffirming their adherence to the purposes and principles of that Protocol and calling upon all States to comply strictly with them,

Guided by the resolutions of the United Nations General Assembly which has condemned all actions contrary to the Geneva Protocol of 17 June 1925 as well as the use in international armed conflicts of any chemical and any biological means of warfare,

Noting the conclusions contained in the report submitted to the United Nations General Assembly and the Disarmament Committee on the grave consequences for mankind that might result from the use of chemical and bacteriological (biological) weapons,

Convinced that an agreement on bacteriological (biological) weapons will facilitate progress towards the achievement of agreement on effective measures for the complete prohibition of chemical weapons, on which negotiations will be continued,

Anxious to contribute to the realization of the purposes and principles of the Charter of the United Nations,

Have agreed as follows:

ARTICLE I

Each State Party to this Convention undertakes not to develop, produce, stockpile or otherwise acquire:

(1) microbiological or other biological agents or toxins of such types and in such quantitites as are not designed for the prevention of disease or for other peaceful purposes;

(2) auxiliary equipment or means of delivery designed to facilitate the use of such agents or toxins for hostile purposes.

ARTICLE II

Each State Party to this Convention undertakes to destroy within a period of three months after the entry into force of the Convention—observing all the necessary precautions—or to divert to peaceful uses all previously accumulated weapons in its possession as well as the equipment and means of delivery mentioned in article I of the Convention.

ARTICLE III

Each State Party to the Convention undertakes not to assist, encourage or induce any particular State, group of States or international organizations to take action contrary to the provisions of this Convention.

ARTICLE IV

Each State Party to the Convention shall be internationally responsible for compliance with its provisions by legal or physical persons of that State.

ARTICLE V

Each State Party to the Convention undertakes to take as soon as possible, in accordance with its constitutional procedures, the necessary legislative and administrative measures for prohibiting the development, production and stockpiling of the weapons, equipment and means of delivery mentioned in article I of the Convention, and for destroying them.

ARTICLE VI

The States Parties to the Convention undertake to consult one another and to co-operate in solving any problems which may arise in the application of the provisions of this Convention.

ARTICLE VII

1. Each State Party to the Convention which finds that actions of any other State Party constitute a breach of the obligations assumed under the provisions of this Convention may lodge a complaint with the Security Council of the United Nations. Such a complaint should include all possible evidence confirming its validity, as well as a request for its consideration by the Security Council. The Council shall inform the States Parties to the Convention of the result of the investigation.

2. Each State Party to the Convention undertakes to co-operate in carrying out any investigations which the Security Council may undertake, in accordance with the provisions of the United Nations Charter, on the basis of the complaint received by the Council.

ARTICLE VIII

Nothing in this Convention shall be interpreted as in any way limiting or detracting from the obligations assumed by any State under the Geneva Protocol of 17 June 1925 on the Prohibition of the Use in War of Asphyxiating, Poisonous or Other Gases, and of Bacteriological Methods of Warfare, an instrument which embodies generally recognized rules of international law.

ARTICLE IX

Each State Party to this Convention undertakes to conduct negotiations in good faith on effective measures for prohibiting the development, production and stockpiling of chemical weapons and for their destruction, and on appropriate measures concerning equipment and means of delivery specifically designed for the production or use of chemical weapons as means of warfare.

ARTICLE X

1. The States Parties to the Convention undertake to facilitate, and have the right to participate in, the fullest possible exchange of equipment, materials and scientific and technological information for the use of bacteriological (biological) agents and toxins for peaceful purposes.

2. This Convention shall be implemented in a manner designed to avoid hampering the economic or technological development of States Parties to the Convention or international co-operation in the field of peaceful bacteriological (biological) activities, including the international exchange of bacteriological (biological) agents and toxins and equipment for the processing, use or production of bacteriological (biological) agents and toxins for peaceful purposes in accordance with the provisions of this Convention.

ARTICLE XI

Any State Party may propose amendments to this Convention. Amendments shall enter into force for each State Party accepting the amendments upon their acceptance by a majority of the States Parties to the Convention and thereafter for each remaining State Party on the date of acceptance by it.

ARTICLE XII

1. This Convention shall be of unlimited duration.

2. Five years after the entry into force of this Convention, a conference of States Parties to the Convention shall be held at Geneva, Switzerland, to review the operation of this Convention, so as to be sure that the purposes of the preamble and the provisions of the Convention, including the provisions concerning negotiations on chemical weapons, are being realized. Such review shall take into account any new scientific and technological developments relevant to this Convention.

ARTICLE XIII

1. This Convention shall be open to all States for signature. Any State which does not sign the Convention before its entry into force in accordance with paragraph 3 of this article may accede to it at any time.

2. This Convention shall be subject to ratification by signatory States. Instruments of ratification and instruments of accession shall be deposited with the Governments of .... which are hereby designated the Depositary Governments.

3. This Convention shall enter into force after the deposit of the instruments of ratification by .... Governments, including the Governments designated as Depositaries of the Convention.

4. For States whose instruments of ratification or accession are deposited subsequent to the entry into force of this Convention, it shall enter into force on the date of the deposit of their instruments of ratification or accession.

5. The Depositary Governments shall promptly inform all signatory and acceding States of the date of each signature, the date of deposit of each instrument of ratification or of accession and the date of the entry into force of this Convention, and shall transmit other notices to them.

6. This Convention shall be registered by the Depositary Governments pursuant to Article 102 of the Charter of the United Nations.

ARTICLE XIV

This Convention, of which the Chinese, English, French, Russian and Spanish texts are equally authentic, shall be deposited in the archives of the Depositary Governments. Duly certified copies of this Convention shall be transmitted by the Depositary Governments to the Governments of the signatory and acceding States.

In witness whereof the undersigned, duly authorized, have signed this Convention.

DONE in ... copies at ..., this ... day of ..., ...

Appendix 4.

Revised draft convention on the prohibition of the development, production and stockpiling of bacteriological (biological) and toxin weapons and on their destruction, submitted by Bulgaria, Byelorussian SSR, Czechoslovakia, Hungary, Mongolia, Poland, Romania, Ukrainian SSR, and the USSR on 5 August 1971[1] (A parallel and identical text was submitted by the USA.[2])

The States Parties to this Convention,

Determined to act with a view to achieving effective progress towards general and complete disarmament including the prohibition and elimination of all types of weapons of mass destruction, and convinced that the prohibition of the development, production and stockpiling of bacteriological (biological) weapons and toxins intended for use as weapons and their elimination will facilitate the achievement of general and complete disarmament under strict and effective international control,

Desiring thereby, for the sake of all mankind, to exclude completely the possibility of bacteriological (biological) agents and toxins being used as weapons,

Convinced of the immense importance and urgent necessity of eliminating from the arsenals of states such dangerous weapons of mass destruction as weapons using bacteriological (biological) agents and toxins,

Desiring to contribute to the strengthening of confidence between peoples and the general improvement of the international atmosphere,

Believing that scientific discoveries in the field of bacteriology (biology) must in the interests of all mankind be used solely for peaceful purposes,

Recognizing nevertheless that in the absence of appropriate prohibitions the development of scientific knowledge throughout the world would increase the risk of the use of bacteriological (biological) methods of warfare,

Convinced that such use would be repugnant to the conscience of mankind and that no effort should be spared to minimize this risk,

[1] Disarmament Conference document CCD/337.
[2] Disarmament Conference document CCD/338.

Recognizing the important significance of the Geneva Protocol of 17 June 1925 for the Prohibition of the Use in War of Asphyxiating, Poisonous or Other Gases, and of Bacteriological Methods of Warfare, and conscious also of the contribution which the said Protocol has already made, and continues to make, to mitigating the horrors of war,

Reaffirming their adherence to the purposes and principles of that Protocol and calling upon all States to comply strictly with them,

Recalling resolutions of the United Nations General Assembly, which has condemned all actions contrary to the principles and purposes of the Geneva Protocol of 17 June 1925,

Convinced that an agreement on the prohibition of bacteriological (biological) and toxin weapons will facilitate progress towards the achievement of agreement on effective measures to prohibit the development, production and stockpiling of chemical weapons, on which negotiations will be continued,

Anxious to contribute to the realization of the purposes and principles of the Charter of the United Nations,

Have agreed as follows:

ARTICLE I

Each State Party to this Convention undertakes not to develop, produce, stockpile or otherwise acquire or retain:

(1) Microbial or other biological agents or toxins of types and in quantities that have no justification for prophylactic or other peaceful purposes;

(2) Weapons, equipment or means of delivery designed to use such agents or toxins for hostile purposes or in armed conflict.

ARTICLE II

Each State Party to this Convention undertakes to destroy, or to divert to peaceful purposes, as soon as possible but not later than . . . months after the entry into force of the Convention all agents, toxins, weapons, equipment and means of delivery specified in Article I of the Convention, which are in its possession or under its jurisdiction or control. In implementing the provisions of this Article all necessary safety precautions shall be observed to protect the population and the environment.

ARTICLE III

Each State Party to this Convention undertakes not to transfer to any recipient whatsoever, directly, or indirectly, and not in any way to assist, encourage, or induce any State, group of States or international organizations to manufacture or otherwise acquire any agent, toxin, weapon, equipment or means of delivery specified in Article I of the Convention.

ARTICLE IV

Each State Party to this Convention shall, in accordance with its constitutional processes, take any necessary measures to prohibit and prevent development, production, stockpiling, acquisition or retention of the agents, toxins, weapons, equipment and means of delivery specified in Article I of the Convention, within the territory of such State, under its jurisdiction or under its control anywhere.

ARTICLE V

The States Parties to the Convention undertake to consult one another and to co-operate in solving any problems which may arise in the application of the provisions of this Convention.

ARTICLE VI

(1) Each State Party to the Convention which finds that actions of any other State Party constitute a breach of the obligations assumed under the provisions of this Convention may lodge a complaint with the Security Council of the United Nations. Such a complaint should include all possible evidence confirming its validity, as well as a request for its consideration by the Security Council. The Security Council shall inform the States Parties to the Convention of the result of the investigation.

(2) Each State Party to the Convention undertakes to co-operate in carrying out any investigations which the Security Council may undertake, in accordance with the provisions of the United Nations Charter, on the basis of the complaint received by the Council.

ARTICLE VII

Nothing in this Convention shall be interpreted as in any way limiting or detracting from the obligations assumed by any State under the Geneva Protocol of 17 June 1925 for the Prohibition of the Use in War of Asphyxiating, Poisonous or Other Gases, and of Bacteriological Methods of Warfare.

ARTICLE VIII

Each State Party to this Convention undertakes to conduct negotiations in good faith on effective measures for prohibiting the development, production and stockpiling of chemical weapons and for their destruction and on appropriate measures concerning the equipment and means of delivery specifically designed for the production or use of chemical weapons for warfare.

ARTICLE IX

(1) The States Parties to the Convention undertake to facilitate, and have the right to participate in, the fullest possible exchange of equipment, materials and scientific and technological information for the use of bacteriological (biological) agents and toxins for peaceful purposes.

(2) This Convention shall be implemented in a manner designed to avoid hampering the economic or technological development of States Parties to the Convention or international co-operation in the field of peaceful bacteriological (biological) activities, including the international exchange of bacteriological (biological) agents and toxins and equipment for the processing, use or production of bacteriological (biological) agents and toxins for peaceful purposes in accordance with the provisions of this Convention.

ARTICLE X

Any State Party may propose amendments to this Convention. Amendments shall enter into force for each State Party accepting the amendments upon their acceptance by a majority of the States Parties to the Convention and thereafter for each remaining State Party on the date of acceptance by it.

ARTICLE XI

Five years after the entry into force of this Convention, or earlier if it is requested by a majority of Parties to the Convention by submitting a proposal to this effect to the Depositary Governments, a conference of States Parties to the Convention shall be held at Geneva, Switzerland, to review the operation of this Convention, with a view to assuring that the purposes of the preamble and the provisions of the Convention, including the provisions concerning negotiations on chemical weapons, are being realized. Such review shall take into account any new scientific and technological developments relevant to this Convention.

ARTICLE XII

(1) This Convention shall be of unlimited duration.

(2) Each State Party to this Convention shall in exercising its national sovereignty have the right to withdraw from the Convention if it decides that extraordinary events, related to the subject matter of this Convention, have jeopardized the supreme interests of its country. It shall give

US/Socialist draft convention

notice of such withdrawal to all other States Parties to the Convention and to the United Nations Security Council three months in advance. Such notice shall include a statement of the extraordinary events it regards as having jeopardized its supreme interests.

ARTICLE XIII

(1) This Convention shall be open to all States for signature. Any State which does not sign the Convention before its entry into force in accordance with paragraph 3 of this Article may accede to it at any time.

(2) This Convention shall be subject to ratification by signatory States. Instruments of ratification and instruments of accession shall be deposited with the Governments of . . . . . . which are hereby designated the Depositary Governments.

(3) This Convention shall enter into force after the deposit of the instruments of ratification by . . . . Governments, including the Governments designated as Depositaries of the Convention.

(4) For States whose instruments of ratification or accession are deposited subsequent to the entry into force of this Convention, it shall enter into force on the date of the deposit of their instruments of ratification or accession.

(5) The Depositary Governments shall promptly inform all signatory and acceding States of the date of each signature, the date of deposit of each instrument of ratification or of accession and the date of the entry into force of this Convention, and of other notices.

(6) This Convention shall be registered by the Depositary Governments pursuant to Article 102 of the Charter of the United Nations.

ARTICLE XIV

This Convention, the Chinese, English, French, Russian and Spanish texts of which are equally authentic, shall be deposited in the archives of the Depositary Governments. Duly certified copies of this Convention shall be transmitted by the Depositary Governments to the Governments of the signatory and acceding States.

In witness whereof the undersigned, duly authorized, have signed this Convention.

Done in . . . . copies at . . . . . ., this . . . . day of . . . . ., ...

Appendix 5.

List of states which have signed, ratified, acceded or succeeded to the 1925 Geneva Protocol

Protocol for the Prohibition of the Use in War of Asphyxiating, Poisonous or other Gases, and of Bacteriological Methods of Warfare. Signed at Geneva on 17 June 1925

The Undersigned Plenipotentiaries, in the name of their respective Governments:

Whereas the use in war of asphyxiating, poisonous or other gases, and of all analogous liquids, materials or devices, has been justly condemned by the general opinion of the civilized world; and

Whereas the prohibition of such use has been declared in Treaties to which the majority of Powers of the world are Parties; and

To the end that this prohibition shall be universally accepted as a part of International Law, binding alike the conscience and the practice of nations;

Declare:

That the High Contracting Parties, so far as they are not already Parties to Treaties prohibiting such use, accept this prohibition, agree to extend this prohibition to the use of bacteriological methods of warfare and agree to be bound as between themselves according to the terms of this declaration.

The High Contracting Parties will exert every effort to induce other States to accede to the present Protocol. Such accession will be notified to the Government of the French Republic, and by the latter to all signatory and acceding Powers, and will take effect on the date of the notification by the Government of the French Republic.

The present Protocol, of which the French and English texts are both authentic, shall be ratified as soon as possible. It shall bear to-day's date.

The ratifications of the present Protocol shall be addressed to the Government of the French Republic, which will at once notify the deposit of such ratification to each of the signatory and acceding Powers.

The instruments of ratification of and accession to the present Protocol will remain deposited in the archives of the Government of the French Republic.

The present Protocol will come into force for each signatory Power as from the date of deposit of its ratification, and, from that moment, each Power will be bound as regards other Powers which have already deposited their ratifications.

In witness whereof the Plenipotentiaries have signed the present Protocol.

Done at Geneva in a single copy, the seventeenth day of June, One Thousand Nine Hundred and Twenty-Five.

Note

Some states, former non-self-governing territories, acceded to the Geneva Protocol without referring to the obligations previously undertaken on their behalf by the colonial power. In these cases, the date of the notification by the Government of France, the depositary government, is indicated as the date of entry into force of the accession for the countries concerned, in accordance with paragraph 2 of the operative part of the Protocol.

Other states, former non-self-governing territories, officially informed the Government of France that they consider themselves bound by the Geneva Protocol by virtue of its ratification by the power formerly responsible for their administration. In such cases of continuity of obligations under the Geneva Protocol, the date of the country's communication, addressed to the French Government, is indicated, and in the absence of a statement to the contrary the succession is regarded as applying also to reservations attached to the ratification of the Protocol.

States which, upon attaining independence, made general statements of continuity to the treaties concluded by the power formerly responsible for their administration, but have not notified the Government of France that their statements specifically applied to the Geneva Protocol, are not listed here.

To determine the actual number of parties to the Geneva Protocol, account should also be taken of the facts that Estonia, Latvia and Lithuania, which signed and ratified the Protocol, no longer have independent status; both the Federal Republic of Germany and the German Democratic Republic are bound by ratification on behalf of Germany; both the People's Republic of China and the Republic of China (Taiwan) are bound by accession on behalf of China.

A. List of signatories and ratifications

Signatory	Deposit of ratification
Austria	9 May 1928
Belgium	4 Dec. 1928[1]
Brazil	28 Aug. 1970
British Empire	9 April 1930[2]

Bulgaria	7 March 1934[3]
Canada	6 May 1930[4]
Chile	2 July 1935[5]
Czechoslovakia	16 Aug. 1938[6]
Denmark	5 May 1930
Egypt	6 Dec. 1928
El Salvador	
Estonia	28 Aug. 1931[7]
Ethiopia	20 Sept. 1935[8]
Finland	26 June 1929
France	10 May 1926[9]
Germany	25 April 1929[10]
Greece	30 May 1931
India	9 April 1930[11]
Italy	3 April 1928
Japan	21 May 1970
Latvia	3 June 1931
Lithuania	15 June 1933
Luxembourg	1 Sept. 1936
Netherlands	31 Oct. 1930[12]
Nicaragua	
Norway	27 July 1932
Poland	4 Feb. 1929
Portugal	1 July 1930[13]
Romania	23 Aug. 1929[14]
Serbs, Croats and Slovenes, Kingdom of the (Yugoslavia)	12 April 1929[15]
Siam (Thailand)	6 June 1931
Spain	22 Aug. 1929[16]
Sweden	25 April 1930
Switzerland	12 July 1932
Turkey	5 Oct. 1929
USA	
Uruguay	
Venezuela	8 Feb. 1928

B. List of accessions and successions

Country	Notification
Argentina	12 May 1969
Australia	24 May 1930[17]

Central African Republic	31 July 1970
Ceylon	20 Jan. 1954
China	24 Aug. 1929[18]
Cuba	24 June 1966
Cyprus	21 Nov. 1966[19]
Dominican Republic	8 Dec. 1970
Ecuador	16 Sept. 1970
Gambia	11 Oct. 1966[20]
Ghana	3 May 1967
Holy See	18 Oct. 1966
Hungary	11 Oct. 1952
Iceland	2 Nov. 1967
Indonesia	13 Jan. 1971[21]
Iraq	8 Sept. 1931[22]
Irish Free State (Ireland)	29 Aug. 1930[23]
Israel	20 Feb. 1969[24]
Ivory Coast	27 July 1970
Jamaica	28 July 1970[25]
Kenya	6 July 1970
Lebanon	17 April 1969
Liberia	17 June 1927
Malagasy Republic	2 Aug. 1967
Malaysia	10 Dec. 1970
Malawi	14 Sept. 1970
Maldives	19 Dec. 1966[26]
Malta	25 Sept. 1970[27]
Mauritius	27 Nov. 1970[28]
Mexico	28 May 1932
Monaco	6 Jan. 1967
Mongolia	6 Dec. 1968[29]
Morocco	13 Oct. 1970
Nepal	9 May 1969
New Zealand	24 May 1930[30]
Niger	18 March 1967[31]
Nigeria	15 Oct. 1968[32]
Pakistan	13 April 1960[33]
Panama	4 Dec. 1970
Paraguay	22 Oct. 1933[34]
Persia (Iran)	5 Nov. 1929
Rwanda	21 March 1964[35]
Saudi Arabia	27 Jan. 1971

Sierra Leone	20 March 1967
South Africa	24 May 1930[36]
Syria	17 Dec. 1968[37]
Tanzania	22 April 1963
Togo	5 April 1971
Tonga	28 July 1971
Trinidad and Tobago	9 Oct. 1970[38]
Tunisia	12 July 1967
Uganda	24 May 1965
Upper Volta	3 March 1971
USSR	15 April 1928[39]
Yemen (Arab Republic of)	17 March 1971

---

[1] (1) The said Protocol is only binding on the Belgian Government as regards States which have signed or ratified it or which may accede to it. (2) The said Protocol shall *ipso facto* cease to be binding on the Belgian Government in regard to any enemy State whose armed forces or whose allies fail to respect the prohibitions laid down in the Protocol.

[2] The British Plenipotentiary declared when signing: "my signature does not bind India or any British Dominion which is a separate Member of the League of Nations and does not separately sign or adhere to the Protocol".

(1) The said Protocol is only binding on His Britannic Majesty as regards those Powers and States which have both signed and ratified the Protocol or have finally acceded thereto. (2) The said Protocol shall cease to be binding on His Britannic Majesty towards any Power at enmity with Him whose armed forces, or the armed forces of whose allies, fail to respect the prohibitions laid down in the Protocol.

[3] The said Protocol is only binding on the Bulgarian Government as regards States which have signed or ratified it or which may accede to it. The said Protocol shall *ipso facto* cease to be binding on the Bulgarian Government in regard to any enemy State whose armed forces or whose allies fail to respect the prohibitions laid down in the Protocol.

[4] (1) The said Protocol is only binding on His Britannic Majesty as regards those States which have both signed and ratified it, or have finally acceded thereto. (2) The said Protocol shall cease to be binding on His Britannic Majesty towards any State at enmity with Him whose armed forces, or whose allies *de jure* or in fact fail to respect the prohibitions laid down in the Protocol.

[5] (1) The said Protocol is only binding on the Chilean Government as regards States which have signed and ratified it or which may definitely accede to it; (2) The said Protocol shall *ipso facto* cease to be binding on the Chilean Government in regard to any enemy State whose armed forces, or whose allies, fail to respect the prohibitions which are the object of this Protocol.

[6] The Czechoslovak Republic shall *ipso facto* cease to be bound by this Protocol towards any State whose armed forces, or the armed forces of whose allies, fail to respect the prohibitions laid down in the Protocol.

[7] (1) The said Protocol is only binding on the Estonian Government as regards States which have signed or ratified it or which may accede to it. (2) The said Protocol shall *ipso facto* cease to be binding on the Estonian Government in regard to any enemy State whose armed forces or whose allies fail to respect the prohibitions laid down in the Protocol.

[8] The document deposited by Ethiopia, a signer of the Protocol, is registered as an accession. The date given is therefore the date of notification by the French Government.

⁹ (1) The said Protocol is only binding on the Government of the French Republic as regards States which have signed or ratified it or which may accede to it. (2) The said Protocol shall *ipso facto* cease to be binding on the Government of the French Republic in regard to any enemy State whose armed forces or whose allies fail to respect the prohibitions laid down in the Protocol.

¹⁰ On 2 March 1959, the Embassy of Czechoslovakia transmitted to the French Ministry for Foreign Affairs a document stating the applicability of the Protocol to the German Democratic Republic.

¹¹ (1) The said Protocol is only binding on His Britannic Majesty as regards those States which have both signed and ratified it, or have finally acceded thereto. (2) The said Protocol shall cease to be binding on His Britannic Majesty towards any Power at enmity with Him whose armed forces, or the armed forces of whose allies, fail to respect the prohibitions laid down in the Protocol.

¹² Including Netherlands Indies, Surinam and Curaçao.
As regards the use in war of asphyxiating, poisonous or other gases, and of all analogous liquids, materials or devices, this Protocol shall *ipso facto* cease to be binding on the Royal Netherlands Government with regard to any enemy State whose armed forces or whose allies fail to respect the prohibitions laid down in the Protocol.

¹³ (1) The said Protocol is only binding on the Government of the Portuguese Republic as regards States which have signed and ratified it or which may accede to it. (2) The said Protocol shall *ipso facto* cease to be binding on the Government of the Portuguese Republic in regard to any enemy State whose armed forces or whose allies fail to respect the prohibitions which are the object of this Protocol.

¹⁴ (1) The said Protocol only binds the Romanian Government in relation to States which have signed and ratified or which have definitely acceded to the Protocol; (2) The said Protocol shall cease to be binding on the Romanian Government in regard to all enemy States whose armed forces or whose allies *de jure* or in fact do not respect the restrictions which are the object of this Protocol.

¹⁵ The said Protocol shall cease to be binding on the Government of the Serbs, Croats and Slovenes in regard to any enemy State whose armed forces or whose allies fail to respect the prohibitions which are the object of this Protocol.

¹⁶ Declares as binding *ipso facto,* without special agreement with respect to any other Member or State accepting and observing the same obligation, that is to say, on condition of reciprocity, the Protocol for the Prohibition of the Use in War of Asphyxiating, Poisonous and other Gases and of Bacteriological Methods of Warfare, signed at Geneva, June 17, 1925.

¹⁷ Subject to the reservations that His Majesty is bound by the said Protocol only towards those Powers and States which have both signed and ratified the Protocol or have acceded thereto, and that His Majesty shall cease to be bound by the Protocol towards any Power at enmity with Him whose armed forces, or the armed forces of whose allies, do not respect the Protocol.

¹⁸ On 13 July 1952, the People's Republic of China issued a statement recognizing as binding upon it the accession to the Protocol in the name of China.
The People's Republic of China considers itself bound by the Protocol on condition of reciprocity on the part of all the other contracting and acceding powers.

¹⁹ In a note of this date Cyprus declared that it was bound by the Protocol which had been made applicable to it by the British Empire.

²⁰ In a declaration of this date, Gambia confirmed its participation in the Protocol which had been made applicable to it by Great Britain.

²¹ In an official declaration of this date, addressed to the French Government, the Government of Indonesia reaffirmed its acceptance of the Geneva Protocol which had been ratified on its behalf by the Netherlands on 31 October 1930, and stated that it remained signatory to that Protocol.

²² On condition that the Iraq Government shall be bound by the provisions of the Protocol only towards those States which have both signed and ratified it or have acceded thereto, and that they shall not be bound by the Protocol towards any State at enmity with them, whose armed forces, or the forces of whose allies, do not respect the provisions of the Protocol.

²³ The Government of the Irish Free State does not intend to assume, by this ac-

cession, any obligation except towards the States having signed and ratified this Protocol or which shall have finally acceded thereto, and should the armed forces or the allies of an enemy State fail to respect the said Protocol, the Government of the Irish Free State would cease to be bound by the said Protocol in regard to such State.

[24] The said Protocol is only binding on the State of Israel as regards States which have signed and ratified or acceded to it. The said Protocol shall cease *ipso facto* to be binding on the State of Israel as regards any enemy State whose armed forces, or the armed forces of whose allies, or the regular or irregular forces, or groups or individuals operating from its territory, fail to respect the prohibitions which are the object of this Protocol.

[25] On this date Jamaica declared to the depositary government that it considered itself bound by the provisions of the Protocol on the basis of the ratification by the United Kingdom in 1930.

[26] In a declaration of this date Maldives confirmed its adherence to the Protocol.

[27] By a notification of this date the Government of Malta informed the French Government that it considers itself bound by the Geneva Protocol as from 21 September 1964, the provisions of the Protocol having been extended to Malta by the Government of the United Kingdom, prior to the former's accession to independence.

[28] By a notification of this date the Government of Mauritius informed the French Government that it considers itself bound by the Geneva Protocol as from 12 March 1968, the date of its accession to independence.

[29] In the case of violation of this prohibition by any State in relation to the People's Republic of Mongolia or its allies, the Government of the People's Republic of Mongolia shall not consider itself bound by the obligations of the Protocol towards that State.

[30] Same reservations as Australia. (See footnote 17.)

[31] In a letter of this date Niger declared that it was bound by the adherence of France to the Protocol.

[32] The Protocol is only binding on Nigeria as regards States which are effectively bound by it and shall cease to be binding on Nigeria as regards States whose forces or whose allies' armed forces fail to respect the prohibitions which are the object of the Protocol.

[33] By a note of this date Pakistan informed the depositary government that it was a party to the Protocol, by virtue of Paragraph 4 of the Annex to the Indian Independence Act of 1947.

[34] This is the date of receipt of the instrument of accession. The date of the notification by the French Government "for the purpose of regularization" is 13 January 1969.

[35] In a declaration of this date Rwanda recognized that it was bound by the Protocol which had been made applicable to it by Belgium.

[36] Same reservation as Australia. (See footnote 17.)

[37] The accession by the Syrian Arab Republic to this Protocol and the ratification of the Protocol by its Government does not in any case imply recognition of Israel or lead to the establishment of relations with the latter concerning the provisions laid down in this Protocol.

[38] By a note of this date the Government of Trinidad and Tobago notified the French Government that it considers itself bound by the Geneva Protocol, the provisions of which had been made applicable to Trinidad and Tobago by the British Empire prior to the former's accession to independence.

[39] (1) The said Protocol only binds the Government of the Union of Soviet Socialist Republics in relation to the States which have signed and ratified or which have definitely acceded to the Protocol. (2) The said Protocol shall cease to be binding on the Government of the Union of Soviet Socialist Republics in regard to any enemy State whose armed forces or whose allies *de jure* or in fact do not respect the prohibitions which are the object of this Protocol.

On 2 March 1970, the Byelorussian Soviet Socialist Republic stated that "it recognizes itself to be a Party" to the Geneva Protocol of 1925 (United Nations doc. A/8052, Annex III).

References

Chapter 1: 1920–1924

1. League of Nations. Procès-Verbal of the Fifth Session of the Council, Document 20/29/11.
2. League of Nations. Minutes of the First Session of the Permanent Advisory Commission for Military, Naval and Air Questions, Document C.P.C./1st SESS/PV-2; *and* Minutes of the Second Session, Document C.P.C./2nd SESS/PV-1.
3. League of Nations. Procès-Verbal of the Tenth Session of the Council, Annex 112a and 112b, Document 20/29/16.
4. League of Nations. Annex 137c to the Procès-Verbal of the Eleventh Session of the Council.
5. League of Nations. Minutes of the Twelfth Session of the Council, Document 21/29/1.
6. League of Nations. Minutes of the Third Committee; Report to the Assembly, Document A.158.1921.
7. League of Nations. Second Assembly. Resolution No. 19, Document A.178.121.
8. League of Nations. Annex 2 to the Records of the Third Assembly/Meetings of the Committees/, Documents A.31.1922; C.631.1922; CTA.173.
9. League of Nations. Minutes of the Third Committee and Records of the Third Assembly, Plenary Meetings, Volume II, Annex 24, Document A.124. 1922.IX.
10. League of Nations. Records of the Third Assembly, Plenary Meetings, Volume I. Text of the Debates.
11. League of Nations. Report of the Temporary Mixed Commission, August 15th 1923. Document A.35.1923.IX.
12. League of Nations. Official Journal No. 8, August 1923, Annex 523, Document C.445.1923.IX.
13. League of Nations. Official Journal, Special Supplement No. 16, Records of the Fourth Assembly, Meetings of the Committees, Minutes of the Third Committee, and Annex 1 and 10 /Part II/; Document A.III.1923.IX /Second Part/.
14. League of Nations. Records of the Fourth Assembly, Text of the Debates.
15. League of Nations. Official Journal, February 1924, Minutes of the Twenty-Seventh Session of the Council. Annex 585; Document C.741/1/.1923.IX.
16. League of Nations. Temporary Mixed Commission for the Reduction of Armaments, Documents C.T.A. 246 and C.T.A./Experts/1–9.
17. League of Nations. Report of the Temporary Mixed Commission for the Reduction of Armaments, July 30th, 1924. Document A.16.1924.IX.
18. League of Nations. Official Journal, Special Supplement No. 26, Records of

the Fifth Assembly, Meetings of the Committees, Minutes of the Third Committee.
19. League of Nations. Official Journal, Special Supplement No. 23, Records of the Fifth Assembly, Plenary Meetings, Text of the Debates.
20. League of Nations. Official Journal No. 10, October 1924, Minutes of the Thirtieth Session of the Council.

Chapter 2: 1925

1. League of Nations. Official Journal, Special Supplement no. 23, Records of the Fifth Assembly, Plenary Meetings, Text of the Debates.
2. League of Nations. Official Journal, 5th Year, No. 10, October 1924, Minutes of the Thirtieth Session of the Council.
3. League of Nations. Document A.13.1925.IX. Verbatim Reports of the Conference for the Supervision of the International Trade in Arms and Ammunition and in Implements of War. Second Plenary Meeting.
4. League of Nations. Document A.13.1925.IX. First and Second Meetings of the General Committee of the Conference.
5. League of Nations. Document A.13.1925.IX. Second Meeting of the Legal Committee of the Conference.
6. League of Nations. Document A.13.1925.IX. Eighth Meeting of the Legal Committee of the Conference.
7. League of Nations. Document A.13.1925.IX. Ninth Meeting of the Legal Committee of the Conference. Report No. 1 of the Legal Committee, Document C.C.I.A./55.
8. League of Nations. Annex 3 to the Minutes of the Advisory Committees of the Conference. Document A.13.1925.IX. Part III.
9. League of Nations. Minutes of the Thirteenth, Fourteenth, Fifteenth and Sixteenth Meetings of the Military, Naval and Air Technical Committee. Document A.13.1925.IX, Part III, Annex 4.
10. League of Nations. Document A.13.1925.IX, Part III, Annex 15; C.C.I.A. 60.
11. League of Nations. Document A.13.1925.IX, Verbatim Report of the General Committee, Seventeenth Meeting; C.C.I.A./C.G./C.R.17.
12. League of Nations. Document A.13.1925.IX, Verbatim Report of the General Committee, Twentieth Meeting; C.C.I.A./74; C.C.I.A./80.
13. League of Nations. Document C.C.I.A./90.
14. League of Nations. Document A.13.1925.IX, Verbatim Report of the General Committee, Twenty-Second Meeting.
15. League of Nations. Document A.13.1925.IX, Verbatim Report of the Plenary Meetings of the Conference, Sixth Meeting.
16. League of Nations. Document C.C.I.A./105; A.13.1925.IX; Verbatim Report of the Plenary Meetings of the Conference, Seventh Meeting; *and* Documents A.16.1925.IX; A.109.1925.IX.

Chapter 3: 1926–1930

1. League of Nations. Documents C.428.M.178.1931.IX; C.761.1925.IX; C.D.C.16.

References

2. League of Nations. Minutes of the First Session of the Preparatory Commission for the Disarmament Conference, Third, Fifth, Sixth, Seventh, Ninth Meetings. Document C.P.D.1.
3. League of Nations. Minutes of the Sixth Meeting of the Drafting Committee, Annex VII, Document C.P.D.1 and Report of the Preparatory Commission to the Council on the Work of its First Session. Document C.301.1926.IX.
4. League of Nations. Documents C.739. M.278. 1926.IX./C.P.D.28/; *and* 1926. IX.16.
5. League of Nations. Documents C.738. M.277. 1926.IX; /C.P.D.29/; /C.P.D./C."B"/ *and* 1926.IX.15.
6. League of Nations. Preparatory Commission for the Disarmament Conference, Sub-Commission A /Third Session/. Summary of Proceedings. Document C.P.D./C."A"/Third Session.
7. League of Nations. Minutes of the Third Session of the Preparatory Commission for the Disarmament Conference. Documents C.310.M.109. 1927. IX./C.P.D.I/c//; *and* 1927.IX.5.
8. League of Nations. Minutes of the Sixth Session /First Part/ of the Preparatory Commission for the Disarmament Conference. Document C.195. M.74.1929.IX./C.P.D.1/g/.
9. League of Nations. Annex 2 to the Minutes of the Sixth Session /First Part/ of the Preparatory Commission for the Disarmament Conference. Document C.P.D./138.
10. League of Nations. Eighth Meeting of the Sixth Session /First Part/ of the Preparatory Commission for the Disarmament Conference. Document 1929. IX.3.
11. League of Nations. Annex 5 to the Minutes of the Fifth Session of the Preparatory Commission for the Disarmament Conference. Document C.P.D.117.
12. League of Nations. Annex 6 to the Minutes of the Sixth Session /First Part/ of the Preparatory Commission for the Disarmament Conference. Document C.P.D./151.
13. League of Nations. Annex 7 to the Minutes of the Sixth Session /First Part/ of the Preparatory Commission for the Disarmament Conference. Document C.P.D./150.
14. League of Nations. Annex 8 to the Minutes of the Sixth Session /First Part/ of the Preparatory Commission for the Disarmament Conference. Document C.P.D./155.
15. League of Nations. Ninth Meeting of the Sixth Session /First Part/ of the Preparatory Commission for the Disarmament Conference. Document 1929. IX.3.
16. League of Nations. Tenth Meeting of the Sixth Session /First Part/ of the Preparatory Commission for the Disarmament Conference. Document 1929. IX.3.
17. League of Nations. Eleventh Meeting of the Sixth Session /First Part/ of the Preparatory Commission for the Disarmament Conference. Document 1929.IX.3.

18. League of Nations. Twentieth Meeting of the Sixth Session /Second Part/ of the Preparatory Commission for the Disarmament Conference. Documents C.4.M.4.1931.IX/C.P.D.I/i//; 1931.IX.1.
19. League of Nations. Twenty-Third Meeting of the Sixth Session /Second Part/ of the Preparatory Commission for the Disarmament Conference. Document 1931.IX.1.
20. League of Nations. Documents C.690.M.289.1930.IX; C.P.D.295/I/.

Chapter 4: 1931 – November 1932

1. League of Nations. Official Journal, 12th Year, No. 7, July 1931.
2. League of Nations. Records of the Conference for the Reduction and Limitation of Armaments. Document 1932.IX.60; Third and Fourth Plenary Meetings and Conf. D. 56; *and* Conf. D. 95.
3. League of Nations. Fifth Plenary Meeting of the Conference for the R. and L. of A; *and* Conf. D. 85.
4. League of Nations. Sixth Plenary Meeting of the Conference for the R. and L. of A; *and* Conf. D. 81.
5. League of Nations. Seventh Plenary Meeting of the Conference for the R. and L. of A; *and* Conf. D. 82; *and* Conf. D. 87.
6. League of Nations. Eighth Plenary Meeting of the Conference for the R. and L. of A; *and* Conf. D. 90.
7. League of Nations, Eighth Plenary Meeting of the Conference for the R. and L. of A; *and* Conf. D. 91.
8. League of Nations. Tenth Plenary Meeting of the Conference for the R. and L. of A; *and* Conf. D. 78.
9. League of Nations. Tenth Plenary Meeting of the Conference for the R. and L. of A; *and* Conf. D. 89.
10. League of Nations. Tenth Plenary Meeting of the Conference for the R. and L. of A; *and* Conf. D. 84.
11. League of Nations. Eleventh Plenary Meeting of the Conference for the R. and L. of A; *and* Conf. D. 86.
12. League of Nations. Eleventh Plenary Meeting of the Conference for the R. and L. of A; *and* Conf. D. 93.
13. League of Nations. Twelfth Plenary Meeting of the Conference for the R. and L. of A; *and* Conf. D. 93.
14. League of Nations. Thirteenth Plenary Meeting of the Conference for the R. and L. of A; *and* Conf. D. 79.
15. League of Nations. Thirteenth Plenary Meeting of the Conference for the R. and L. of A; *and* Conf. D. 80.
16. League of Nations. Fourteenth Plenary Meeting of the Conference for the R. and L. of A.
17. League of Nations. Minutes of the General Commission of the Conference for the R. and L. of A; Document 1932.IX.64.
18. League of Nations. Series of Publications: 1932.IX.32; Conf. D. 106.
19. League of Nations. Series of Publications: 1932.IX.39; Conf. D. 112.
20. League of Nations. Series of Publications: 1932.IX.40; Conf. D. 113.

References

21. League of Nations. Series of Publications: 1923.IX.46; Conf. D. 120. /Conf. D/A.C.B.16(I)/.
22. League of Nations. Conf. D.136(I); /Conf.D/C.G.31(1)/.
23. League of Nations, Records of the Conference for the R. and L. of A; Series C. Minutes of the Bureau, Volume I. Document 1935.IX.2; *and* Conf. D/Bureau/18.
24. League of Nations. Conference for the R. and L. of A; Conf. D/Bureau/14.
25. League of Nations. Conference for the R. and L. of A; Conf. D/Bureau/21.
26. League of Nations. Series of Publications: 1932.IX.55; Conf. D.142 (Conf. D/Bureau/24).
27. League of Nations. Series C. Minutes of the Bureau, Conference for the R. and L. of A; Volume I; Series of Publications: 1935.IX.2.
28. League of Nations. Conference for the R. and L. of A; Conf. D/Bureau/30 (1).

Chapter 5: November 1932–1933

1. League of Nations. Series of Publications: 1932.IX.65; Conf.D.152 /Conf. D./A.C.B./37(1)/.
2. League of Nations. Conference for the R. and L. of A; Conf.D./A.C.B./ P.V.
3. League of Nations. Minutes of the Bureau, Conference for the R. and L. of A; Series C, Vol. I. Thirty-fifth and Thirty-sixth meetings; Document 1935.IX.2.
4. League of Nations. Conference for the R. and L. of A; Conf. D/Bureau/41.
5. League of Nations. Conference for the R. and L. of A; Conf.D/Bureau/45.
6. League of Nations. Series of Publications: 1933.IX.2; Conf.D.157(1).
7. League of Nations. Minutes of the General Commission, Conference for the R. and L. of A; Vol. II; Document 1933.IX.10.
8. League of Nations. Conference for the R. and L. of A; Conf.D.163(1).
9. League of Nations. Conference for the R. and L. of A; Conf.D/C.G.123.
10. League of Nations. Conference for the R. and L. of A; Conf.D/C.G.136.
11. League of Nations. Series of Publications: 1936.IX.3; Conf.D.171(1).

Chapter 6: 1935–1938

1. League of Nations. Official Journal, February 1936; C.506.M.275.1935.VII.
2. League of Nations. Official Journal, February 1936; C.8.M.7.1936.VII.
3. League of Nations, Official Journal, February 1936; C.16.M.14.1936.VII.
4. League of Nations. Official Journal, February 1936; C.51.M.21.1936.VII.
5. League of Nations. Official Journal, April 1936; Annex 1592; C.155.M.93.1936.VII.
6. League of Nations. Official Journal, April 1936; Annex 1592; C.119.M.58.1936.VII; and C.143.M.82.1936.VII.
7. League of Nations. Document C.133.M.72.1936.VII, Annexes I and IV.

References

8. League of Nations. Official Journal, April 1936; Annex 1592; C.152.M.90. 1936.VII.
9. League of Nations. Official Journal, April 1936; 91st Session of the Council, Ninth Meeting.
10. League of Nations. Official Journal, April 1936; Annex 1592; C.159.M.97. 1936.VII.
11. League of Nations. Official Journal, April 1936; Annex 1592; C.161.M.99. 1936.VII.
12. League of Nations. Document C.176.M.112.1936.VII.
13. League of Nations, Document C.17.M.112.1936.VII, Annex 3.
14. League of Nations. Official Journal, April 1936; 91st Session of the Council, Tenth Meeting.
15. League of Nations. Official Journal, April 1936; 91st Session of the Council, Eleventh Meeting.
16. League of Nations. Appendix to Document C.248.M.144.1936.VII.
17. League of Nations. Official Journal, June 1936; 92nd Session of the Council; Annex 1597; C.208.M.130.1936.VII.
18. League of Nations. Series of Publications: 1936.VII.4; C.201.M.126.1936.VII.
19. *Ibid.*, Appendix 3.
20. *Ibid.*, Appendix 4.
21. *Ibid.*, Appendix 6.
22. *Ibid.*, Appendix 7.
23. *Ibid.*, Appendix 8.
24. *Ibid.*, Appendix 10.
25. League of Nations. Series of Publications: 1936.VII.5; C.207.M.129.1936.VII.
26. League of Nations. Series of Publications: 1936.VII.6; C.225.M.137.1936.VII.
27. League of Nations. Appendix I to Document C.280.M.170.1936.VII.
28. League of Nations. Document C.279.M.169.1936.VII.
29. League of Nations, Official Journal, Special Supplement No. 151, Records of the Sixteenth Assembly. Eighteenth Plenary Meeting.
30. League of Nations, Official Journal, Special Supplement No. 145. Co-ordination Committee. Second Meeting (Private).
31. League of Nations. Official Journal, Special Supplement No. 145. Co-ordination Committee. Fourth Meeting.
32. League of Nations. Series of Publications: General Questions. 1935.10. Co-ordination Committee 106(I).
33. League of Nations. Official Journal, Special Supplement No. 146. Co-ordination Committee. Minutes of the Second Session. Fourth Meeting.
34. League of Nations. Official Journal, Special Supplement No. 150. Co-ordination Committee/10.
35. League of Nations. Official Journal, Special Supplement No. 150. Co-ordination Committee/23 (g).
36. League of Nations. Official Journal, Special Supplement No. 150. Co-ordination Committee/130(1).

References

37. League of Nations. Records of the Sixteenth Session of the Assembly. Twenty-fifth Plenary meeting.
38. League of Nations. Official Journal, May–June 1938, Annex 1702; Document C.166.M.93.1938.VII.
39. League of Nations. Official Journal, May–June 1938, 101st Session of the Council, Second Meeting.
40. League of Nations. Official Journal, May–June 1938, 101st Session of the Council, Eighth Meeting.
41. League of Nations. Official Journal, November 1938, 102nd Session of the Council, Fourth Meeting.
42. League of Nations. Official Journal, November 1938, 103rd Session of the Council, Second Meeting.

Chapter 7: 1946–1953

1. United Nations. General Assembly. Resolution 1(I).
2. U.N. G.A. Resolution 41(I).
3. U.N. document. S/C. 3/SC. 3/7.
4. U.N. document. S/C. 3/SC. 3/SR. 3.
5. U.N. document. S/C. 3/SC. 3/7/Rev. 1.
6. U.N. documents. S/C. 3/SC. 3/SR. 4, *and* S/C. 3/24.
7. U.N. document. S/C. 3/SR. 12.
8. U.N. document. S/C. 3/SR. 13.
9. U.N. G.A. Official records; Third Session. Supplement no. 1/(A/565).
10. U.N. G.A. Resolution 502 (VI)
11. U.N. document. S/2142.
12. U.N. document. S/2684.
13. U.N. document. S/2684/Add. 1.
14. U.N. document. S/2802.
15. U.N. document. A/C. 1/L. 28.
16. U.N. document. S/2979.
17. U.N. documents. DC/PV. 2, 3 and 6, *and* DC/4 and Corr. 1.
18. U.N. document. DC/PV. 3.
19. U.N. document. DC/PV. 4.
20. U.N. document. DC/PV. 5.
21. U.N. document. DC/PV. 6.
22. U.N. document. DC/PV. 8.
23. U.N. document. DC/6.
24. U.N. document. DC/PV. 11.
25. U.N. document. DC/PV. 16.
26. U.N. documents. DC/PV. 19, *and* DC/15.
27. U.N. document. DC/PV. 20.
28. U.N. documents. DC/13, *and* DC/13/Rev. 1, *and* DC/PV. 19.
29. U.N. document. DC/PV. 21.
30. U.N. documents. DC/PV. 24, *and* DC/14.
31. U.N. Security Council. Official records, Seventh Year, 577th meeting. *and* U.N. document. S/2663.
32. U.N. S.C. Official Records, Seventh Year, 578th meeting.
33. U.N. S.C. Official Records, Seventh Year, 579th meeting.

References

34. U.N. S.C. Official Records, Seventh Year, 581st meeting.
35. U.N. S.C. Official Records, Seventh Year, 582nd meeting.
36. U.N. S.C. Official Records, Seventh Year, 583rd meeting.
37. U.N. S.C. Official Records, Seventh Year, 584th meeting.
38. U.N. document. S/2674/Rev. 1.
39. U.N. S.C. Official Records, Seventh Year, 585th meeting.
40. U.N. S.C. Official Records, Seventh Year, 586th meeting.
41. U.N. S.C. Official Records, Seventh Year, 587th meeting, *and* U.N. document. S/2671.
42. U.N. S.C. Official Records, Seventh Year, 590th meeting, *and* U.N. document. S/2688.
43. U.N. document. S/2707.
44. U.N. G.A. Resolution 704 (VII), *and* 424th plenary meeting, document. A/PV. 424.
45. U.N. documents. A/C. 1/SR. 590, *and* A/C. 1/L. 35.
46. U.N. documents. A/C. 1/SR. 590, *and* A/C. 1/L. 36 and Rev.
47. U.N. document. A/C. 1/SR. 591.
48. U.N. document. A/C. 1/SR. 593.
49. U.N. document. A/C. 1/SR. 592.
50. U.N. G.A. Resolution 706 (VII).
51. U.N. document. A/PV. 428.
52. U.N. document. A/C. 1/L. 39.
53. U.N. document. A/C. 1/SR. 596.
54. U.N. document. A/C. 1/SR. 602.
55. U.N. document. A/2426.
56. U.N. documents. A/2231, *and* A/C. 1/SR. 648, *and* A/C. 1/L. 66.
57. U.N. document. A/C. 1/L. 67.
58. U.N. document. A/C. 1/SR. 649.
59. U.N. document. A/C. 1/SR. 650.
60. U.N. document. A/C. 1/SR. 651.
61. U.N. documents. A/C. 1/SR. 653, *and* A/2535, *and* A/PV. 456, *and* U.N. G.A. Resolution 714 (VIII).
62. U.N. G.A. Resolution 715 (VIII), *and* U.N. document. A/PV. 460.
63. U.N. documents. DC/SC. 118, *and* DC/53, Annex 7.
64. U.N. document. A/PV. 470.

Chapter 8: 1954–1967

1. U.N. document. DC/44.
2. U.N. document. DC/PV. 35.
3. U.N. documents. DC/53, *and* DC/SC. 1/4.
4. U.N. documents. DC/SC. 1/34; DC/71, Annex 23.
5. U.N. document. DC/SC. 1/35.
6. U.N. document. DC/SC. 1/44.
7. U.N. G.A. Official Records, Fourteenth Session, 798th plenary meeting, *and* U.N. document. A/C. 1/820.
8. U.N. document. A/4219.
9. U.N. document. A/RES/1378 (XIV).

References

10. Disarmament Conference document. TNCD/3.
11. Disarmament Conference document. TNCD/4.
12. Disarmament Conference document. TNCD/6/ Rev. 1; U.N. document. A/4505.
13. Disarmament Conference document. TNCD/7.
14. U.N. document. A/4503.
15. U.N. document. A/C. 1/L. 250.
16. U.N. document. A/C. 1/L. 259 and Add. 1–2.
17. U.N. document. A/C. 1/L. 251.
18. U.N. document. A/4879.
19. U.N. document. A/4891.
20. U.N. document. A/RES/1653 (XVI).
21. Disarmament Conference document. ENDC/2.
22. Disarmament Conference document. ENDC/30; DC/214/Add. 1.
23. Disarmament Conference document. ENDC/PV. 35.
24. Disarmament Conference document. ENDC/PV. 36.
25. Disarmament Conference document. ENDC/PV. 37.
26. Disarmament Conference document. ENDC/40/Rev. 1.
27. U.N. documents. A/C. 1/867; DC/213/Add. 1, *and* Disarmament Conference document. ENDC/2/Rev. 1 and Corr. 1.
28. U.N. document. S/5839.
29. U.N. document. S/5847.
30. U.N. document. S/5852.
31. U.N. document. S/5852/Add. 1.
32. U.N. document. S/5896.
33. U.N. document. S/5894.
34. U.N. document. S/6260.
35. U.N. document. S/6270.
36. U.N. document. A/AC. 91/SR. 13.
37. U.N. Disarmament Commission. Official Records, 72nd meeting.
38. U.N. Disarmament Commission. Official Records, 73rd meeting.
39. U.N. document. A/C. 1/L. 374.
40. U.N. document. A/C. 1/PV. 1451.
41. U.N. document. A/C. 1/PV. 1452.
42. U.N. document. A/C. 1/PV. 1454.
43. U.N. document. A/C. 1/PV. 1457.
44. U.N. document. A/C. 1/L. 381.
45. U.N. document. A/C. 1/L. 381/Rev. 1.
46. U.N. document. A/C. 1/L. 374/Rev. 1.
47. U.N. document. A/C. 1/L. 382/Rev. 1.
48. U.N. document. A/C. 1/L. 382/Rev. 2.
49. U.N. document. A/C. 1/PV. 1461.
50. U.N. document. A/C. 1/PV. 1462.
51. U.N. document. A/PV. 1484.
52. U.N. document. A/RES/2162 B (XXI).
53. U.N. documents. S/7842, *and* S/7793, *and* S/7816, *and* S/7821.
54. Disarmament Conference document. ENDC/PV. 288.
55. U.N. document. A/C. 1/L. 411.

References

56. U.N. document. A/C. 1/PV. 1547.
57. U.N. document. A/C. 1/L. 411/Rev. 1.
58. U.N. document A/C. 1/L. 412.
59. U.N. document. A/C. 1/PV. 1547.
60. U.N. document. A/C. 1/PV. 1548.
61. Annex to a letter from the Permanent Representative of the United Arab Republic in Geneva, of 1 June 1970.
62. U.N. documents S/9224, *and* S/9263, *and* S/9374.
63. US Department of State. Press release No. 335.
64. Supplement to the *American Journal of International Law,* Volume 49, 1955.
65. Report of Tripartite Conference of Berlin (17 July – 2 August, 1945), also referred to as the "Potsdam Declaration". Supplement to the *American Journal of International Law,* Volume 39, 1945.
66. Annex A of the Agreement relating to the prohibited and limited industries in the United States, United Kingdom and French-occupied areas of Germany, signed at Frankfurt, on 14 April 1949.
67. U.N. document. A/C.1/988.
68. U.N. document. A/C.1/989.
69. U.N. document. A/C.1/991.
70. Major Peace Treaties of Modern History, Volume IV, New York, 1967.

Chapter 9: 1968–1969

1. U.N. document. A/C. 1/PV. 1561.
2. U.N. document. A/7134, *and* Disarmament Conference document ENDC/227.
3. *UN Monthly Chronicle,* August–September 1968.
4. Disarmament Conference document. ENDC/PV. 381.
5. Disarmament Conference document. ENDC/PV. 385.
6. Disarmament Conference document. ENDC/231.
7. Disarmament Conference document. ENDC/PV. 387.
8. Disarmament Conference document. ENDC/PV. 389.
9. Disarmament Conference document. ENDC/PV. 390.
10. Disarmament Conference document. ENDC/PV. 389.
11. Disarmament Conference documents. ENDC/PV. 383, *and* 391.
12. Disarmament Conference document. ENDC/236.
13. U.N. document. A/7201/Add. 1.
14. U.N. document. A/C. 1/L. 444.
15. U.N. document. A/C. 1/L. 445.
16. U.N. document. A/C. 1/PV. 1615.
17. U.N. document. A/C. 1/PV. 1635.
18. U.N. document. A/C. 1/PV. 1613.
19. U.N. document. A/C. 1/PV. 1612.
20. U.N. document. A/C. 1/PV. 1634.
21. U.N. document. A/C. 1/PV. 1614.
22. U.N. document. A/C. 1/PV. 1608.
23. U.N. document. A/PV. 1750.

References

24. U.N. document. A/C. 1/PV. 1630.
25. U.N. document. A/C. 1/PV. 1606.
26. U.N. document. A/PV. 1750.
27. U.N. document. A/RES/2454A (XXIII).
28. Disarmament Conference documents. ENDC/PV.396, *and* 404.
29. Disarmament Conference document. ENDC/PV. 397.
30. U.N. documents. A/7575/Rev. 1, *and* S/9292/Rev. 1.
31. Health aspects of chemical and biological weapons. Geneva: World Health Organization, 1970.
32. Disarmament Conference document. ENDC/236.
33. U.N. document. A/RES/2603B (XXIV).

Part II: Developments in 1969–1970

1. Report of the U.N. Secretary-General. U.N. documents. A/7575/Rev. 1, *and* S/9292/Rev. 1.
2. Health aspects of chemical and biological weapons. Geneva: World Health Organization, 1970.
3. US Arms Control and Disarmament Agency, 9th Annual Report to Congress.
4. Office of the White House Press Secretary. Press release. 14 February 1970.
5. UN document. A/C. 1/PV. 1716.
6. Disarmament Conference document. CCD/PV.460.
7. Disarmament Conference document. CCD/PV. 480.
8. Disarmament Conference document. CCD/PV. 458.
9. Disarmament Conference document. CCD/PV. 481.
10. U.N. document. A/RES/2603 B(XXIV).
11. *New York Times,* 20 August 1970.
12. U.N. document. A/C. 1/L. 498; *and* Disarmament Conference documents. CCD/PV. 453 *and* CCD/PV. 474.
13. Disarmament Conference document. CCD/PV. 471.
14. *Hansard,* (Commons) vol. 795: col. *17–18.*
15. *Hansard,* (Lords) vol. 308: nr. 44.
16. U.N. document. A/RES/2603 A(XXIV).
17. US *Congressional Record.* Vol. 116, nr. 149.
18. Disarmament Conference document. CCD/PV. 491.
19. Disarmament Conference document. CCD/PV. 457.
20. Disarmament Conference document. CCD/295.
21. Disarmament Conference document. CCD/310.
22. Disarmament Conference document. CCD/PV.494.
23. Disarmament Conference document. CCD/PV.466.
24. Disarmament Conference document. CCD/283.
25. Disarmament Conference document. ENDC/255.
26. Disarmament Conference document. ENDC/255/Rev. 1.
27. Disarmament Conference document. CCD/286.
28. Disarmament Conference document. CCD/290.
29. Disarmament Conference document. CCD/255/Rev. 2.

30. U.N. document. A/7655.
31. Disarmament Conference document. CCD/285 and Corr. 1.
32. Disarmament Conference document. CCD/300.
33. Disarmament Conference document. CCD/308.
34. Disarmament Conference document. CCD/293.
34. Disarmament Conference document. CCD/293.
35. Disarmament Conference document. CCD/311.
36. Disarmament Conference document. CCD/303.
37. Disarmament Conference document. CCD/302.
38. Disarmament Conference document. CCD/299.
39. Disarmament Conference document. CCD/PV.464.
40. Disarmament Conference document. CCD/314.
41. Disarmament Conference document. CCD/PV.456.
42. Disarmament Conference document. CCD/301.
43. Disarmament Conference document. CCD/PV.463.
44. Disarmament Conference document. CCD/PV.486.
45. Disarmament Conference document. CCD/PV.465.
46. Disarmament Conference document. CCD/288.
47. Disarmament Conference document. CCD/289.
48. Disarmament Conference document. CCD/304.
49. U.N. document. A/8136.
50. WHO document. WHA 23.53.
51. US Department of State *Bulletin,* Vol. 16, part 2.
52. U.N. document. A/8001/ Add. 1.
53. U.N. document. A/RES/2662(XXV).
54. U.N. document. A/C.1/PV.1754.
55. U.N. document. A/C.1/PV.1765.
56. U.N. document. A/C.1/PV.1750.
57. U.N. document. A/C.1/PV.1753.
58. Chemical-biological warfare: US policies and international effects. US House of Representatives, Committee on Foreign Affairs, Subcommittee on National Security Policy and Scientific Developments. Washington, 1970.
59. U.N. document. A/C.1/PV.1755.
60. U.N. document. A/C.1/PV.1757.
61. US White House press release on herbicides of 26 December 1970.
62. U.N. documents. A/PV.1928 *and* A/8187.
63. U.N. document. A/C.1/PV.1752.
64. U.N. document. A/PV.1856.
65. U.N. document. A/8052.
66. Disarmament Conference document. CCD/287.
67. Disarmament Conference document. CCD/334.

Postscript
1. Disarmament Conference document CCD/332
2. Disarmament Conference document CCD/PV.499
3. Disarmament Conference document CCD/320
4. Disarmament Conference document CCD/PV.504
5. Disarmament Conference document CCD/PV.502

References

6. Disarmament Conference document CCD/PV.509
7. Disarmament Conference document CCD/324
8. Disarmament Conference document CCD/PV.505
9. Disarmament Conference document CCD/319
10. Disarmament Conference document CCD/325; CCD/325/Rev. 1
11. Disarmament Conference document CCD/PV.508
12. Disarmament Conference document CCD/PV.507
13. Disarmament Conference document CCD/PV.510
14. Daily Bulletin No. 74, United States Mission, Geneva, 1971
15. Disarmament Conference document CCD/PV.513
16. Disarmament Conference document CCD/PV.511
17. Disarmament Conference document CCD/PV.512
18. Disarmament Conference document CCD/PV.514
19. Disarmament Conference document CCD/PV. 516
20. US Congressional Record—Senate, 9 March 1971
21. Disarmament Conference document CCD/328
22. Disarmament Conference document CCD/PV.519
23. US Congressional Record—Senate, 8 June 1971
24. Disarmament Conference document CCD/333
25. Disarmament Conference document CCD/PV.522
26. Disarmament Conference document CCD/337
27. Disarmament Conference document CCD/338
28. Disarmament Conference document CCD/PV.527
29. *Le Monde* 10 August 1971
30. Disarmament Conference document CCD/339
31. Disarmament Conference document CCD/PV.528
32. Disarmament Conference document CCD/341
33. Disarmament Conference document CCD/PV.531
34. Disarmament Conference document CCD/347
35. Disarmament Conference document CCD/346
36. Disarmament Conference document CCD/353
37. *Le Monde* 24 September 1971

Guide to CB disarmament proposals in the League of Nations and the United Nations, 1920–71[1]

Note

Only essential proposals are listed. Those of a procedural character, or concerning the ways of drafting treaty obligations, or containing appeals for adherence to existing international rules, have been omitted.

If identical proposals were made by different countries or bodies, and at various periods, only the original movers are indicated; similar proposals, differing in details, are often grouped together.

Proposals covering several aspects of CB disarmament have been re-arranged and placed under separate headings, but some overlap proved unavoidable.

Under each heading, the proposals, as a rule, follow a chronological order.

Response to proposals is given whenever it is considered significant.

The texts have been abridged; the wording however is close to the original.

Abbreviations

Arms Trade Conference	Conference for the Supervision of the International Trade in Arms and Ammunition and in Implements of War, convened in Geneva on 4 May 1925. The main bodies set up by the Conference were: General, Military and Legal Committees.
B	Biological
C	Chemical
CBW	Chemical and biological warfare
Disarmament Conference	Conference for the Reduction and Limitation of Armaments, convened in Geneva on 2 February 1932.

[1] Up to August 1971.

Guide to proposals

ENDC and CCD	Eighteen-Nation Committee on Disarmament, set up in 1961, and composed of: Brazil, Bulgaria, Burma, Canada, Czechoslovakia, Ethiopia, France, India, Italy, Mexico, Nigeria, Poland, Romania, Sweden, UAR, UK, USA and USSR. The Committee was enlarged in 1969 by the inclusion of eight countries. The name of the conference it holds in Geneva is now "The Conference of the Committee on Disarmament" (CCD).
Geneva Protocol	Protocol for the Prohibition of the Use in War of Asphyxiating, Poisonous or Other Gases, and of Bacteriological Methods of Warfare, signed in Geneva on 17 June 1925.
L. of N.	League of Nations
L. of N. Assembly	Plenary meetings of the League of Nations Assembly and meetings of the Third Committee of the Assembly, dealing with disarmament questions.
PAC	Permanent Advisory Commission for Military, Naval and Air Questions, constituted by the League of Nations Council on 19 May 1920.
Permanent Disarmament Commission	A body to be established by a League of Nations disarmament convention to supervise the implementation of the obligations undertaken by the parties.
Preparatory Commission	Commission set up by the League of Nations Council resolution of 12 December 1925, and entrusted with the preparation for the Conference for the Reduction and Limitation of Armaments (Disarmament Conference).
Socialist Draft BW Convention	Draft Convention on the prohibition of the development, production and stockpiling of bacteriological (biological) weapons and toxins and on their destruction, submitted on 30 March 1971 to the Conference of the Committee on Disarmament by Bulgaria, Byelorussia, Czechoslovakia, Hungary, Mongolia, Poland, Romania, Ukraine and the USSR. The draft was revised on 5 August 1971.
Socialist Draft CBW Convention	Draft Convention on the prohibition of the development, production and stockpiling of chemical and bacteriological (biological) weapons, and on the destruction of such weapons, submitted on 19 September 1969 to the

Guide to proposals

	UN General Assembly by Bulgaria, Byelorussia, Czechoslovakia, Hungary, Mongolia, Poland, Romania, Ukraine and the USSR. The draft was revised on 23 October 1970.
Special Committee's Report I	Report of 31 May 1932 by the Special Committee on chemical and bacteriological weapons, set up at the Disarmament Conference.
Special Committee's Report II	Report of 13 December 1932 by the Special Committee on chemical, incendiary and bacterial weapons, set up at the Disarmament Conference.
Sub-Commission A	A body of the Preparatory Commission for the Disarmament Conference, composed of a military, naval and air expert for each of the countries represented on the Preparatory Commission. Its terms of reference included items of a technical nature.
Sub-Commission B	A body of the Preparatory Commission for the Disarmament Conference, composed of a representative of each delegation. Its terms of reference included all non-military questions related mainly to the economic and financial aspects of disarmament.
TMC	Temporary Mixed Commission established by the League of Nations Council on 25 February 1921, to prepare proposals for the reduction of armaments.
TMC Report	Temporary Mixed Commission's report of 30 July 1924, on chemical and bacteriological warfare.
UK Draft Disarmament Convention	Draft Disarmament Convention submitted on 16 March 1933 by the United Kingdom to the Disarmament Conference. Part IV of the Draft contained provisions concerning chemical, incendiary and bacterial warfare.
UK Draft BW Convention	Draft Convention for the prohibition of biological methods of warfare, submitted on 10 July 1969 by the United Kingdom to the Conference of the Committee on Disarmament. It was revised on 26 August 1969 and on 18 August 1970. The latter revision, following a US amendment, included toxins in the prohibitions under the Draft Convention.

Guide to proposals

UN	United Nations
UN General Assembly	Plenary meetings of the United Nations General Assembly and meetings of the First Political Committee of the Assembly, dealing with disarmament questions.
US Draft BW Convention	Draft Convention on the prohibition of the development, production and stockpiling of bacteriological (biological) and toxin weapons and on their destruction, submitted on 5 August 1971 by the USA to the Conference of the Committee on Disarmament. The text was identical to the revised Socialist Draft BW Convention tabled simultaneously.
Washington Treaty	Treaty of 6 February 1922, relating to the use of submarines and noxious gases in warfare. Article 5 of the Treaty prohibited the use in war of asphyxiating, poisonous or other gases, and all analogous liquids, materials or devices.
WHO	World Health Organization, a United Nations specialized agency.

Contents of the Guide

I. Study of CBW problems 365
II. Prohibition of use of CB weapons 366
III. Scope of CBW prohibition and definitions 368
IV. Revision of the 1925 Geneva Protocol 374
V. Limitations on production and possession of CB weapons 374
VI. Prohibition of production of CB weapons 376
VII. Prohibition or limitation of CBW means of delivery 379
VIII. Prohibition of transfers between countries 381
IX. Destruction of stocks 384
X. Prohibition of other forms of preparation and training for use of CB weapons 386
XI. Restraint on CBW research and development 388
XII. Supervision of observance of the prohibitions 390
XIII. Investigation of complaints 395
XIV. Action in case of violations 397
XV. Protection against CB weapons 401
XVI. Arousing public opinion about CBW 402

Guide to proposals

I. Study of CBW problems

1. 1922 South Africa
   To collect information on the possible effects of CBW.
   Response TMC Report of 1924
   Reference pp. 48–49, 55.

2. 1948 UN Secretary-General
   To study some problems involved in the control of B and lethal C weapons.
   Reference p. 195.

3. 1967 Malta
   To prepare a report on the nature and probable effects of existing CB weapons and on the economic and health implications of the possible use of such weapons with particular reference to states that are not in a position to establish comprehensive methods of protection.
   Reference p. 247.

4. 1968 United Kingdom
   To prepare a report on the nature and possible effects of C weapons and on the implications of their use.
   Reference p. 254.

5. 1968 Poland
   To prepare a report on the effects of CB weapons.
   Reference p. 254.

6. 1968 ENDC
   The UN Secretary-General to appoint a group of experts to study the effects of the possible use of CB means of warfare.
   Response UN General Assembly resolution requesting the UN Secretary-General to prepare a report in accordance with the above recommendation.
   UN Secretary-General's Report of 1969
   WHO Report of 1969.
   Reference pp. 260, 264, 266, 274.

7. 1970 UN Secretary-General's Report on "Respect for Human Rights in Armed Conflicts"
   To make a study of napalm weapons and the effects of their use, so as to facilitate action by the United Nations with a view to curtailing or abolishing such uses of the weapons in question as might be established as inhumane.
   Reference p. 284.

Guide to proposals

II. Prohibition of use of CB weapons

1. 1922 Colombia and Australia

   Colombia: To adopt a convention condemning the use of poisonous gas; the text to be identical with that of Article 5 of the 1922 Washington Treaty.

   Australia: To recommend adherence to the Washington Treaty.

   Response L. of N. Assembly resolution requesting the Council to recommend adhesion to the Washington Treaty.

   Reference p. 47.

2. 1925 USA, France, Poland

   To adopt a resolution (USA) or a protocol (France), based on Article 5 of the Washington Treaty, and open it for signature by States.

   Poland: To extend the prohibition to the use of bacteriological methods of warfare.

   Response The signing at the Arms Trade Conference of the Geneva Protocol for the prohibition of the use in war of asphyxiating, poisonous or other gases, and of bacteriological methods of warfare.

   Reference pp. 66–69.

3. 1925 Switzerland

   To codify the existing international principle concerning the prohibition of use of poisonous and similar gases; to lay down definite rules for the application of the principle.

   Response Arms Trade Conference asked the L. of N. Secretary-General to draw the attention of the Committee of Jurists for the Codification of International Law to the Geneva Protocol and to the clause in the Washington Treaty relating to the prohibition of chemical warfare.

   Reference pp. 65, 69.

4. 1927 Belgium, Poland, Kingdom of the Serbs, Croats and Slovenes, Czechoslovakia and Romania

   To include in the Convention for the Limitation and Reduction of Armaments an undertaking to abstain from the use in war of asphyxiating, poisonous or similar gases, and of all analogous liquids, substances or processes, and to abstain from the use of all bacteriological methods of warfare.

   Reference p. 89.

5. 1929 Belgium

   To undertake, subject to reciprocity, to abstain from the use in war of asphyxiating, poisonous or similar gases, and of all analogous liquids, substances or processes. To undertake unreservedly to abstain from the use of all bacteriological methods of warfare.

   Response Proposal adopted by the Preparatory Commission and included in the Draft Convention for the Limitation and Reduction of Armaments.

   Reference pp. 94, 100.

Guide to proposals

6. 1932 Norway

    To undertake unreservedly to abstain from the use in war of asphyxiating, poisonous or similar gases, and of all analogous liquids, substances or processes.

    To undertake also unreservedly to abstain from the use of all bacteriological methods of warfare.

    Response Chairman of the Special Committee on CBW: The prohibition must be absolute. The right of resorting to the prohibited means of warfare to be excluded even if an undertaking not to use these weapons could not be given by all countries; even if a State were the victim of an unlawful war. However control must be satisfactory and, above all, the penalties must be effective.

    Reference pp. 120–23.

7. 1933 UK Draft Disarmament Convention

    The following provision to be accepted as an established rule of international law:

    The use of chemical, incendiary or bacterial weapons as against any state, whether or not a party to the Convention, and in any war, whatever its character, is prohibited. This provision does not, however, deprive any party which has been the victim of the illegal use of chemical or incendiary weapons of the right to retaliate, subject to such conditions as may hereafter be agreed.

    Response The delegations of the Little Entente to the Disarmament Conference: the prohibition must be absolute.

    Reference pp. 167, 171.

8. 1968 United Kingdom

    To declare that the use of microbiological methods of warfare of any kind and in any circumstances should be treated as contrary to international law and a crime against humanity;

    To undertake never to engage in such methods of warfare in any circumstances.

    Reference p. 256.

9. 1969 UK Draft BW Convention

    To undertake, never, in any circumstances, by making use for hostile purposes of microbial or other biological agents, to engage in biological methods of warfare.

    Reference p. 295.

10. 1969, 1970 Italy

    To withdraw the reservation (made by a number of countries) to the Geneva Protocol exempting parties from prohibition of use of CB weapons against non-parties.

    Reference p. 284.

Guide to proposals

11. 1970 Japan

    States should undertake never, under any circumstances, to engage in CB warfare.

    Reference p. 285.

12. 1968–1970 Sweden, Ethiopia, Nepal, Ghana, UN Secretary-General

    To remove all the reservations to the Geneva Protocol, limiting the applicability of the prohibition to parties and to first-use.

    Response The United States renounced B weapons unconditionally, and proposed to ratify the Protocol with a reservation permitting the use of C weapons if an enemy or its allies were to employ them first.

    The USSR stated that the adoption of a convention for complete elimination of CB weapons, with the participation of a wide range of states, would make the reservations to the Protocol pointless.

    Reference pp. 258, 280, 284–85.

III. Scope of CBW prohibition and definitions

1. 1924 TMC Report

    Agents used in chemical warfare could be classified as: irritant (lachrymatory, sneeze-producing and blistering) agents; suffocating or asphyxiating agents; and toxic agents.

    Reference p. 50.

2. 1926 Kingdom of the Serbs, Croats and Slovenes

    To include incendiary material in the category of chemical weapons.

    Response The Soviet Draft Convention on the Reduction of Armaments, of 1928, provided for the abolition of flame-throwers along with other "methods of and appliances for chemical aggression."

    Special Committee's Report I found that incendiary weapons possessed a specific means of action which assimilated them to chemical weapons: projectiles specifically intended to cause fires and appliances designed to attack persons by fire should be prohibited.

    Special Committee's Report II: It is possible to prohibit projectiles and bombs specifically incendiary.

    UK Draft Disarmament Convention of 1933:

    The prohibition of the use of incendiary weapons shall apply to:

    (1) The use of projectiles specifically intended to cause fires.

    The prohibition shall not apply to:
    - (a) Projectiles specially constructed to give light or to be luminous and generally to pyrotechnics not intended to cause fires, or to projectiles of all kinds capable of producing incendiary effects accidentally;
    - (b) Incendiary projectiles designed specifically for defence against aircraft, provided that they are used exclusively for that purpose.

(2) The use of appliances designed to attack persons by fire, such as flame-projectors.

Reference pp. 73, 92, 116–17, 152–53, 168.

3. 1930 United Kingdom, France

To consider the use of lachrymatory gases in war as covered by the prohibition under the Geneva Protocol.

Reponse US objection: the agents in question are used by many governments in peacetime against their own population.

Reference pp. 102–104.

4. 1932 Italy

To abolish the use in time of war of chemical weapons of all kinds and particularly to prohibit all asphyxiating, toxic, lachrymatory or similar gases, all liquids or other substances or devices producing results similar to the above-mentioned gases and bacteriological methods of all kinds.

Reference p. 111.

5. 1932 Special Committee's Report I

There should be prohibited the use for the purpose of injuring an adversary, of all natural or synthetic noxious substances, whatever their state, whether solid, liquid or gaseous, whether toxic, asphyxiating, lachrymatory, irritant, vesicant, or capable in any way of producing harmful effects on the human or animal organism, whatever the method of their use.

This definition does not extend to explosives, provided that the latter have not been designed or used with the object of producing noxious substances, or to smoke or fog, provided they do not produce harmful effects under normal conditions of use.

There should be prohibited all methods for the projection, discharge or dissemination in any manner, in places inhabited or not, of pathogenic microbes in whatever phase they may be (virulent or capable of becoming so), or of filter-passing viruses, or of infected substances, whether for the purpose of bringing them into immediate contact with human beings, animals or plants, or for the purpose of affecting any of the latter in any indirect manner—for example, by polluting the atmosphere, water, foodstuffs, or any other objects.

Response US delegation to the Disarmament Conference: there is no question of use of tear gas in time of war, but it would be difficult to give up the preparation and employment of this gas for local police purposes. The Special Committee's formulations were included in the UK Draft Disarmament Convention.

Reference pp. 113, 116–17, 134, 167–68.

6. 1932 Special Committee's Report II

Lachrymatory substances should not be treated separately from poisonous as far as the prohibition of use in war is concerned.

Guide to proposals

Some supervision should be exercised over the stocks of lachrymatory substances and apparatus for their use in police operations in peacetime.

Response The US government was willing to forego the use of lachrymatory substances for military purposes in wartime, but the use of those substances for police purposes and for protecting private property should be permitted; special regulations could be introduced to prevent abuse.

Reference pp. 153–54.

7. 1947 USA

To include in the definition of weapons of mass destruction "lethal chemical and biological weapons".

Response Australia: To delete the word "lethal" in the above text.

The US definition was adopted in 1948 by the UN Commission for Conventional Armaments

Reference pp. 193–95.

8. 1954 Federal Republic of Germany

Chemical weapons to be defined as any equipment or apparatus expressly designed to use, for military purposes, the asphyxiating, toxic, irritant, paralysant, growth-regulating, anti-lubricating or catalysing properties of any chemical substance; chemical substances, having such properties and capable of being used in the equipment or apparatus referred to above, shall be included in this definition.

Biological weapons to be defined as any equipment or apparatus expressly designed to use, for military purposes, harmful insects or other living or dead organisms, or their toxic products; insects, organisms and their toxic products of such nature and in such amounts as to make them capable of being used in the equipment or apparatus referred to above, shall be included in this definition.

Response These definitions of CB weapons were included in Annex II to Protocol No. III signed by the members of the Western European Union (Belgium, France, the FRG, Italy, Luxembourg, Netherlands and the UK).

Reference pp. 224–25.

9. 1955 France and United Kingdom

A definition of armaments should include bombs and equipment for discharging or disseminating CB substances. A definition of essential components of armaments—to include major components manufactured elsewhere than at the place of assembly of the whole armament, such as the filling of CB weapons.

Reference p. 227.

10. 1966 Hungary

To declare that the use of CB weapons for the purpose of destroying human beings and the means of their existence constitutes and international crime.

Response USA: No rule of international law prohibits the use in combat against an enemy, for humanitarian purposes, of agents that are commonly used to control riots; similarly, the Geneva Protocol does not apply to herbicides.
Reference pp. 238–39, 243.

11. 1967 Malta

    To consider as chemical weapons toxic chemical agents used for hostile purposes which produce their effects directly as a result of their chemical properties rather than as a result of blast, heat or other physical effects of a chemical reaction.

    To consider as biological weapons all microorganisms including viruses, or their toxic products intentionally used for hostile purposes.
    Reference p. 248.

12. 1968 Sweden

    To give the Geneva Protocol a broad interpretation and to consider all existing CB weapons as belonging in one set and the prohibition to use any of them as valid without exceptions.
    Reference p. 258.

13. 1968 USA

    Biological means of warfare should be understood to mean disease-causing living microorganisms, be they bacteria, or viruses, used as deliberate weapons of war.
    Reference p. 264.

14. 1969 Group of UN consultant experts on CB weapons

    Chemical agents of warfare are chemical substances—whether gaseous, liquid or solid—which might be employed because of their direct toxic effects on man, animals and plants.

    Bacteriological agents of warfare are living organisms, whatever their nature, or infective material derived from them, which are intended to cause disease or death in man, animals, or plants, and which depend for their effects on their ability to multiply in the person, animal or plant attacked.
    Reference pp. 267–68.

15. 1969 UN Secretary-General

    To make a clear affirmation that the prohibition contained in the Geneva Protocol applies to the use in war of all CB agents (including tear-gas and other harassing agents), which now exist or which may be developed in the future.

    Response UN General Assembly resolution, initiated by Sweden, stated that the Geneva Protocol embodied the generally recognized rules of international law prohibiting the use in international armed conflicts of all

Guide to proposals

CB methods of warfare, regardless of any technical developments. It declared as contrary to those rules the use of chemical and biological agents as defined by the group of UN consultant experts (see above). Australia, Portugal and USA voted against the resolution.
Reference pp. 273, 286.

16. 1970 United Kingdom

CS and other such gases not significantly harmful to man in other than wholly exceptional circumstances are not prohibited under the Geneva Protocol.
Reference p. 287.

17. 1970 Canada

The use or the prohibition of use in war of tear gas and other riot-control agents presents practical problems which require detailed study and resolution.
Reference p. 287.

18. 1970 USA

The use in war of riot-control agents and chemical herbicides is not prohibited; smoke, flame and napalm are also not covered by the Geneva Protocol.
Reference pp. 283–84.

19. 1970 Netherlands

The use of all CB agents of warfare—including tear gases—should be prohibited.
It is necessary to establish a clear rule for the future, which would exclude the use of herbicides and defoliants for warlike purposes.
Reference pp. 287–89.

20. 1970 Norway

To achieve an effective ban on the use of CB weapons and on their development, production and stockpiling, including a ban on the use in warfare of tear gases.
A ban on the use of CB weapons should include a ban on the use of herbicides in warfare.
Reference pp. 288–89.

21. 1970 Finland

To request the International Court of Justice to provide an advisory opinion as to the scope of the rules and prohibitions of the Geneva Protocol.
Reference p. 289.

22. 1970 USA

To include toxins in the prohibition of biological means of warfare
Reference p. 297.

Guide to proposals

23. 1970 UK Draft BW Convention (revised)

    To prohibit the use, production, acquisition and research of microbial or other biological agents or toxins causing death, damage or disease to man, animals or crops, as well as ancillary equipment or vectors the purpose of which is to facilitate the use of such agents or toxins for hostile purposes.
    Reference pp. 297–98, 323.

24. 1970 Socialist Draft CBW Convention (revised)

    To prohibit the development, production, stockpiling or other acquisition of CB weapons, or equipment or vectors specially designed for the use of CB weapons as means of warfare.
    Reference p. 300.

25. 1970 Yugoslavia

    In enacting national laws prohibiting research, development, production and stockpiling of agents for CB weapons, an exception could be made, in line with the provisions of a treaty on the complete prohibition of CB weapons, for types and quantitites of agents used for riot-control purposes within the country.
    Reference p. 303.

26. 1971 Sweden

    Not to develop, test, produce, stockpile or otherwise acquire (and transfer) CB weapons.
    Not to produce, test and stockpile (and transfer) agents. The agents to be divided into two categories: Category *a:* super-toxic agents, whether chemical, toxins or biological which have a practically exclusive use as potential means of warfare. They would include all substances more toxic than 1 mg per kg body weight (i.a. chemical components of nerve gases and mustards, as well as all toxins). Category *b:* all remaining chemical agents which can be used as means of warfare but also have recognized peaceful uses (i.a. hydrogen cyanide, phosgene, tear gases and defoliants); and most biological agents in so far as they are produced for non-military purposes.
    To prohibit ancillary equipment and vectors.
    Different verification procedures to be applied to category *a* and to category *b* and equipment. (See under XII.20).
    Reference p. 312.

27. 1971 Netherlands

    To use a chemical formula in connection with a toxicity level of 0.5 mg/kg in order to define organophosphorus compounds likely to have nerve-gas properties, which would be subject to unconditional prohibition.
    Reference p. 313.

28. 1971 Socialist Draft BW Convention

    To prohibit the development, production, stockpiling or other acquisition of microbiological or other B agents or toxins, as well as of auxiliary

equipment or means of delivery designed to facilitate the use of such agents or toxins for hostile purposes.
Reference p. 314.

29. 1971 Sweden

All toxins that have potential use as warfare agents, whatever their origin or mode of preparation, should be prohibited.
Reference p. 318.

30. 1971 Revised Socialist Draft BW Convention and US Draft BW Convention

Not to develop, produce, stockpile or otherwise acquire or retain microbial or other biological agents or toxins, and weapons, equipment or means of delivery designed to use such agents or toxins for hostile purposes or in armed conflict.
Reference p. 337.

IV. Revision of the 1925 Geneva Protocol

1. 1967 Malta[1]

   To consider the problems relating to the use of CB weapons with a view to revision, updating or replacement of the Geneva Protocol.
   Reference p. 247.

V. Limitations on production and possession of CB weapons

1. 1920 United Kingdom[2]

   To impose for each country a limitation on the quantity of poisonous gases.
   Response PAC conclusion: it would be useless to seek to restrict the use of gases in wartime by prohibiting or limiting their manufacture in peacetime.
   Reference pp. 43–44.

2. 1926–1930 France

   If substances utilizable for CBW have a normal utility in peacetime, to restrict their manufacture to commercial requirements.
   Reference p. 94.

3. 1932 Chairman of the Special Committee on CBW

   To fix for each state the maximum quantity of chemical substances capable of use for military purposes, which may be stored in industrial establish-

---

[1] Similar proposals were made previously by some Western countries; Malta submitted it formally as a UN draft resolution.
[2] The UK raised the problem in a general way. The proposal was formulated in a questionnaire examined by the PAC.

ments of every kind in its territory. (That would involve making an estimate of the needs of consumers and the normal reserves necessary to meet those needs.)

Response Special Committee's Report II: It is not possible to limit the chemical production capacity of states nor the quantity of chemical products in stock. The latter limitation would only be possible in the case of substances exclusively used for chemical warfare.

Reference pp. 125, 158.

4. 1932 Special Committee's Report II

    A State wishing to use lachrymatory substances for police purposes should inform the Permanent Disarmament Commission, specifying the substances, the implements which it proposed to employ and their number. The Commission would examine whether there is any disproportion between the arms notified and police requirements. In those countries where industrial firms manufacture or sell implements or devices charged with lachrymatory substances for the protection of private property, the state should be responsible for its nationals.

    Response UK Draft Disarmament Convention. (See under V.5.)

    Reference pp. 154–55.

5. 1933 UK Draft Disarmament Convention

    The parties should inform the Permanent Disarmament Commission of the lachrymatory substances intended to be used by their authorities for police operations as well as of the number of the various appliances by means of which they are to be utilized.

    Response US amendment: The parties shall inform the Permanent Disarmament Commission annually of the nature of the lachrymatory substances used by their governmental agencies or instrumentalities for police operations, as well as of the number and character of the various appliances by which the said lachrymatory substances are utilized.

    Reference pp. 169, 172.

6. 1952 USA

    At appropriate stages of a comprehensive disarmament programme, agreed measures should become effective providing for progressive curtailment of production, progressive dismantling of plants and progressive destruction of stockpiles of bacterial weapons and related appliances.

    Reference p. 205.

7. 1960 USA

    In the second stage of general and complete disarmament:
    Quantities of all kinds of armaments of each state, including CB weapons and all means for their delivery, to be reduced to agreed levels and the resulting excesses destroyed or converted to peaceful uses.

    Reference p. 230.

Guide to proposals

8. 1962 USA

In the second stage of general and complete disarmament:
Reduction, by agreed categories, of stockpiles of CB weapons of mass destruction to levels fifty per cent below those existing at the beginning of that stage.
Reference pp. 232–33.

VI. Prohibition of production of CB weapons

1. 1927, 1929 Belgium, Poland, Kingdom of the Serbs, Croats and Slovenes, Czechoslovakia and Romania

Not to permit the manufacture of substances utilizable for CBW when they are manufactured with a view to such use.
France
   To prevent the manufacture of substances utilizable for CBW, so far as these have no normal utility in peacetime.
Reference pp. 89, 94.

2. 1928 USSR

The industrial undertakings engaged in the production of the means of chemical aggression or bacteriological warfare to discontinue production from the date of the entry into force of the Protocol (supplementary to the 1925 Geneva Protocol).
Reference pp. 91–92.

3. 1932 Italy

Not to manufacture chemical substances (asphyxiating, toxic, lachrymatory or similar gases, all liquids or other substances or devices producing results similar to the above-mentioned gases) and bacteriological substances, specifically intended to harm the belligerents or the civil population, with the exception of chemical and bacteriological substances capable of being utilized for peaceable industrial and scientific purposes, and for such purposes only.
Reference p. 111.

4. 1932 Norway

To prevent the manufacture of materials or apparatus capable of use for CB warfare. If such substances or apparatus meet a normal peaceful requirement, the parties will keep their manufacture within the limits of commercial requirements, and will supply the Permanent Disarmament Commission with annual statistics.
Reference p. 121.

5. 1932 Chairman of the Special Committee on CBW

To prohibit the manufacture or storage of chemical substances exclusively suited to the conduct of chemical warfare. To prohibit the manufacture or storage of chemical substances capable of being used both for peaceful and for military purposes with the intention of using them in war.

Response Special Committee's Report II: It would be possible to prohibit the manufacture and possession of substances exclusively used for chemical warfare, but such prohibition would be of only limited value. In any event, there should be no total prohibition, because a certain amount of such substances would have to be prepared to study protection. It is not possible to prohibit the manufacture and possession of substances capable of employment both for peaceful and warlike purposes.

Reference pp. 124, 150–53.

6. 1933 UK Draft Disarmament Convention

It shall be prohibited:

To manufacture or be in possession of substances exclusively suited to chemical or incendiary warfare. The quantitites of chemical substances necessary for protective experiments, therapeutic research and laboratory work shall be excepted. The parties shall inform the Permanent Disarmament Commission of the quantitites of the said substances necessary for their protective experiments. The manufacture of and trade in these substances may not be undertaken without government authorization.

To manufacture or be in possession of substances suitable for both peaceful and military purposes with intent to use them in violation of the prohibition contained in the Convention.

Response US objection to the phrase "intent to use them in violation of the prohibition".

Reference pp. 168–69, 171.

7. 1953 UN General Assembly resolution

Prohibition of atomic, hydrogen, bacterial, chemical and all such other weapons of war and mass destruction.

Reference p. 220.

8. 1959 United Kingdom

To ban the manufacture of CB weapons at the third (final) stage of a programme for comprehensive disarmament.

Reference p. 228.

9. 1959 USSR

At the third (last) stage of the programme of general and complete disarmament: Entry into force of the prohibition on the production, possession and storage of chemical and bacteriological weapons.

Reference p. 228.

10. 1960 USSR

Joint studies to be undertaken in the first stage of a general disarmament process of the measures to be implemented in the second stage relating to

the discontinuance of the manufacture of nuclear, chemical and biological weapons. In the second stage—a complete prohibition of nuclear, chemical, biological and other weapons of mass destruction, with the cessation of manufacture of such weapons.
Reference p. 229.

11. 1960 USA

In the third stage of general and complete disarmament:
No manufacture of any armaments except for agreed types and quantities for use by the international peace force and agreed remaining national contingents.
Reference p. 230.

12. 1961 USA

In the first stage of a programme for general disarmament, to establish within an international disarmament organization (to be set up within the framework of the United Nations) a chemical, biological, radiological (CBR) experts commission for the purpose of examining and reporting on the feasibility and means for accomplishing a verifiable reduction and eventual elimination of CBR weapons stockpiles and the halting of their production;
In the second stage, depending upon the findings of the experts commission, to cease the production of CBR weapons.
Reference p. 231.

13. 1962 and 1965 USSR

At the second stage of general and complete disarmament:
To discontinue completely the production of all kinds of chemical, biological and radiological weapons and of all means and devices for their combat use, transportation and storage. All plants, installations and laboratories that are wholly or partly engaged in the production of such weapons, to be destroyed or converted to production for peaceful purposes.
Reference p. 232.

14. 1962 USA

In the second stage of general and complete disarmament:
Cessation of all production of chemical and biological weapons of mass destruction.
Dismantling or conversion to peaceful uses of all facilities engaged in the production or field testing of CB weapons of mass destruction.
Reference p. 233.

15. 1969 UN Secretary-General

To halt the development, production and stockpiling of all CB agents for purposes of war and to achieve their effective elimination.
Reference p. 273.

Guide to proposals

16. 1969–1970 UK Draft BW Convention

    To prohibit production or other acquisition of microbial or other B agents or toxins of types and in quantities that have no justification for prophylactic or other peaceful purposes.

    Reference pp. 295, 297–98.

17. 1969–1970 Socialist Draft CBW Convention

    Not to produce, stockpile or otherwise acquire CB weapons.

    Reference p. 298.

18. 1970 Sweden

    To prohibit unconditionally B agents. To prohibit unconditionally single purpose chemical agents such as nerve gases, as well as toxins. To prohibit conditionally (prohibition with partial restraints) chemical agents with dual purpose, having a legitimate use in peaceful activities.

    Reference p. 292.

19. 1970 Mexico

    To renounce the manufacture and stockpiling of B weapons through unilateral declarations; the renunciation would acquire a contractual character when overall agreement on CBW is achieved.

    Reference p. 282.

20. 1971 Socialist Draft BW Convention

    Not to produce, stockpile or otherwise acquire microbiological or other biological agents or toxins of such types and in such quantities as are not designed for the prevention of disease or for other peaceful purposes.

    Reference p. 314.

21. 1971 Revised Socialist Draft BW Convention and US Draft BW Convention

    Not to produce, stockpile or otherwise acquire or retain microbial or other biological agents or toxins of types and in quantities that have no justification for prophylactic or other peaceful purposes; as well as weapons designed to use such agents or toxins.

    Reference p. 337.

VII. Prohibition or limitation of CBW means of delivery

1. 1926 United Kingdom

    To limit the instruments enabling the use of gas, thus limiting its "utilizability".

    Reference p. 72.

2. 1929 Germany

    To introduce a general restriction in the use of the most important weapons by which chemical substances can be employed in war—the air weapon.
    Not to launch weapons of offence of any kind from the air by means of

aircraft, not to employ unpiloted aircraft controlled by wireless or otherwise carrying explosive or incendiary gaseous substances.

Response Proposal formally rejected.

Reference pp. 89–90, 95, 102.

3. 1932 Italy

   Not to manufacture appliances for the utilization of chemical weapons and bacteriological methods of warfare.

   Reference p. 111.

4. 1932 Chairman of the Special Committee on CBW

   To prohibit the manufacture or storage of appliances exclusively suited to the conduct of chemical warfare (special apparatus for the projection of gases or propulsion of gas-shells).

   To prohibit the manufacture or storage of appliances capable of being used both for peaceful and for military purposes with the intention of using them in war.

   Response France: The appliances peculiar to chemical warfare differ very little from other military appliances. The difficulty is even greater with prohibiting material which could be used for peaceful and warlike purposes, alike.

   Switzerland: There can be no real distinction between material intended exclusively for war purposes and material which might be used for both peaceful and military purposes.

   Special Committee's Report II: It would be possible to prohibit the manufacture and possession of apparatus exclusively used for chemical warfare, but such a prohibition would be of only limited value. It is not possible to prohibit the manufacture and possession of apparatus capable of employment for both peaceful and warlike purposes.

   Reference pp. 124, 135–36, 152.

5. 1933 UK Draft Disarmament Convention

   It shall be prohibited:

   To manufacture or be in possession of appliances exclusively suited to chemical or incendiary warfare.

   To manufacture or be in possession of appliances suitable for both peaceful and military purposes with intent to use them in violation of the prohibition contained in the Convention.

   Response US objection to the phrase "intent to use them in violation of the prohibition".

   Reference pp. 168–69, 171.

6. 1962 USSR

   Production of all means and devices for the combat use of CB weapons to be completely discontinued at the second stage of general and complete disarmament, together with the cessation of production of CB weapons.

   Reference p. 232.

7. 1962 USA

    To eliminate all means of delivery of weapons of mass destruction—an objective of a treaty on general and complete disarmament.

    Reference p. 232.

8. 1969–1970 UK Draft BW Convention

    Not to produce or otherwise acquire ancillary equipment or vectors the purpose of which is to facilitate the use of microbial or other biological agents or toxins for hostile purposes.

    Reference p. 295.

9. 1970 Socialist Draft CBW Convention

    Not to produce, stockpile or otherwise acquire equipment or vectors specially designed for the use of CB weapons as means of warfare.

    Reference p. 300.

10. 1971 Socialist Draft BW Convention

    Not to produce, stockpile or otherwise acquire auxiliary equipment or means of delivery designed to facilitate the use of microbiological or other biological agents or toxins for hostile purposes.

    Reference p. 314.

11. 1971 Revised Socialist Draft BW Convention and US Draft BW Convention

    Not to produce, stockpile or otherwise acquire or retain weapons, equipment or means of delivery designed to use microbial or other biological agents or toxins for hostile purposes or in armed conflict.

    Reference p. 337.

VIII. Prohibition of transfers between countries

1. 1925 USA[3]
    1. To prohibit the export of any asphyxiating, poisonous or other gases, and all analogous liquids, intended or designed for use in connection with operations of war.
    2. To control the traffic in poisonous gases by prohibiting the exportation of all asphyxiating, toxic or deleterious gases and all analogous liquids, materials and devices manufactured and intended for use in warfare under adequate penalties, applicable in all places where the parties exercise jurisdiction or control.
    3. To prohibit the export of all asphyxiating, toxic or deleterious gases, and all analogous liquids, exclusively designed or intended for use in connection with operations of war.
    4. To couple the prohibition of the traffic in means of chemical warfare with a universal prohibition of the use of poison gases.

[3] The texts are listed in the order of submission.

Guide to proposals

Response Military Committee of the Arms Trade Conference: The prohibition of export of CB arms is, in most cases, practically impossible, and would, moreover, be of no effect until all nations undertook to abstain from CB warfare of all kinds.
Reference pp. 59–61, 64.

2. 1925 Brazil

To prohibit the export of appliances used in chemical warfare; this measure should be completed by the prohibition to manufacture anything connected with chemical warfare. The two measures, in turn, ought to be followed by a universal agreement entirely outlawing chemical warfare.
Reference p. 63.

3. 1925 Poland

To prohibit the export of bacteriological means of warfare.
Response See under VIII.1.
Reference pp. 60, 64.

4. 1927 Belgium, Poland, Kingdom of the Serbs, Croats and Slovenes, Czechoslovakia and Romania

Not to permit the importation or the exportation of substances utilizable for chemical or bacteriological warfare, when they are imported or exported with a view to such use.
France
    To prevent the importation or the exportation of substances utilizable for chemical or bacteriological warfare, so far as these have no normal utility in peacetime. If such substances have a normal utility in peacetime, to restrict their importation or exportation to commerical requirements.
Reference pp. 89, 94.

5. 1932 Italy

Not to import chemical and bacteriological appliances of any kind specifically intended for warlike purposes.
Reference p. 111.

6. 1932 Norway

To prevent the import or export of materials or apparatus capable of use for CB warfare. If such substances or apparatus meet a normal peacetime requirement, the parties should keep their import and export within the limits of commercial requirements and supply the Permanent Disarmament Commission with annual statistics.
Reference p. 121.

Guide to proposals

7. 1932 Chairman of the Special Committee on CBW

    To prohibit the import of appliances (special apparatus for the projection of gases or propulsion of gas-shells) and chemical substances, exclusively suited to the conduct of chemical warfare.

    To prohibit the import of appliances and chemical substances capable of being used both for peaceful and for military purposes with the intention of using them in war.

    Response Special Committee's Report II: It would be possible to prohibit the import of apparatus and substances exclusively used for chemical warfare, but such a prohibition would be of only limited value. It is not possible to prohibit the import and export of apparatus and substances capable of employment for peaceful and warlike purposes.

    Reference pp. 124, 152.

8. 1933 UK Draft Disarmament Convention

    It shall be prohibited:

    To import or export appliances or substances exclusively suited to chemical or incendiary warfare. (The quantities of chemical substances necessary for protective experiments, thereapeutic research and laboratory work shall be excepted).

    To import or export appliances or substances suitable for both peaceful and military purposes with intent to use them in violation of the prohibition contained in the Convention.

    Response US objection to the phrase "intent to use them in violation of the prohibition".

    Reference pp. 168–69, 171.

9. 1969–1970 UK Draft BW Convention

    Not to acquire, or assist in or permit the acquisition of microbial or other biological agents or toxins of types and in quantitites that have no justification for prophylactic or other peaceful purposes; and ancillary equipment or vectors the purpose of which is to facilitate the use of such agents or toxins for hostile purposes.

    Reference pp. 295, 297–98.

10. 1970 Sweden

    To prohibit unconditionally transfers between countries of all biological agents of warfare and of an increasing number of chemical agents.

    Reference p. 292.

11. 1969–1970 Socialist Draft CBW Convention

    Not to acquire CB weapons, or equipment or vectors specially designed for the use of CB weapons as means of warfare.

    Not to assist, encourage or induce any state, group of states or international organizations to acquire CB weapons.

    Reference pp. 298, 300.

Guide to proposals

12. **1971 Socialist Draft BW Convention**
Not to acquire, or assist, encourage or induce any particular state, group of states or international organizations to acquire microbiological or other biological agents or toxins of such types and in such quantities as are not designed for the prevention of disease or for other peaceful purposes; and auxiliary equipment or means of delivery designed to facilitate the use of such agents or toxins for hostile purposes.
Reference pp. 314, 332.

13. **1971 Revised Socialist Draft BW Convention and US Draft BW Convention**
Not to transfer to any recipient whatsoever, directly, or indirectly, any agent, toxin, weapon, equipment or means of delivery specified in the convention.
Reference pp. 337–38.

IX. Destruction of stocks

1. **1928 USSR**
To destroy within three months of the date of the entry into force of the Protocol (supplementary to 1925 Geneva Protocol): all methods of and appliances for chemical aggression (all asphyxiating gases used for warlike purposes, as well as all appliances for their discharge, such as gas-projectors, pulverisers, balloons, flame throwers and other devices) and for bacteriological warfare, whether in service with troops or in reserve or in process of manufacture.
Reference p. 92.

2. **1932 Italy**[4]
To destroy, within a period of X months, as from the entry into force of the convention for the limitation and reduction of armaments, all quantities of chemical substances (asphyxiating, toxic, lachrymatory or similar gases, all liquids or other substances or devices producing results similar to the above-mentioned gases) and bacteriological substances, constituting reserve depots or material for experiment, as well as the plant serving for their manufacture and all appliances serving for their utilization. Plant capable of direct employment by the chemical and pharmaceutical industry for non-military purposes may be retained on condition that it is strictly utilized for the needs of peaceful industries.
To destroy within a period of X months as from the entry into force of the convention, all artillery or hand ammunition and projectiles of all kinds loaded with chemical and bacteriological substances of the above-mentioned categories and intended for discharge by aircraft.
Reference pp. 111, 120–21.

[4] A similar proposal was tabled by Norway.

3. 1932 Chairman of the Special Committee on CBW

   Any stocks of chemical substances exclusively suited to the conduct of chemical warfare should be destroyed.

   Reference p. 124.

4. 1959 USSR

   At the third (last) stage of the programme of general and complete disarmament—all stockpiles of CB weapons in the possession of states should be removed and destroyed.

   Reference p. 228.

5. 1960 USSR

   Joint studies to be undertaken in the first stage of a general disarmament process of the measures to be implemented in the second stage relating, among others, to the destruction of stockpiles of CB weapons.
   In the second stage—destruction of all stockpiles of such weapons with on-site inspection by representatives of the control organization.

   Reference p. 229.

6. 1960 USA

   In the third stage of general and complete disarmament, all armaments, including weapons of mass destruction and vehicles for their delivery to be destroyed or converted to peaceful uses.

   Reference p. 230.

7. 1961 USA

   In the second stage of a programme for general disarmament, depending on the findings of an experts commission (see under VI.12), to reduce progressively existing stocks and destroy the resulting excess quantities or convert them to peaceful uses.
   In the third stage, to destroy or convert to peaceful purposes all armaments, except for those of agreed types and quantities to be used by the UN peace force and those required to maintain internal order.

   Reference p. 231.

8. 1962, 1965 USSR

   At the second stage of general and complete disarmament:
   To eliminate from the arsenals of states and destroy (neutralize) all kinds of chemical, biological and radiological weapons, whether directly attached to the troops or stored in various depots and storage places. To destroy simultaneously all instruments and facilities for the combat use of such weapons, special facilities for their transportation and special devices and facilities for their storage and conservation.

   Reference pp. 232, 234.

Guide to proposals

9. 1969–1970 UK Draft BW Convention

   To destroy, or divert to peaceful purposes, within three months after the Convention comes into force for each party, any stocks in its possession of such agents or toxins or ancillary equipment or vectors as have been produced or otherwise acquired for hostile purposes.
   Reference pp. 295, 323.

10. 1969–1970 Socialist Draft CBW Convention

    To destroy within a period of ...—observing all the necessary precautions—or to divert to peaceful uses all previously accumulated CB weapons, as well as equipment and vectors specially designed for the use of CB weapons as a means of warfare.
    Reference pp. 298, 300, 327.

11. 1971 Sweden

    To destroy C agents by means of reactive chemicals or through decomposition by heating, pyrolisis or combustion. To destroy B agents by combustion, in autoclave or by means of disinfectants.
    Reference p. 313.

12. 1971 Socialist Draft BW Convention

    To destroy within three months after the entry into force of the convention—observing all the necessary precautions—or to divert to peaceful uses all previously accumulated B weapons, as well as the equipment and means of delivery.
    Reference p. 314.

X. Prohibition of other forms of preparation and training for use of CB weapons

1. 1927, 1929 Germany

   To abstain from any preparation in peacetime of the use of chemical and bacteriological methods of warfare: stockpiling materials for chemical warfare with the intention of using them in war; preparing bombs with chemical gases; training soldiers for chemical warfare etc.
   Reference pp. 89–91.

2. 1929 France

   To abstain from any preparation in peacetime with a view to the use in war of the methods of CB warfare, and undertake effectual steps to prevent

Guide to proposals

private persons from making preparations for the use of such methods in war.
Reference p. 94.

3. 1932 Italy

   Not to maintain or train personnel specialized in the use of chemical and bacteriological appliances of all kinds; not to publish even for purely theoretical purposes regulations or instructions dealing with the use of the said appliances.
   Reference p. 111.

4. 1932 Norway

   To abstain from any preparation in time of peace for the use in wartime of all apparatus and appliances for CB warfare, and in particular from any training of troops, and to prevent all preparation on the part of private persons or enterprises for the employment of such methods of warfare.
   Reference p. 121.

5. 1932 Chairman of the Special Committee on CBW

   Not to issue appliances and substances capable of use for military and peaceful purposes to the armed forces; not to keep them in military establishments (arsenals, fortifications, barracks, etc.), not to store them for army use. (Should the armed forces need a certain quantity of such appliances or substances for use otherwise than as fighting weapons, e.g. for disinfection purposes, the states should say what quantity they would wish to leave at the disposal of the armed forces.)
   To prohibit training for CB warfare (defensive training aimed exclusively at protecting people against the effects of chemical and bacteriological warfare must be authorized).

   Response France: There is no difference between instructing a unit in the release of non-toxic smoke for purposes of cover and training the same unit in the release of toxic gases. No special training is needed for releasing gases.

   Special Committee's Report II: The prohibition of the possession by armed forces of certain substances capable both of peaceful and military utilization would be ineffective.
   It is possible to prohibit the training of armed forces in the use of chemical weapons, but the practical effect of such a prohibition would be very small.
   It is not possible in practice to prevent preparation for bacteriological warfare.
   Reference pp. 124–25, 135, 151–52.

6. 1933 UK Draft Disarmament Convention

   All preparations for chemical, incendiary or bacterial warfare should be prohibited in time of peace as in time of war. It should in particular be

prohibited to instruct or train armed forces in the use of chemical, incendiary or bacterial weapons and means of warfare, or to permit any instruction or training for such purposes.
Reference pp. 168–69.

7. 1970 Sweden

    To prohibit preparation of instructions and manuals as well as training for CB warfare.
    Reference p. 292.

8. 1970 Yugoslavia

    To cease the training of troops in the use of CB weapons and to delete from army manuals all relevant instructions except for sections dealing with protection.
    Reference p. 303.

XI. Restraint on CBW research and development

1. 1920 United Kingdom[5]

    To prohibit experiments in the laboratories.
    Response PAC conclusion: The prohibition of laboratory experiments is out of the question.
    Reference pp. 43–44.

2. 1926 Belgium

    To prohibit preparation of poisonous gases in laboratories.
    Response Sub-Commission A of the Preparatory Commission: It is impossible to prevent the study of poisonous substances in the laboratories.
    Reference pp. 72, 78.

3. 1926 Sub-Commission B of the Preparatory Commission

    To abolish all subsidies to official laboratories and private institutions promoting research in the matter of poisonous gases for purely military purposes.
    Reference p. 78.

4. 1959 USSR

    In the third (last) stage of a programme for general and complete disarmament, scientific research for military purposes and the development of weapons and military equipment should be prohibited.
    Reference p. 228.

---

[5] The UK raised the problem in a general way. The proposal was formulated in a questionnaire examined by the PAC.

Guide to proposals

5. 1962 USA

In the second stage of general and complete disarmament, to stop field testing of CB weapons of mass destruction; to dismantle or convert to peaceful uses all facilities engaged in such testing.

In the third stage, subject to agreed requirements for non-nuclear armaments of agreed types for national forces and UN peace force, all applied research and development of armaments should stop and the facilities for such purposes—dismantled or converted to peaceful uses.

Reference pp. 232–33.

6. 1969–1970 UK Draft BW Convention

Not to conduct, assist or permit research aimed at production of microbial or other biological agents or toxins that have no justification for prophylactic or other peaceful purposes, as well as ancillary equipment or vectors the purpose of which is to facilitate the use of such agents or toxins for hostile purposes.

Reference pp. 295, 297–98, 323.

7. 1969–1970 Socialist Draft CBW Convention

Not to develop CB weapons or equipment or vectors specially designed for the use of CB weapons as means of warfare.

Reference pp. 298, 300–301, 327.

8. 1970 Sweden

To exempt from prohibition research with regard to chemical and biological agents. To prohibit unconditionally development of warfare agents and of devices for their dissemination. To prohibit simultaneously the testing of chemical and biological warfare agents.

Reference p. 292.

9. 1970 Yugoslavia

To prohibit research and development of agents for CB weapons. To abolish proving grounds for the testing of CB weapons.

Reference p. 303.

10. 1971 Socialist Draft BW Convention

Not to develop microbiological or other B agents or toxins that are not designed for the prevention of disease or for other peaceful purposes, as well as auxiliary equipment or means of delivery designed to facilitate the use of such agents or toxins for hostile purposes.

Reference p. 314.

Guide to proposals

XII. Supervision of observance of the prohibitions

1. 1926 Sub-Commission B of the Preparatory Commission

    To institute between industries in different countries agreements, sanctioned by the states concerned, providing for the rationing of manufacture; the agreements would cover both the nature of the products and the quantities manufactured. This would also make it possible to exercise stricter supervision as regards prohibition to manufacture certain products which appear to be of use only for military purposes.

    Response Chairman of the Special Committee on CBW: Supervision would gain from the existence of such agreements, but one can hardly contemplate creating them in order to facilitate supervision.

    Reference pp. 78, 126.

2. 1926 Belgium

    To set up general supervision with regard to poisonous gases.

    Response Sub-Commission A of the Preparatory Commission: It is impossible to prevent or hinder the manufacture of poisonous gases in peacetime.

    Reference pp. 72, 78.

3. 1928, 1929 USSR

    In enterprises capable of being utilized for the manufacture of means of chemical and bacteriological warfare, a permanent labour control should be organized by the workers' committees of the factories or by other organs of the trade unions operating in the respective enterprises with a view to limiting the possibility of breaches.

    To supervise the observance of the convention on the reduction of armaments, investigations would be carried out on the spot by an international commission in the event of reasonable suspicion of a breach.

    Reference pp. 92–93.

4. 1932 Denmark

    National cartels for chemical and bacteriological manufacture to organize an international cartel responsible for ensuring that such private manufacture shall not be employed for preparation of CB warfare.

    Response See under XII. 1.

    Reference pp. 109, 111–12, 126.

5. 1932 Chairman of the Special Committee on CBW

    To establish a specialized section for chemical warfare in the Permanent Disarmament Commission.

    It would be the duty of the section to collect all the information that could be found in official trade statistics or private statistics carrying some weight, relating to the manufacture of chemical substances in the territory of each signatory state, and their import into that territory.

Guide to proposals

Having made a preliminary study of this information, the section may ask the governments of the countries concerned to supply it with any further information, if necessary.

Any signatory State may, on the strength of any information received by it, apply to the section for explanations regarding chemical substances manufactured in the territory of another state or imported into that territory. The section would make a preliminary examination of such applications in order to establish whether it was worthwhile to obtain fuller information. If it were so decided, the state whose position had been impugned would be asked for explanations.

When in possession of the information required, the section would decide whether the matter is to de dropped or laid before the Permanent Commission. In the latter case, the Commission would take steps within the limits of the powers conferred upon it by the disarmament convention.

Response Special Committee's Report II: It is not possible to enforce the prohibition against the preparation of chemical warfare by examining the commercial statistics of activities of chemical industries in all-countries. It is conceivable in theory, but impossible in practice, to exercise control by entrusting national or international bodies with the duty of inspecting chemical factories and of making public the character of the products manufactured, the existing stocks and the production capacity.

Reference pp. 127, 158.

6. 1932 Special Committee's Report II

No manufacture of or trade in poisonous substances suitable exclusively for the conduct of chemical warfare, should be permissible without government authorization.

Reference pp. 158–59.

7. 1952 France

Measures of international control should include:
Publication and verification of laboratory research, control of media of bacterial dissemination and emission and, as far as possible, control of the most dangerous special chemical products capable of being used to produce the most destructive type of emission.

Reference p. 203.

8. 1955 United Kingdom

All states should supply the control organ, set up under a disarmament convention, with all the information required on plants making CB weapons; the control organ would have the right to analyze and check the information and, at a later stage of the disarmament programme, to establish resident inspection posts at those plants or inspect them through routine visits.

Reference p. 227.

Guide to proposals

9. 1955 France and United Kingdom

    The list of objects to be controlled in the first stage of disarmament should include chemical and bacteriological armaments; all military establishments and installations which use or store chemical or bacteriological armaments; all documents such as estimates and accounts, necessary to verify expenditure on CB armaments. In the second stage, factories and other installations in which armaments are being manufactured or assembled or in which essential components are being manufactured, or which can be readily adapted to the manufacture of such armaments or components, would be subject to control.
    Reference p. 227.

10. 1959 USSR

    Upon the completion of general and complete disarmament, which should include the disbandment of all services of the armed forces and the destruction of all types of weapons, including CB weapons, an international control organ would have free access to all objects of control.
    Reference pp. 228–29.

11. 1960 USSR

    Representatives of the control organization should conduct on-site inspection of the destruction of all existing stockpiles of CB weapons.
    Reference p. 229.

12. 1962 USSR

    Elimination and destruction, as well as discontinuance of production of CB weapons and of related instruments and facilities (at the second stage of a general and complete disarmament programme) to be implemented under the control of inspectors of an international disarmament organization.
    Reference p. 232.

13. 1962 USA

    In the second stage of general and complete disarmament, an international disarmament organization should verify the cessation of production and field testing and the reduction of stockpiles of CB weapons, as well as the dismantling or conversion to peaceful uses of all facilities engaged in the production or field testing of CBW, so as to provide assurance that retained levels of CB weapons did not exceed agreed levels and that activities subject to limitations were not conducted at undeclared locations.

    In the third stage, upon elimination of all armaments, subject to agreed requirements for non-nuclear armaments of agreed types, the international disarmament organization would carry out verification and provide assurance that retained armaments were of the agreed types and did not exceed agreed levels; it would verify at declared locations the cessation of all applied research, development, production and testing of armaments as well as the

dismantling or conversion to peaceful uses of all facilities for such purposes; it would also provide assurance that the banned activities were not conducted at undeclared locations.
Reference pp. 232–33.

14. 1969–1970 Socialist Draft CBW Convention

    Each party to a convention prohibiting CBW should be internationally responsible for compliance with the provisions of the convention by legal and physical persons exercising their activities in and outside its territory; it should take the necessary legislative and administrative measures to prohibit the development, production and stockpiling of CB weapons and to destroy such weapons. The parties would consult one another and cooperate in solving any problems which may arise in the application of the provisions of the convention.
    Reference pp. 298–99.

15. 1970 Czechoslovakia

    Self-supervision of a CBW prohibition could be carried out by national bodies having an international reputation, such as academies of science.
    Reference p. 303.

16. 1970 Japan

    To trace in each state the flow of materials which may be used for the production of the most dangerous agents, so as to deter the use of highly poisonous organophosphorus agricultural chemicals as chemical warfare agents.

    Statistics of certain chemical substances concerning the amount of their production, preferably on a factory basis, exportation and importation as well as consumption for different purposes, might be used as part of the data forming the evidence for possible complaints of violations. In *ad hoc* inspections gaschromatography could be applied to microanalyze substances from the chemical plant concerned existing in liquid wastes, the soil and dust in and around the premises, on the production devices or on the workers' clothes. To limit the scope of the chemical items to be accounted for, a certain level of lethal dose by hypodermic injection might be employed as a criterion. (A tentative list of substances to be reported on, was suggested.) A roster of experts to be made available for investigations should be provided for and kept by the UN Secretary-General.
    Reference pp. 304–306.

17. 1970 Mongolia

    To establish special government agencies, on the pattern prescribed in the Single Convention on Narcotic Drugs, for the purpose of ensuring compliance with a CBW convention; the agencies might include representatives of important research institutes, medical and veterinary services, departments responsible for chemical industries, etc. In addition to a national system of compulsory registration of CB agents, to introduce control of the im-

port and export of agents which could be converted into weapons, as well as control of the manufacture, import and export of equipment and apparatus that could be used for the production of CB weapons.
Reference p. 303.

18. 1970 Sweden

To introduce international open exchange of information on pertinent peaceful scientific, technical and other activities; periodic reporting on chemical and biological agents, applying to both qualitative and quantitative factors, an international organ having the duty of receiving, storing, analyzing and distributing the information contained in the reports.
Reference p. 305.

19. 1970 Yugoslavia

To enact a law prohibiting research for weapons purposes and the development, production or stockpiling of agents for CB weapons; to enact a law for compulsory publication of the names of institutions and facilities engaged in or which, by their nature, could engage in the activities prohibited under the treaty, as well as data concerning the production of materials or agents which could be used for the production of CB weapons; to promulgate a decision to eliminate existing stockpiles and to abolish proving grounds for the testing of the prohibited weapons, and all installations related exclusively to such weapons. All such national legislative measures should be preceded by the placing under civilian administration or control of all institutions engaged in the research, development or production of chemical and biological weapons.

An international organ collecting information on CB agents should be able, at the request of states, to carry out preliminary investigations in order to ascertain whether a violation of the treaty had occurred.

To ensure appropriately regulated access (in accordance with the concept of verification by challenge) to institutions which prior to the ban were engaged in research, development, production and testing of CB weapons, as well as to institutions which by their nature could be engaged in such activities; lists of these facilities would have to be declared by governments. Control from the air by satellites and other devices for remote detection.
Reference pp. 303, 305–306.

20. 1971 Sweden

The most rigorous methods of control to be applied to chemicals more toxic than 1 mg per kg body weight, toxins and biological agents without any recognized peaceful use.

The production of these compounds would in principle be prohibited. Any deviation from this general rule would have to be reported to an international agency, the report giving the reasons for the production (scientific use, protective measures etc.). In case of any large-scale production (i.e., over 1 kg) or in case of suspected undeclared production, the international agency might be entitled to conduct an on-site inspection, either on the invitation of the producing or suspected party, or obligatory.

Guide to proposals

The compounds comprising all remaining chemical and biological agents which can be used as means of warfare, but also have peaceful uses, as well as the ancillary equipment and vectors would be controlled by national means only, such national control being possibly complemented by statistical reporting by the parties to an international agency; they would further be subject, if suspicion were aroused, to the sequence of processes foreseen in the complaints procedure. (See under XIII. 8.)

There should be some form of international surveillance of the destruction of stocks. In the case of diversion of CB warfare agents to peaceful purposes, international involvement could take the form of transfers of stocks for research or health protection for the benefit of developing countries.

Reference pp. 312–13.

21. 1971 Socialist Draft BW Convention

    Each state party to the convention should be internationally responsible for compliance with its provisions by legal or physical persons of that state; it should take the necessary legislative and administrative measures for prohibiting the development, production and stockpiling of B weapons, equipment and means of delivery, and for destroying them. The parties would consult one another and co-operate in solving any problems which may arise in the application of the provisions of the convention.

    Reference pp. 314–15, 333.

XIII. Investigation of complaints

1. 1926 Belgium, British Empire, Czechoslovakia, Finland, France, Netherlands, Poland, Romania, Kingdom of the Serbs, Croats and Slovenes, Spain and Sweden.

    In cases in which it is possible to institute an enquiry on the sole basis of documents—to set up a special commission, competent and impartial, to draw up a report after having examined the complaint and the documents. In cases of preparation for aggression and in all cases in which the time required for the employment of the preceding method would be incompatible with the nature of infraction, direct enquiry would have to be carried out on the spot. It would be essential to fix very short time limits and to secure the element of surprise.

    Response Chile, Italy, Japan and USA: enquiries would in general prove fruitless and illusory.

    Reference pp. 84–86.

2. 1926 Sub-Commission B of the Preparatory Commission

    Not to send too many commissions in case of investigation on the spot; to have recourse, if possible, to a single investigator of assured competence.

    Reference p. 87.

Guide to proposals

3. 1932 Chairman of the Special Committee on CBW

   The Permanent Disarmament Commission should have the duty to establish the fact of use of CB weapons. It would have the right to carry out for this purpose any preliminary enquiries, both in the territory subject to the authority of the complainant states and in the territory subject to the authority of the state against which a complaint was made.
   Reference p. 133.

4. 1932 Special Committee's Report II:

   The establishment of facts of infringement should be extremely rapid; it should afford guarantees of impartiality, and be carried out by persons of recognized qualifications and of high moral standing. The duty of collecting evidence would be entrusted to an international commission for urgent initial investigation, constituted in peacetime. Failing this, the complainant state should apply to the doyen of the diplomatic corps who would appoint members of the investigation commission.
   The state involved, and all the other signatory states, should take necessary steps to enable the commissioners to perform their duty.
   Reference pp. 160–61.

5. 1933 UK Draft Disarmament Convention

   The Permanent Disarmament Commission shall examine the complaints put forward by any party which may allege that the prohibition to prepare for chemical, incendiary or bacterial warfare has been violated.
   Any party claiming that chemical, incendiary or bacterial weapons have been used against it shall notify the Permanent Disarmament Commission. It shall, at the same time, notify the authority designated for the purpose by the Permanent Disarmament Commission or, failing such authority, the doyen of the diplomatic corps accredited to it, with a view to the immediate constitution of a commission of investigation.
   If the above-mentioned authority has received the necessary powers, it shall itself act as a commission of investigation.
   The commission of investigation shall proceed with all possible speed to the enquiries necessary to determine whether chemical, incendiary or bacterial weapons have been used.
   It shall report to the Permanent Disarmament Commission.
   The Permanent Disarmament Commission shall invite the party against which the complaint has been made to furnish explanations.
   It may send commissioners to the territory under the control of that party for the purposes of proceeding to an enquiry, to determine whether chemical, incendiary or bacterial arms have been used.
   The Permanent Commission may also carry out any other enquiry with the same object.
   The parties involved in the above-mentioned operations, and, in general, all the parties to the Convention, shall take the necessary measures to facilitate these operations, particularly as regards the rapid transport of persons and correspondence.

Guide to proposals

According to the result of the above-mentioned operations, the Permanent Commission, acting with all possible speed, shall establish whether chemical, incendiary or bacterial weapons have been used.
Reference pp. 169–70.

6. 1969–1970 UK Draft BW Convention

Any party to the Convention, believing that B weapons have been used against it, would be entitled to lodge a complaint with the UN Secretary-General and request that the complaint be investigated and that a report on the result of the investigation be submitted to the Security Council; any party believing that another party has acted in breach of other undertakings under the convention, would be entitled to lodge a complaint with the Security Council and request that the complaint be investigated. The complaints would have to be supported by all evidence at the disposal of the complaining party. Each party would undertake to cooperate in any investigation that may be carried out.
Reference pp. 295–97, 323.

7. 1969–1970 Socialist Draft CBW Convention

The parties to the Convention would be entitled to lodge complaints of possible violations with the UN Security Council. The complaints should include all possible evidence confirming their validity, and a request for consideration. Each party would undertake to co-operate in carrying out any investigations which the Security Council may undertake, in accordance with the UN Charter, on the basis of the complaint received. (The possibility of on-site inspection is envisaged under this procedure.)
Reference pp. 299, 303, 328.

8. 1970, 1971 Sweden

The complaints procedure should take the form of a system of successive steps, including consultations between the parties and other fact-finding measures.
A system of "verification by challenge" would permit a party under suspicion of having violated its engagements to free itself from that suspicion through the supply of relevant information, not excluding invitation to inspection. These steps should precede resort to the UN Security Council.
Reference p. 305–306.

9. 1971 Socialist Draft BW Convention
See under XIII. 7.

XIV. Action in case of violations

1. 1926 Belgium, Bulgaria, Czechoslovakia, Finland, France, Poland, Romania and Kingdom of the Serbs, Croats and Slovenes.

    All states in possession of a chemical industry should undertake:
    To put at the disposal of any state attacked by gas the raw materials, chemical products and means of operation necessary for reprisals;

Guide to proposals

To engage themselves in joint reprisals, so far as distance permits, by the use of chemical means against the state which has committed an act of aggression by the use of gas.

Response Netherlands and Germany were opposed to the employment of chemical weapons as sanction.

Reference pp. 81–82.

2. 1929 Romania and Kingdom of the Serbs, Croats and Slovenes.

To place at the disposal of any state which is the victim of aggression by means of poisonous or bacteriological substances such raw materials, products and appliances as may be necessary to meet this aggression;

To undertake to participate, as far as distance will allow, in collective reprisals by employing chemical and bacteriological means against the state guilty of an aggression by such means.

Reference p. 94.

3. 1932 Yugoslavia

If in case of hostilities one of the parties transgressed the obligation not to use CB weapons (even for legitimate defence), the Council of the League of Nations should pronounce its outlawry from the civilized world. The signatory states would be obliged to render military assistance to the state victim of the transgression.

Reference p. 112.

4. 1932 France

The only retaliation which could be admitted must be decided upon by the community of states; it is necessary to make preparations for international punitive action.

Reference p. 138.

5. 1932 Chairman of the Special Committee on CBW

The declaration of the Permanent Disarmament Commission establishing the fact of use of CB weapons would entail immediate action on the part of the third states. It would be their right and duty to bring pressure to bear on the offending state (diplomatic representations, rupture of diplomatic, economic and financial relations, blockade). Third states would at the earliest possible moment decide, if necessary, on the punitive or other action to be taken.

The state victim of the breach would have the right to retaliate within the fighting area. The possibility of retaliating by the actual use of the prohibited weapon would apply only to chemical and incendiary weapons and not to the bacteriological weapon. The use of the latter, if duly established, would give, however, the other belligerent the right to employ other prohibited weapons by way of reprisal.

Response France: the scale of sanctions laid down above would not be effective. If individual retaliation on the part of the attacked state were permitted, it would be impossible to prevent preparation of chemical warfare.
Japan and Spain opposed the employment of gas as retaliatory measure.
Reference pp. 129–33, 137–39.

6. 1932 Spain

   If a state signatory to a convention has recourse to the use of chemical, incendiary or bacteriological weapons, it shall be, *ipso facto,* considered as having committed an act of war against the other parties to the convention; and the other states shall take repressive action against the state violating the prohibition. The action shall be progressively accentuated with a view to inducing the state in question to forego the use of chemical, incendiary or bacteriological weapons, or preventing it from continuing the use of them, in the last resort employing military sanctions to enforce respect for the obligations under the convention.
   The right of retortion against the use of chemical, incendiary or bacteriological weapons is formally forbidden. Any state having recourse to reprisals will thereby place itself outside the convention.
   Reference p. 139.

7. 1932 Sub-Committee of the Disarmament Conference, charged with defining sanctions.

   The Permanent Disarmament Commission having established the fact of use of CB weapons:
   Third states shall individually bring pressure to bear upon the guilty state to induce it to give up the use of those weapons or to deprive it of the possibility of continuing to use them.
   A consultation shall be held among third states to determine what joint steps shall be taken and to decide on the joint punitive action. The decisions to be taken by a majority, the minority shall not be bound.
   Third states situated in a given region may pledge themselves to undertake jointly, and as rapidly as possible, severe punitive actions against the delinquent state, and create beforehand a joint police force.
   The state against which chemical, incendiary or bacteriological weapons have been employed shall in no circumstances retaliate by the use of the same weapons.
   Response Italy questioned the efficacy of sanctions of a regional character. Universal action, such as a blockade by all the states in the world, would be more powerful.
   Reference pp. 140–41.

8. 1932 Special Committee's Report II

   The attacked state should be given scientific, medical and technical assistance in repairing, attenuating or preventing the effects of the use of prohibited weapons. Preparations for the granting of assistance might be entrusted to an international information and documentation service for protection against chemical weapons, to be established.

To stop supplies to the guilty state of raw materials, products and appliances necessary for chemical and incendiary warfare.

The period between the submission of a complaint and the establishment of the fact of a breach should be used to make preparations with a view to the possible application of penalties.

Reference pp. 161–63.

9. 1933 UK Draft Disarmament Convention

The party which has been the victim of the illegal use of chemical or incendiary weapons is not deprived of the right to retaliate, subject to such conditions as may hereafter be agreed.

Response France: Collective sanctions are necessary in case of any breach of the Convention.

Reference pp. 167, 171.

10. 1955 France and United Kingdom

Any alleged breach would be considered by an executive committee, consisting of China, France, USSR, UK and USA, which would be permanent members of the committee, and of ten other states which would be elected for two-year terms. If the executive committee considered that an irregularity covered by Chapter VII of the UN Charter had occurred, it should immediately inform the UN Security Council or General Assembly and suspend such of the prohibitions, limitations and reductions provided for in the disarmament treaty as it may consider necessary, pending a decision under the Charter by the Security Council or General Assembly. The executive committee would equally have the right to suspend such of the prohibitions, limitations and reductions as it might consider necessary if it decided on the basis of a report from the director-general of the international disarmament organization that a participating state was markedly and consistently late in fulfilling its obligations.

Reference pp. 227–28.

11. 1969–1970 UK Draft BW Convention

Each party should affirm its intention to provide or support appropriate assistance to any other party, if the UN Security Council concludes that biological methods of warfare have been used against that party.

Each party shall have the right to withdraw from the Convention, if it decides that extraordinary events, related to the subject matter of the Convention, have jeopardized the supreme interests of its country.

Reference pp. 295, 323–24.

XV. Protection against CB weapons

1. 1923 Hungary
   To publish information concerning research undertaken with a view to discovering means of safeguarding humanity from the consequences of CBW; the use of such means to be made accessible to all.
   Reference p. 49.

2. 1925 Hungary
   To exclude from the prohibition the methods of defence (gas masks) against asphyxiating, poisonous, or other similar gases employed as a means of warfare.
   Reference pp. 59–60.

3. 1930 Poland
   To conclude a convention on international aid to be afforded to any country chemically or bacteriologically attacked. Such aid would be essentially of humanitarian nature (sanitary, scientific etc.).
   Reference p. 104.

4. 1932 Chairman of the Special Committee on CBW
   Material solely intended to protect combatants and civilians against chemical and bacteriological weapons must not be prohibited. Also training aimed at protecting people against the effects of CBW must be authorized. The use of defensive appliances presupposes the possibility of manufacturing gases for experimental purposes. However, this manufacture would be on such a small scale, being confined to laboratories, that it would not be necessary for the convention to deal with it.
   Response Japan: laboratory tests are not sufficient to work out protection against poisonous gases.
   Special Committee's Report II:
   For defensive purposes, tests in the field are indispensable, but the quantities of poisonous substances required are extremely small compared to those needed for a chemical attack; restrictions might be imposed.
   Reference pp. 124–25, 134–35, 148.

5. 1932 France
   To prohibit private manufacture of defensive appliances and experimental material for those appliances; to provide for strict government supervision and international supervision.
   Reference p. 135.

6. 1932 Special Committee's Report II
   To set up an international information service, attached to the Permanent Disarmament Commission, for the collection of material regarding protection against chemical weapons, so as to enable all countries to keep abreast

of methods of preparing the defence of civilians. Certain protective devices, e.g. masks, could be subjected to technical testing by an international body. Such tests might lead to the introduction of standard devices.

The organization of a suitable health service in time of peace represents the most effective means of defence against bacterial infection.

Reference pp. 149–50.

7. 1933 UK Draft Disarmament Convention

Not to restrict the freedom of the parties in regard to material and installations intended exclusively to ensure individual or collective protection against the effects of chemical, incendiary or bacterial weapons, or to training with a view to individual or collective protection against the effects of the said weapons.

Reference p. 169.

8. 1971 United Kingdom

To provide assistance, primarily of medical and relief nature, to a party-victim of a biological attack, at the request of the victim.

Reference p. 318.

XVI. Arousing public opinion about CBW

1. 1921 South Africa

    To address an appeal to scientists of all countries to publish their discoveries on poison gas and similar subjects, in order to minimize the likelihood of their being used in war.

    Response TMC conclusion: The proposed appeal is not a practical measure.

    Reference pp. 45–46.

2. 1926 Sub-Commission B of the Preparatory Commission

    To establish as a crime in common law and to punish with suitable penalties any exercises or training by military persons or civilians in the use of poisons and bacteria and, in particular, any exercises or training by air squadrons.

    Reference p. 87.

3. 1932 Special Committee's Report II

    To introduce penal legislation in each country providing for punishment of the authors of the preparation of a prohibited form of warfare—for example, chemists and bacteriologists convicted of preparing chemical or bacteriological weapons.

    Reference p. 159.

4. 1932 Some members of the Special Committee-Report II

    A stigma involving the prohibition to practice their profession to be attached to those engaged in work aimed at the preparation of CB warfare.

Response France: The studies of scientists may serve a two-fold purpose. Research in poisonous substances may be in the interest of mankind. Italy: A chemist's first duty is to his own country.

Reference p. 159.

5. 1970 Poland

To include in the textbooks dealing with chemistry and biology an indication that the use of CB agents for any warlike purposes is a violation of international law and is liable to prosecution.

Reference p. 303–304.

6. 1970 Mexico

Individuals should become active participants in the denunciation of treaty violations in the interests of the international community.

Reference p. 304.

Addendum

Draft convention on the prohibition of the development, production and stockpiling of bacteriological (biological) and toxin weapons and on their destruction, submitted by Bulgaria, Canada, Czechoslovakia, Hungary, Italy, Mongolia, Netherlands, Poland, Romania, USSR, UK and USA on 28 September 1971[1]

The States Parties to this Convention,
    Determined to act with a view to achieving effective progress towards general and complete disarmament including the prohibition and elimination of all types of weapons of mass destruction, and convinced that the prohibition of the development, production and stockpiling of chemical and bacteriological (biological) weapons and their elimination, through effective measures, will facilitate the achievement of general and complete disarmament under strict and effective international control,
    Recognizing the important significance of the Geneva Protocol of 17 June 1925 for the Prohibition of the Use in War of Asphyxiating, Poisonous or Other Gases, and of Bacteriological Methods of Warfare, and conscious also of the contribution which the said Protocol has already made, and continues to make, to mitigating the horrors of war,
    Reaffirming their adherence to the principles and objectives of that Protocol and calling upon all States to comply strictly with them,
    Recalling that the General Assembly of the United Nations has repeatedly condemned all actions contrary to the principles and objectives of the Geneva Protocol of 17 June 1925.
    Desiring to contribute to the strengthening of confidence between peoples and the general improvement of the international atmosphere,
    Desiring also to contribute to the realization of the purposes and principles of the Charter of the United Nations,
    Convinced of the importance and urgency of eliminating from the arsenals of states, through effective measures, such dangerous weapons of mass destruction as those using chemical or bacteriological (biological) agents,
    Recognizing that an agreement on the prohibition of bacteriological (bio-

---

[1] Disarmament Conference document CCD/353.

logical) and toxin weapons represents a first possible step towards the achievement of agreement on effective measures also for prohibition of the development, production and stockpiling of chemical weapons, and determined to continue negotiations to that end,

Determined, for the sake of all mankind, to exclude completely the possibility of bacteriological (biological) agents and toxins being used as weapons,

Convinced that such use would be repugnant to the conscience of mankind and that no effort should be spared to minimize this risk,

Have agreed as follows:

ARTICLE I

Each State Party to this Convention undertakes never in any circumstances to develop, produce, stockpile or otherwise acquire or retain:

(1) Microbial or other biological agents, or toxins whatever their origin or method of production, of types and in quantities that have no justification for prophylactic, protective or other peaceful purposes;

(2) Weapons, equipment or means of delivery designed to use such agents or toxins for hostile purposes or in armed conflict.

ARTICLE II

Each State Party to this Convention undertakes to destroy, or to divert to peaceful purposes, as soon as possible but not later than nine months after the entry into force of the Convention all agents, toxins, weapons, equipment and means of delivery specified in Article I of the Convention, which are in its possession or under its jurisdiction or control. In implementing the provisions of this Article all necessary safety precautions shall be observed to protect populations and the environment.

ARTICLE III

Each State Party to this Convention undertakes not to transfer to any recipient whatsoever, directly or indirectly, and not in any way to assist, encourage, or induce any State, group of States or international organizations to manufacture or otherwise acquire any of the agents, toxins, weapons, equipment or means of delivery specified in Article I of the Convention.

ARTICLE IV

Each State Party to this Convention shall, in accordance with its constitutional processes, take any necessary measures to prohibit and prevent development, production, stockpiling, acquisition or retention of the agents, toxins, weapons, equipment and means of delivery specified in Article I of the Convention, within the territory of such State, under its jurisdiction or under its control anywhere.

ARTICLE V

The States Parties to the Convention undertake to consult one another and to co-operate in solving any problems which may arise in relation to the objective of, or in the application of the provisions of, this Convention. Consultation and co-operation pursuant to this Article may also be undertaken through appropriate international procedures within the framework of the United Nations and in accordance with its Charter.

ARTICLE VI

(1) Any State Party to the Convention which finds that any other State Party is acting in breach of obligations deriving from the provisions of this Convention may lodge a complaint with the Security Council of the United Nations. Such a complaint should include all possible evidence confirming its validity, as well as a request for its consideration by the Security Council.

(2) Each State Party to the Convention undertakes to co-operate in carrying out any investigation which the Security Council may initiate, in accordance with the provisions of the United Nations Charter, on the basis of the complaint received by the Council. The Security Council shall inform the States Parties to the Convention of the results of the investigation.

ARTICLE VII

Each State Party to the Convention undertakes to provide or support assistance, in accordance with the United Nations Charter, to any Party to the Convention which so requests, if the Security Council decides that such party has been exposed to danger as a result of violation of this convention.

ARTICLE VIII

Nothing in this Convention shall be interpreted as in any way limiting or detracting from the obligations assumed by any State under the Geneva Protocol of 17 June 1925 for the Prohibition of the Use in War of Asphyxiating, Poisonous or Other Gases, and of Bacteriological Methods of Warfare.

ARTICLE IX

Each State Party to this Convention affirms the recognized objective of effective prohibiton of chemical weapons and, to this end, undertakes to continue negotiations in good faith with a view to reaching early agreement on effective measures for the prohibition of their development, production and stockpiling and for their destruction, and on appropriate measures concerning equipment and means of delivery specifically designed for the production or use of chemical agents for weapons purposes.

ARTICLE X

(1) The States Parties to the Convention undertake to facilitate, and have the right to participate in, the fullest possible exchange of equipment, materials and scientific and technological information for the use of bacteriological (biological) agents and toxins for peaceful purposes. Parties to the Convention in a position to do so shall also co-operate in contributing individually or together with other States or international organizations to the further development and application of scientific discoveries in the field of bacteriology (biology) for prevention of disease, or for other peaceful purposes.

(2) This Convention shall be implemented in a manner designed to avoid hampering the economic or technological development of States Parties to the Convention or international co-operation in the field of peaceful bacteriological (biological) activities, including the international exchange of bacteriological (biological) agents and toxins and equipment for the processing, use or production of bacteriological (biological) agents and toxins for peaceful purposes in accordance with the provisions of this Convention.

ARTICLE XI

Any State Party may propose amendments to this Convention. Amendments shall enter into force for each State Party accepting the amendments upon their acceptance by a majority of the States Parties to the Convention and thereafter for each remaining State Party on the date of acceptance by it.

ARTICLE XII

Five years after the entry into force of this Convention, or earlier if it is requested by a majority of Parties to the Convention by submitting a proposal to this effect to the Depositary Governments, a conference of States

Addendum

Parties to the Convention shall be held at Geneva, Switzerland, to review the operation of this Convention, with a view to assuring that the purposes of the preamble and the provisions of the Convention, including the provisions concerning negotiations on chemical weapons, are being realized. Such review shall take into account any new scientific and technological developments relevant to this Convention.

ARTICLE XIII

(1) This Convention shall be of unlimited duration.

(2) Each State Party to this Convention shall in exercising its national sovereignty have the right to withdraw from the Convention if it decides that extraordinary events, related to the subject matter of this Convention, have jeopardized the supreme interests of its country. It shall give notice of such withdrawal to all other States Parties to the Convention and to the United Nations Security Council three months in advance. Such notice shall include a statement of the extraordinary events it regards as having jeopardized its supreme interests.

ARTICLE XIV

(1) This Convention shall be open to all States for signature. Any State which does not sign the Convention before its entry into force in accordance with paragraph 3 of this Article may accede to it at any time.

(2) This Convention shall be subject to ratification by signatory States. Instruments of ratification and instruments of accession shall be deposited with the Governments of . . . . . . which are hereby designated the Depositary Governments.

(3) This Convention shall enter into force after the deposit of the instruments of ratification by twenty-two Governments, including the Governments designated as Depositaries of the Convention.

(4) For States whose instruments of ratification or accession are deposited subsequent to the entry into force of this Convention, it shall enter into force on the date of the deposit of their instruments of ratification or accession.

(5) The Depositary Governments shall promptly inform all signatory and acceding States of the date of each signature, the date of deposit of each instrument of ratification or of accession and the date of the entry into force of this Convention, and of the receipt of other notices.

(6) This Convention shall be registered by the Depositary Governments pursuant to Article 102 of the Charter of the United Nations.

ARTICLE XV

This Convention, the Chinese, English, French, Russian and Spanish texts of which are equally authentic, shall be deposited in the archives of the Depositary Governments. Duly certified copies of this Convention shall be transmitted by the Depositary Governments to the Governments of the signatory and acceding States.

In witness whereof the undersigned, duly authorized, have signed this Convention.

Done in . . . . . copies at . . . ., this . . . . day of . . . ., ...

Country Index

A

Afghanistan, 107, 188
Albania 107, 243, 320
Argentina, 39, 81, 84, 107, 157, 188, 273, 293, 317, 319, 343
Australia, 29, 37, 47, 100, 107, 112, 122, 178, 188, 194–95, 216, 219, 263, 273, 285–86, 288, 308, 343
Austria, 42, 56, 107, 110, 188, 198, 250, 342

B

Belgium, 60, 72–73, 80–83, 85, 89–90, 94–98, 100–101, 107, 112, 133–34, 141, 184, 188, 195, 199, 217, 225, 242, 263, 342
Bolivia, 107, 188
Brazil, 37, 39, 59, 63, 107, 112, 195, 199, 202, 209, 212, 218, 232, 273, 316–17, 319, 342
Bulgaria, 39, 42, 56, 79, 81, 107, 188, 229–30, 232, 239, 298, 314, 326, 331, 336, 343
Burma 39, 230, 232, 319
Byelorussia, 240, 298, 314, 326, 331, 336, 347

C

Cambodia, 230, 235–38
Canada, 39, 100, 103, 107, 188, 202, 217, 227, 229, 232, 241, 261, 281, 284–85, 287, 301, 306, 318, 320, 343
Ceylon, 230, 344
Central African Republic, 344
Chile, 59, 81, 84, 96, 100, 107, 188, 207, 212, 343
China; People's Republic of China; Republic of China (Taiwan), 21, 27, 38, 40, 65, 71, 100, 103, 107, 188–92, 195–99, 201, 206, 210, 212–20, 222, 227, 263, 283, 315, 320, 342, 344
Colombia, 47, 59, 65, 90, 107, 188, 195
Costa Rica, 107
Cuba, 107, 188, 242–43, 344
Cyprus, 344

Czechoslovakia, 39, 79, 81–83, 85–86, 89, 100, 103, 107, 109, 188, 202, 218, 229–30, 232, 298, 303, 314, 320, 326, 331, 336, 343

D

Denmark, 37, 107, 109, 111–12, 147, 157, 188, 343
Dominican Republic, 107, 188, 344

E

Ecuador, 37, 107, 188, 219, 273, 344
Egypt (UAR), 39, 107, 188, 218, 230, 232, 243–46, 251–52, 261, 298, 304, 316–17, 319, 343
Estonia, 107, 188, 342–43
Ethiopia (Abyssinia), 21, 39, 107, 175–80, 182–85, 187–89, 191–92, 204, 232, 253, 263, 285, 319, 343

F

Finland, 80–83, 85–86, 104, 107, 188, 289, 343
France, 18–20, 22–24, 34, 37–39, 45–46, 50, 62, 66, 69, 71, 73, 75, 79–83, 85–89, 94–97, 99–100, 102–103, 105, 107, 111, 113, 119, 122, 134–35, 137–39, 141, 147, 154–55, 157, 159, 165, 171, 188, 195, 199, 203, 207, 210–12, 217–18, 220, 224–25, 227, 229, 232, 242–43, 250, 253, 263, 283, 294, 315, 320, 342–43

G

Gabon, 242–43
Gambia, 344
Germany; German Democratic Republic; Federal Republic of Germany, 18, 20, 24, 27, 34, 37, 41–42, 54–56, 73, 75, 81–82, 84, 88–91, 95, 97, 100, 102, 105, 107, 110, 112–13, 119, 165, 171, 174, 204, 209–10, 222–26, 250, 257, 284, 342–43
Ghana, 230, 285, 344
Greece, 95, 98, 107, 139, 188, 205–208, 212, 220, 343
Guatemala, 107

Country index

H

Haiti, 107, 110, 188
Holy See, 344
Honduras, 107, 188
Hungary, 28, 39, 42, 49, 56, 59, 61–62, 107, 188, 225–26, 238–43, 249–51, 261–62, 273, 298–99, 301, 314, 319–20, 326, 331, 336, 344

I

Iceland, 344
India, 39, 100, 107, 110, 122, 188, 226, 230, 232, 257–58, 261, 316, 319, 343
Indonesia, 217–18, 230, 283, 344
Iran, 95, 97–98, 108, 188, 344
Iraq, 37, 107, 122, 188, 230, 344
Ireland, 60, 68, 100, 103, 107, 122, 188, 263, 344
Israel, 252, 344
Italy, 19, 21, 34, 37, 39, 46, 62–65, 73, 75, 81, 84, 92–93, 95–96, 100, 103, 105, 107–108, 111, 113, 115, 119, 140, 145, 147, 159–60, 166, 175–76, 178–82, 184–88, 191, 198–99, 204, 212, 225, 229–30, 232, 241, 264, 284, 306, 320, 343
Ivory Coast, 344

J

Jamaica, 344
Japan, 19–21, 34, 37, 39, 46, 65, 68, 81, 84, 91, 93, 96, 98, 100, 103, 105, 107, 112–13, 134–35, 138, 147, 154, 189–91, 196, 206, 209, 213, 216, 219, 263, 273, 285, 304–306, 313, 343

K

Kenya, 241, 344
Korea; Democratic People's Republic of Korea; Republic of Korea, 27, 196–97, 199–202, 204, 207–208, 215–16, 218–19, 222, 226, 263

L

Laos, 240
Latvia, 107, 188, 342–43
Lebanon, 344
Liberia, 107, 188, 344
Lithuania, 107, 188, 342–43
Luxembourg, 107, 188, 225, 343

M

Madagascar, 249, 344
Malawi, 344

Malaysia, 344
Maldives, 344
Mali, 249
Malta, 28, 242, 247–50, 252, 254, 262–64, 306, 344
Mauritius, 344
Mexico, 39, 107, 188, 232, 261, 282, 304, 317, 319–20, 344
Monaco, 344
Mongolia, 39, 206, 263, 273, 298–300, 303–304, 314, 319–20, 326, 331, 336, 344
Morocco, 30, 39, 230, 273, 293, 310, 316, 318–19, 344

N

Nepal, 230, 240, 285, 293, 344
Netherlands, 37, 39, 41, 65, 81, 85–86, 95, 97–98, 102, 107, 109, 113, 119, 122, 147, 154, 158, 165, 170, 188, 209, 212, 216, 225, 242, 273, 282, 287–88, 297, 313, 320, 343
New Zealand, 100, 108, 122, 188, 216–17, 344
Nicaragua, 37, 107, 188, 343
Niger, 344
Nigeria, 39, 232, 317, 319, 344
Norway, 47, 66, 107, 109, 119–21, 145, 177, 188, 241, 288, 343

P

Pakistan, 39, 211–12, 218, 273, 317–19, 344
Panama, 108, 188, 344
Paraguay, 37, 108, 344
Peru, 108, 188
Poland, 19, 22, 37, 39, 59–60, 64, 67–68, 71, 79–83, 85–86, 89, 95, 99, 100–101, 104, 108, 113, 139–40, 147, 176, 188, 195, 199, 218, 225, 229–30, 232, 254, 257–58, 261–63, 298–99, 303, 314–15, 318–20, 326, 331, 336, 343
Portugal, 29, 108, 122, 178, 188, 286, 343

R

Romania, 39, 60–62, 79–83, 85–86, 89, 94, 96–97, 101, 103–104, 108, 122, 188, 229–30, 232, 298, 314, 320, 326, 331, 336, 343
Rwanda, 344

S

Salvador, 37, 108
Saudi Arabia, 108, 243–46, 251, 344
Sierra Leone, 345

411

Country index

South Africa, 45, 100, 107, 122, 188, 263, 345
Spain, 37, 81, 85–86, 93, 99, 103, 108, 113, 122, 139, 147, 188, 343
Sweden, 18, 28–29, 39, 63, 81, 85–87, 108–109, 139, 177, 184, 188, 199, 215, 218, 232–34, 247, 257–58, 261, 266, 281, 286, 292, 298, 304–306, 309, 312–13, 316–19, 343
Switzerland, 37, 65, 108–109, 113, 133, 136, 147, 188, 343
Syria, 195, 345

T

Tanzania, 240–41, 345
Togo 345
Tonga, 345
Trinidad and Tobago 262–64, 329, 332
Tunisia, 345
Turkey, 42, 56, 69, 99, 101, 103, 108–109, 112, 188, 207, 212, 343

U

Uganda, 241, 345
Union of Soviet Socialist Republics, 18, 20, 24, 27–28, 30–31, 34, 36–40, 63, 71, 91–92, 96–97, 99, 101–105, 108–109, 113–15, 118–20, 122, 141, 165, 188, 194–95, 199–204, 206–10, 212–15, 218–20, 222–24, 227–32, 234, 237–40, 242, 250, 253, 256–58, 263, 283–84, 293, 298–99, 302, 306, 309–10, 312, 314–16, 320, 326, 331, 336, 345
Ukrainian SSR, 195, 298, 314, 326, 331, 336
United Arab Republic. *See* Egypt
United Kingdom, 18–20, 22, 24–25, 28, 30–31, 34, 37–40, 43, 45–46, 60, 62, 64, 66, 71–72, 74–75, 81, 85, 87–88, 91, 96, 100, 102, 105, 107–108, 113, 119, 121, 134, 136–38, 147, 154–57, 164, 166–67, 170–73, 177–78, 188, 190–91, 194–95, 199, 202, 206, 209–10, 212, 217, 219, 224–25, 227–30, 232, 241–42, 246, 250–52, 254–59, 266, 273, 277, 281, 283, 287, 290–91, 293, 295–302, 309–10, 312–18, 320, 322, 342
United States, 18–20, 23, 26–31, 34, 36–40, 42, 46, 58–64, 66, 71–72, 75, 81, 84, 88, 91, 96, 100, 102–103, 105, 107–108, 110–11, 113–14, 118, 134, 136, 139, 141, 145, 147, 149, 153–54, 156–57, 164–65, 171–72, 174, 193–94, 196–208, 212–19, 221–24, 227, 229–32, 234, 236–43, 246, 249–51, 253, 256–57, 260, 263, 277, 280–81, 283–88, 290–91, 293–94, 297–300, 302, 307–10, 312–18, 320–21, 336, 343
Upper Volta, 345
Uruguay, 59, 316, 188, 218, 343

V

Venezuela, 108, 188, 219, 230, 343
Viet-Nam; Democratic Republic of Viet-Nam; Republic of Viet-Nam, 28, 30, 235–41, 249, 251, 253, 263, 280, 286, 288, 308

Y

Yemen (Arab Republic of), 243–47, 251, 345
Yugoslavia, 22, 39, 61, 73, 79–83, 85–86, 88–89, 94, 96, 99, 101, 103–105, 108–12, 171, 188, 230, 273, 281–82, 303, 305–306, 313, 316–17, 319, 343